LABORATORY EXERCISES IN OCEANOGRAPHY

LABORATORY EXERCISES IN
OCEANOGRAPHY

SECOND EDITION

Bernard W. Pipkin
University of Southern California

Donn S. Gorsline
University of Southern California

Richard E. Casey
University of San Diego

Douglas E. Hammond
University of Southern California

W. H. FREEMAN AND COMPANY • NEW YORK

Cover image: California El Niño of 1982–1983. The photo on the left shows the normal conditions that prevailed in January 1982. The right-hand photo shows El Niño warming in January 1983. See Exercise 11 for further discussion. [Photo by Paul Fiedler, National Marine Fisheries Service/Southwest Fisheries Center, processed at Scripps Satellite Oceanography Facility]

ISBN 0-7167-1810-3

Printed in the United States of America

4567890 SL 99876543210

CONTENTS

PREFACE

The national importance of oceanographic research and the exploration of "inner space" prompted us to devise an elementary course in oceanography that would be suitable for undergraduates interested in this field. Few people are aware that reactions between the atmosphere and oceans are of basic importance in determining world climate. Although people have traveled over, on, and under the ocean surface, there is a general lack of awareness of the vast resources that exist in the sea. Faced with the necessity of feeding 100,000 new mouths daily, we continue to obtain our food, particularly our agricultural crops, largely from the land, even though parts of the ocean are more productive than many marginal farmlands. Exploiting the ocean as a food source would require no more than the introduction of new ideas in food use and some changes in the types of equipment used in contemporary fisheries. Furthermore, numerous inorganic materials, both metallic and nonmetallic, exist in large quantities on the sea floor and are or will soon be available for profitable recovery by existing technology. Consequently, the informed citizen should have a general background in elementary oceanography in order to appreciate the how the law of the sea, which is being developed by the nations of the world, will affect the use of the oceans.

Because most of our largest urban areas are located on our seacoasts, problems of shore erosion, the impact of waves on beaches and steep shores, and the alleviation of such effects have great economic significance. Also, an understanding of these coastal processes is essential for effective coastal land development, harbor construction, and the allocation of lands for recreation areas.

Over the past two decades new data from the oceans and from drilling the oceanic crust have been the basis for a major revolution in our models for the formation of the oceans and continents, the evolution of new life forms, and the locations of major mineral provinces. This new view of our planet has influenced studies of other dense planets of our solar system. For these reasons alone, the study of the oceans and their basins is a basic part of the background of all literate people.

This manual, presently the only one of its kind, has been compiled to meet the needs of a general education course in oceanography. Because it is intended for use in an elementary course, we have simplified some of the illustrative material to emphasize basic points in a given exercise. Although no laboratory course can be truly complete, we have attempted to provide exercises for all of the major domains in oceanography. Our aim has been to provide exercises that can be performed by a student within a period of one or two hours, either with the assistance of an instructor or after minimal preparatory lecture and discussion.

In many of the exercises, data are drawn from actual research projects conducted aboard the research vessel *Velero IV* by scientists from the Allan Hancock Foundation and the Department of Geological Sciences at the University of Southern California. Exposure to genuine oceanographic data gives students a more realistic impression of marine scientific work, and provides them with experience in the manipulation and interpretation of typical observations.

Since most semester or quarter courses will be too short to permit the use of all of the exercises, the instructor will want to select those exercises that provide the best background for the level of the students or that correspond to the purpose of the particular course. A number of the experiments can be accomplished with minimal equipment in a lecture room; some can even be done by the student alone once the instructor has provided some basic instructions. Consequently, even if a science course does not have regularly scheduled laboratory sections, instructors will be able to supplement course material with practical classroom exercises or homework assignments. Exercises can easily be augmented with a variety of films, demonstrations, and equipment.

All of the exercises have been used in our courses over the past decade and have been modified in response to the criticism of students, laboratory instructors, and colleagues. The objectives have been to give students an opportunity to practice simple procedures used by professionals in the field, and thus to provide insight into how science is done. Our procedure has been to cite in class appropriate laboratory exercises as examples illustrating principles we are developing in the lecture. Each exercise ends with a short glossary of terms that have been introduced in the exercise. Each of these terms is printed in **boldface** type the first time it occurs in the text.

A bibliography comprising two sections, one listing general references and the other listing selected relevant articles from past issues of *Scientific American,* is provided at the end of the manual.

This second edition has been completely reviewed and errors found by us and helpful users over the past few years have been corrected. We have removed the supplementary exercises and rearranged some exercises incorporating parts of that material. New laboratory exercises have been included—one on oil spills (Exercises 22), one on paleoceanography (Exercise 23). Updating includes such additions as mention of the finding of a new ecosystem at hydrothermal vents on the East Pacific Rise. We have also added satellite oceanography to Exercise 11 and have used new data for several exercises based on recent national field and laboratory research programs such as the OPUS Project on upwelling and various new deep ocean and coastal ocean surveys.

We thank the several hundred students whose comments have significantly influenced the development of this manual. In addition, Gregg Blake, James Buika, Peter Day, and Brian Edwards were particularly helpful in that they applied their experience as teaching assistants in the course to critical evaluations of the various exercises. Professors James C. Ingle, Jr., of Stanford University; J. Robert Moore of the University of Wisconsin; Peter J. Fischer of California State University, Northridge; John V. Byrne of Oregon State University; and Benno M. Brenninkmeyer of Boston College all provided critical readings of the first edition.

Dr. T. D. Dickey, University of Southern California, did much of the work on the new satellite exercise and also reviewed the other physical oceanographic materials. Dr. Burton Jones, University of Southern California, provided OPUS data and edited the seawater temperature exercise. Thomas Nardin of Exxon Production Research provided the seismic profiles, and a number of authors of various books, papers, and studies graciously permitted us to use their data or illustrations.

We are particularly appreciative of the suggestions and emendations by Harold Thurman of Mt. San Antonio College; Scott Thornton, now of Union Oil Company; Joseph Donaghue of Florida State University; Gilbert Fry, now of Mobil Oil Company; and many teaching assistants over the past four years. Critical review and detailed comments were of great value from W. W. Reynolds of the University of New England; Jon Sloan of California State University, Northridge; Donald Rice of the University of Maryland's Center for Estuarine and Environmental Studies; and John Buck of the University of Connecticut Marine Sciences Institute. To all of these individuals we express our gratitude while acknowledging responsibility for any errors that may remain.

Peter J. Dougherty and Jerry Lyons of W. H. Freeman and Company provided encouragement and the benefit of their experience throughout the revision. Georgia Lee Hadler shepherded the second edition through final editing and production with much appreciated hard labor and critical good sense.

June 1986

Bernard W. Pipkin
Donn S. Gorsline
Richard E. Casey
Douglas E. Hammond

INTRODUCTION

Our knowledge of the oceans is based on observations of the dimensions and characteristics of the ocean water, the materials of the ocean floor, and the life in the seas. All of this information must be classified and condensed into concepts from which theories can be devised about the origin of oceanographic features, their distribution in time and space, and the processes that alter them. In most of the exercises in this manual, the information and questions are based on actual data, collected from the world ocean by workers using a variety of sampling devices and sensors aboard research vessels.

The primary purpose of these exercises is to give you practice and experience in the manipulation, evaluation, and interpretation of ocean data. Each is designed to complement the lecture materials on a particular subject and to give you a graphic description of various marine features. For example, the exercise on sea-floor spreading, or plate tectonics, will demonstrate the enormity of the crustal components involved in this process and the dimensions and magnitude of the magnetic anomalies that form at the mid-ocean spreading centers under the periodically changing magnetic field of the earth. In the exercise on sea floor materials, you will note the distribution of sediment types, and you may even examine samples from the floor itself.

The exercises on marine organisms will give you some understanding of the principles underlying the classification of life forms, and of the factors that influence the activities of these many and varied forms. Other exercises are concerned with the chemistry of the oceans. Your instructor will select those that meet the needs of the particular course you are taking, but you may wish to look through all of them to supplement the lecture material or the discussions in your textbook. Note also that the discussion in each exercise concludes with a glossary of terms. Every term that is introduced in **boldface** type appears in the list of definitions for that exercise. Finally, the readings listed in the bibliography will be useful sources of additional background information.

The communication of ideas, dimensions, qualities, and quantities is one of the most difficult problems confronting modern man. In the sciences, the problem is complicated by the fact that those attempting to solve earth problems or to understand the universe must not only transmit information across national and linguistic boundaries but must do so clearly and accurately. To implement this communication, several scientific languages have been developed. The primary one is mathematics. Consequently, a major requirement in any science course is practice in the mathematical manipulation of data.

Although none of the operations in this manual are very demanding, you will learn to handle simple quantitative information. Another language is the system of symbols used to describe chemical reactions. In the exercises on seawater chemistry you will learn some of these forms of notation.

In conclusion, by the end of the term, you will have gained a broad understanding of the basic operations used by scientists and engineers to reduce observations of existing features, organisms, and phenomena to useful descriptions or classifications. But remember that all classifications and descriptions are the product of the human mind, and each as such presents only one view of our environment. They are useful tools to define problems more clearly so that we may search for answers more effectively. They are neither sacred nor immutable, and, after your work in the laboratory, you may think of better ways of translating observations into useful concepts. Your primary goal is to gain an organized, well-documented understanding of the oceans as well as a general impression of how scientific work is accomplished.

LABORATORY EXERCISES IN
OCEANOGRAPHY

EXERCISE 1

BATHYMETRY

We are all aware of the variety of landforms to be seen on earth, from colossal mountain ranges to deep canyons. However, many more forms, indeed the majority of geographic features on the entire earth, are covered by water. Although direct observation and measurement of these forms are impossible, what information we do have reveals that the most dramatic marvels on land are exceeded by others on the sea floor: there are, for instance, submarine mountain ranges longer and higher than the Rockies, and magnificent canyons into which the Grand Canyon would fit with room to spare.

BATHYMETRY

The term *bathymetry,* from the Greek roots *bathy-* meaning depth, and *-metry* meaning the science of measurement, is defined as the measuring and charting of the topography of the sea floor. At one time **soundings** (depth measurements) were accomplished by dropping a weighted line from a vessel, a task that often took several hours in deep water. Today a more practical electronic device, developed by the U.S. Navy in 1922, enables us to collect a sufficient number of accurately located depths, which in turn disclose details of sea-floor topography. The device, known as an *echo sounder*, or *fathometer,* records the time required for a sound pulse to travel from the ship to the sea floor and back again. Knowing the velocity of sound in seawater (about 4800 feet, or 1463 meters, per second), we can easily calculate depth if the travel time is known (Figure 1-1). In fact, the fathometer is simply a very accurate electronic timer that records the travel times of reflected sound pulses and marks these "echoes" on a graph, or *strip-chart recorder,* at the appropriate depth in feet, meters, or fathoms. The resulting display on the strip chart is known as an *echogram* (Figure 1-2).

In practice, soundings are recorded continuously by the instrument. Periodic notation of time of day, ship speed, and direction of travel are written by the investigator directly on the recording paper, or strip chart. Later the soundings are marked on a chart or base map, the position of each sound being determined by the navigator's chart and the notations on the strip chart. By this procedure a great number of reliable soundings from many different ships have been compiled to construct accurate charts for navigational purposes and scientific analysis. Once the soundings have been noted on the base map, **contours** (isobaths) connecting the points of equal depth can be drawn in. The shape and spacing of the contours will reveal the features of that area of the ocean floor.

A chart or map must also have a **scale.** This can be a calibrated bar that shows the relationship between the

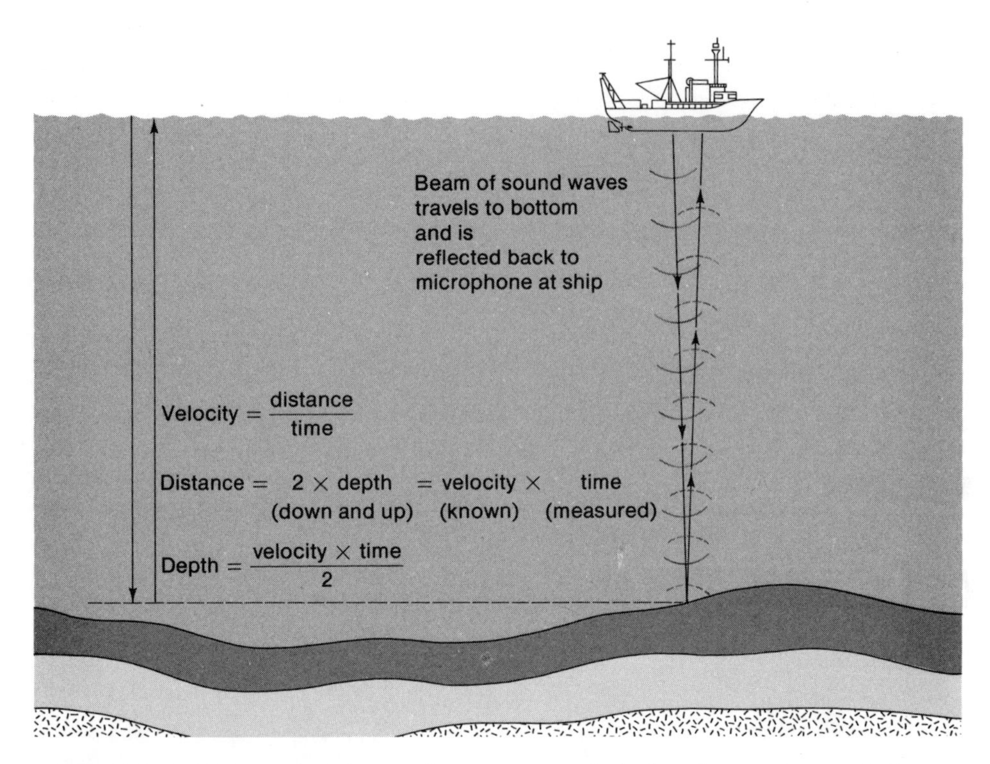

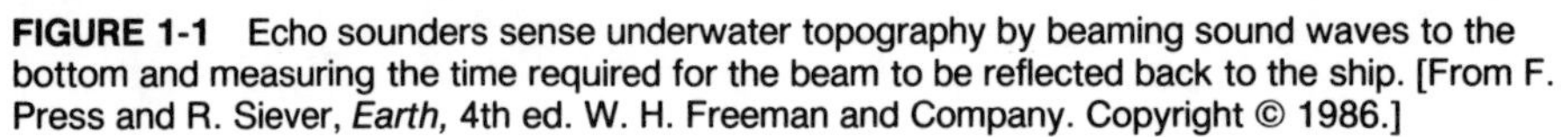

FIGURE 1-1 Echo sounders sense underwater topography by beaming sound waves to the bottom and measuring the time required for the beam to be reflected back to the ship. [From F. Press and R. Siever, *Earth,* 4th ed. W. H. Freeman and Company. Copyright © 1986.]

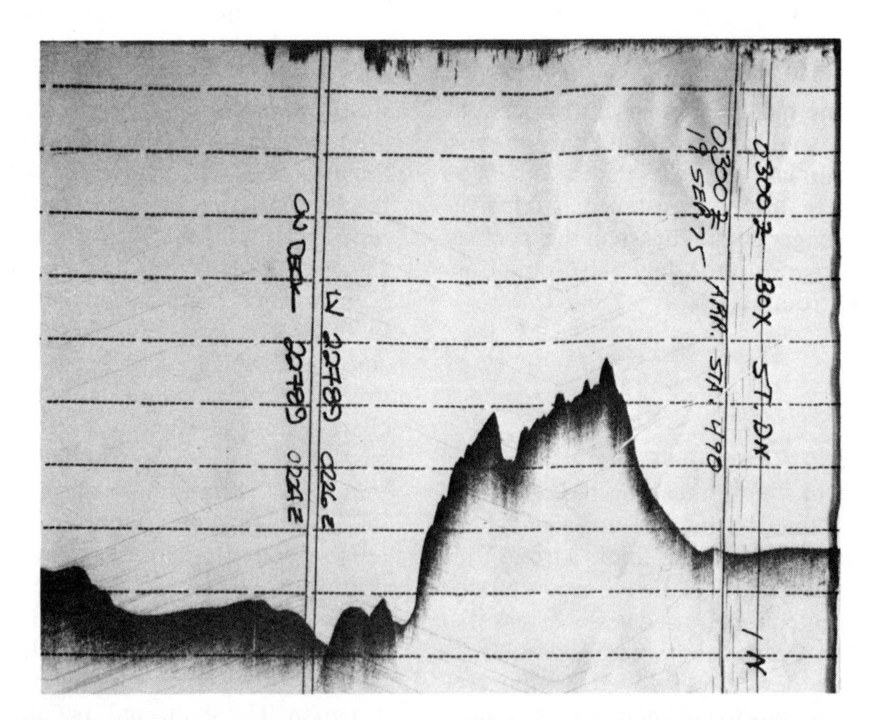

FIGURE 1-2 Echogram of Lasuen Knoll due west of Dana Point, California. Horizontal lines are depth intervals of 20 fathoms; vertical lines with written notations are locations at which bottom sampling was performed. The depth to the top of the seaknoll in this crossing is about 105 fathoms (630 feet), and the bottom materials at Station 490 were fine sand and rock (see Figure 1-5).

units of length used on the particular map and the actual units they represent, such as nautical miles or kilometers. The scale may also be a fractional one: for example, a scale of 1 : 500,000 means that one unit on the map (centimeters, millimeters, or inches) is equal to 500,000 of the same units. Thus 1 centimeter on the map is equal to 500,000 centimeters true distance on the earth, which is equal to 5000 meters or 5 kilometers.

THE CONSTRUCTION OF A BATHYMETRIC CHART

In constructing a bathymetric chart from plotted soundings, observe the following simple rules. Contour lines never cross one another. When they cross a canyon or deeply incised valley, they have a V shape, with the apex of the V pointing upstream. A particular contour line delineates areas shallower than the line from those that are deeper. If a contour is valid, it must separate depth zones completely. For example, if a 100-meter contour is drawn offshore to extend along a coastline, all depths of 100 meters must be on the line, and all the soundings shallower than 100 meters must be between the contour and the shore. If a sounding greater than 100 meters were found inside the area delineated by the contour, the position of the line would have to be adjusted to exclude the deeper area. The same principle applies to the location of increasingly deeper lines. The area between the 100- and 200-meter lines must not contain points less than 100 meters or greater than 200 meters deep. Close spacing between contours indicates a steep slope or an abrupt change in depth, whereas widely spaced contours represent a gentle slope or a gradual change in depth. Finally, a contour line closes upon itself. The contours of smaller features may close within the map area, so that the two ends are joined, whereas contours of larger features will close off the map area.

Figure 1-3a is a sketch of two submarine hills separated by a valley. The hill on the right is low and rounded, extending from 900 meters to 250 meters below sea level. The hill on the left side consists of two peaks with a steep east-facing slope and numerous gulleys on the lower slopes. Figure 1-3b is a bathymetric chart of the same features.

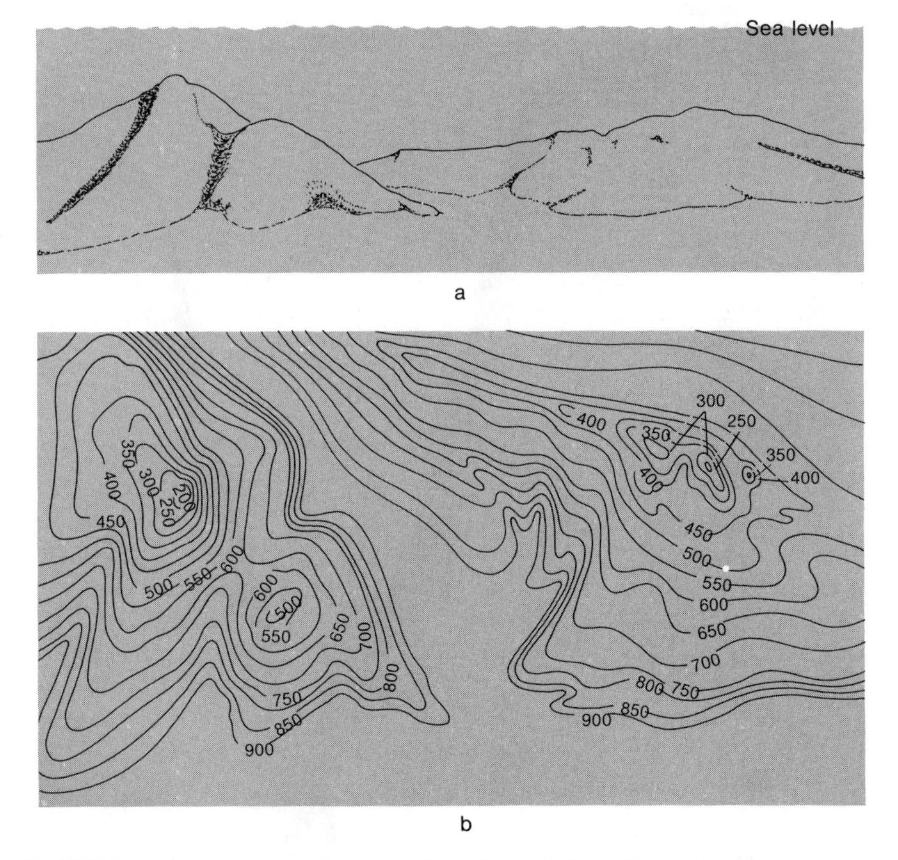

FIGURE 1-3 Representation of an area of sea floor with two hills separated by a valley. Part (a) is a perspective sketch; part (b) is a bathymetry chart in which the features in (a) are represented by contour lines. Depths noted along the contours are in meters.

The most accurate device for measuring and charting submarine topography is a precision depth recorder (PDR). It is similar to a fathometer in construction and placement aboard ship except that the timing of sound pulses is precisely controlled by use of a tuning fork within the instrument. Instruments with timing accuracies of better than one part in 3000 have been developed in the United States and Europe. Such precision is not required for routine navigation purposes but is mandatory for accurate scientific work. Figure 1-4 is part of a PDR profile of the Redondo Submarine Canyon. Vertical lines, such as the two shown on the figure, are marked every 15 minutes as well as whenever there is a change in the ship's course. On this recording the navigator on the bridge fixed the ship's position at 1200 hours and 1209 hours so that the scientist could accurately locate the axis of the submarine canyon, noted at 1203 hours, on a marine chart. Also shown are the ship's course (C 175°) and its speed of 7 knots (S7K). The regularly spaced horizontal lines are 10-fathom (60-foot) depth intervals and they indicate that the north edge of the canyon is at a depth of about 330 feet. This portion of the canyon was recorded on a 0- to -200 fathoms scale; however, as the vessel proceeds farther offshore, where the canyon progressively deepens, the scale must be changed to 200–400 fathoms, 400–600 fathoms, and so on, to accommodate the greater depth.

TOPOGRAPHIC CROSS SECTIONS

A bathymetric chart shows a visualization of the topography of the sea floor as viewed from above. We can derive topography from a contour map or nautical chart by constructing a topographic profile along a preselected line or course. The profile is like a graph

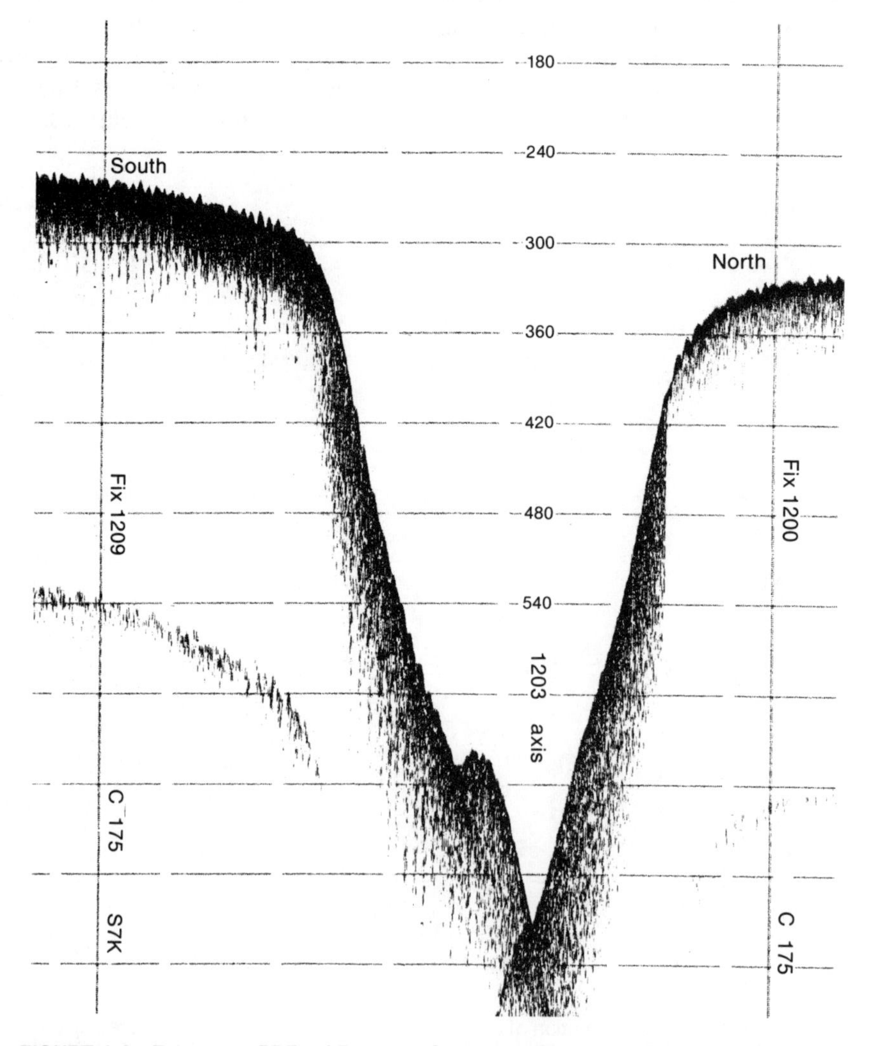

FIGURE 1-4 Echogram (PDR) of Redondo Submarine Canyon off Redondo Beach, California. Horizontal lines at 10-fathom (60-foot) intervals.

with sea-floor depths plotted vertically and distance plotted horizontally. The echograms of Figures 1-2 and 1-4 are topographic profiles on which the data have been plotted electronically.

Commonly, in order to show surface topography in its true relationship to horizontal distance, a profile has the same vertical and horizontal scale. At sea, however, distances are so great relative to **relief** that we must exaggerate the vertical scale. That is, 1 kilometer on the vertical scale of the graph may be five times the length of 1 kilometer on the horizontal scale. In this case the vertical exaggeration (V. E.) is said to be five times (5×). To determine the vertical exaggeration, divide the distance represented by one unit on the horizontal scale by the distance represented by one unit on the vertical grid. For instance, if the horizontal scale is 1:250,000 and the vertical scale is 1:10,000, the vertical exaggeration is

$$\text{V. E.} = \frac{250{,}000 \text{ (distance of one horizontal unit)}}{10{,}000 \text{ (distance of one vertical unit)}}$$
$$= 25\times$$

To construct a cross section, choose a profile line (in this exercise the profile lines are those in Figure 1-2 or Figure 1-4) and lay a piece of paper along that line. Mark the ends of the profile line on the paper, and wherever an isobath intersects the paper, make a pencil tick and write in the depth of that contour. When this step is completed, lay the paper along the base of a graph or grid that is scaled off vertically for depth and horizontally for distance, and make a dot at the depth on the vertical scale that corresponds to each horizontal distance. Connect the points with a smooth curve to complete the profile. If you are exaggerating the vertical scale, indicate this fact. Although in this exercise the lines of section and profile graph are provided, you are not restricted by them. Should you prefer, you may make a profile of any feature shown on either chart with a pencil and graph paper as described.

DEFINITIONS

Contour. A line on a chart, map, or section that connects equal values of a given dimension. In this exercise the contour lines connect points of equal depth below sea level. In marine studies contour lines are normally called *isobaths* meaning "equal depth." A *contour interval* is the vertical distance between adjacent contour lines.

Relief. Vertical distance between the highest and lowest points in an area. High relief means generally rugged topography; low relief describes flat or monotonous terrain.

Scale. A calibrated line or bar that shows the relationship between the units of length used on a particular chart or map and the actual units they represent. The scale may also be a fractional one: for example, a scale of 1:500,000 means that one unit on the map (centimeters, millimeters, or inches) is equal to 500,000 of the same units on the earth.

Sounding. A measurement of water depth.

EXERCISE 1

REPORT

BATHYMETRY

NAME

DATE

INSTRUCTOR

1. Figure 1-5 (on page 8) is part of National Oceanic and Atmospheric Administration (NOAA) Chart 18746. It has been reduced in size by half from a scale of 1:80,000 to 1:160,000. Examine the chart and its soundings and give a brief description of the topography of the sea floor off Newport Beach, California (relief, shelf width, any canyons, and so forth).

2. Contour the soundings in Figure 1-5 using a 50-fathom contour interval. Then draw in the 10- and 20-fathom isobaths to give more detail. Compare the contoured bathymetry with the soundings alone. Which conveys the better picture of the sea floor?

3. On the graph paper in Figure 1-6 (on page 9), construct cross sections along the two profile (AB and XYZ) lines. The upper graph is vertically exaggerated and the lower graph is true scale with no vertical exaggeration. Draw both profiles at the two different vertical scales.

   (a) What is the vertical scale of the upper profile grid?

   (b) Briefly describe and compare the true and exaggerated shapes of the topography off Newport Beach, California.

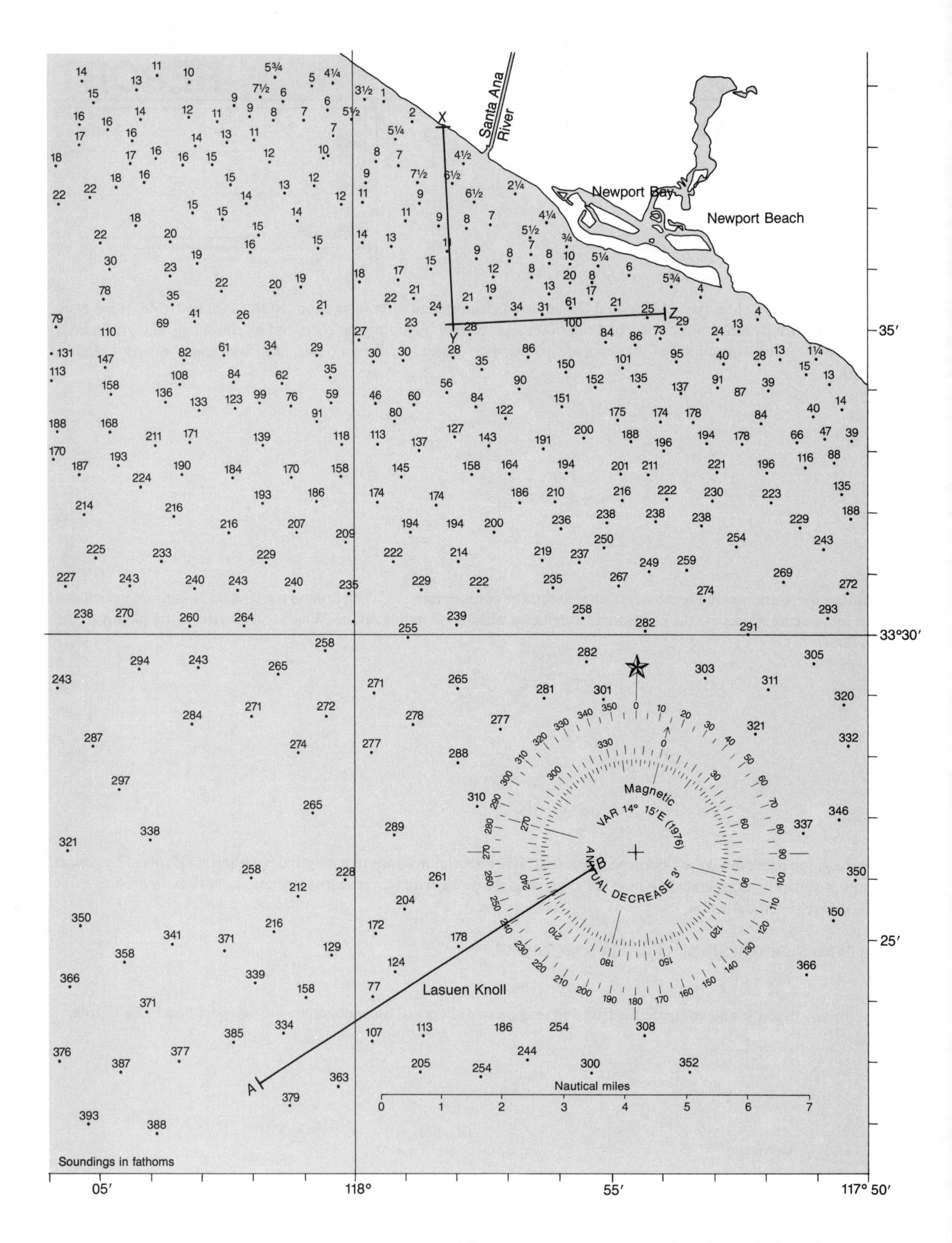

FIGURE 1-5 National Oceanic and Atmospheric Administration Chart 18746, San Pedro Channel, California.

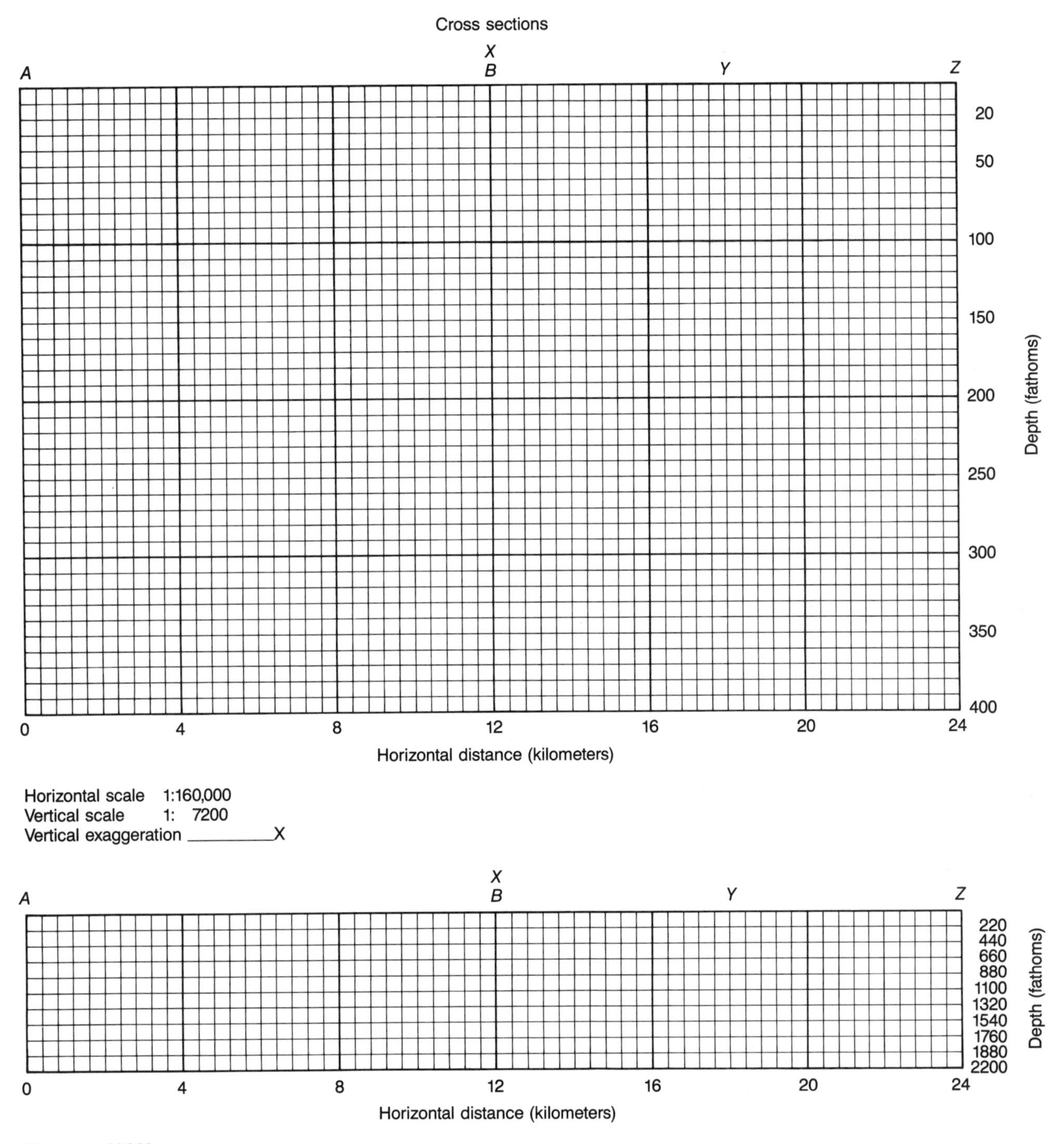

Scale 1:160,000
No vertical exaggeration

FIGURE 1-6

4. Compare the echogram in Figure 1-2 with the profile you drew of Lasuen Knoll (Figure 1-6). Sketch the ship's course on your map in pencil. Note that the depths at the ends of the track are 170 and 195 fathoms and the length of the track is about 3 nautical miles. The track across the hill need not be straight but might consist of several straight segments with turns at the ends.

5. Dredging of the top of Lasuen Knoll found sedimentary rock about 15 million years old (Miocene age). What might be the geologic origin of Lasuen Knoll?

6. The following questions are concerned with the precision-depth-recorder trace of the Redondo Submarine Canyon. To answer them, refer to Figure 1-4 and the text accompanying it.

(a) What is the depth of the canyon?

_______________ feet.

(b) What is the width of the canyon between the notation for 1200 hours and that for 1209 hours?

_______________ nautical miles.

(c) What is the vertical exaggeration of the PDR trace (vertical scale about 1 : 1500, horizontal scale about 1 : 21,000).

(d) Depth *(D)* is equal to the velocity *(V)* of sound in seawater, multiplied by the travel time *(T)* of the sound pulse to the bottom and back divided by 2. The velocity of sound in seawater is generally taken as 4800 feet per second. Using the equation

$$D = V \times \frac{T}{2}$$

calculate the travel time for a sound pulse from the ship to the bottom of the canyon and back again.

_______________ seconds.

(e) What do you suppose is the origin of the faint traces that occur between depths of 90 and 110 fathoms on the south side, and between 100 and 120 fathoms on the north side, of the canyon proper?

EXERCISE 2

ACOUSTIC SEISMIC PROFILING

In the 1940s it was realized that sound could not only be reflected from the sea floor but could also penetrate bottom materials and be reflected from rock or sediment layers hundreds of feet below the floor. Commercial instruments were developed at once, and more than a dozen instrument systems are now available. With the aid of these various seismic profilers, the marine scientist can follow the progress of sound, or *seismic,* waves to determine something about the subsurface structure of the sea floor, an investigation that previously required more costly geophysical techniques or drilling.

SEISMIC-REFLECTION PROFILING

The basic principle of operation for the seismic profiler is the same as for an echo sounder (described in Exercise 1): that is, sound waves are emitted by a sound source, and the return signals are recorded on the receiver. However, in an echo sounder the energy source is installed in the ship's hull, but for seismic-reflection profiling the sound source and receiver must be towed below the surface, a depth less than the true one is usually recorded. (See Figure 2-1.)

A sample record of the **reflections,** or return signals, from this type of array is shown in Figure 2-2, and the travel paths of the sound waves producing the reflections are diagrammed in Figure 2-3. The circled numbers 1–4, designating the reflections and **multiples** in the photograph, correspond to the numbers for each travel path in the diagram. The reflection marked 1 records a wave traveling from the source to the water surface to the receiver, and reflection 2 records one traveling from the source to the bottom to the receiver. Reflection 3 bounces off the bottom and returns to the surface to be reflected to the receiver. Finally, reflection 4 represents a multiple echo, since

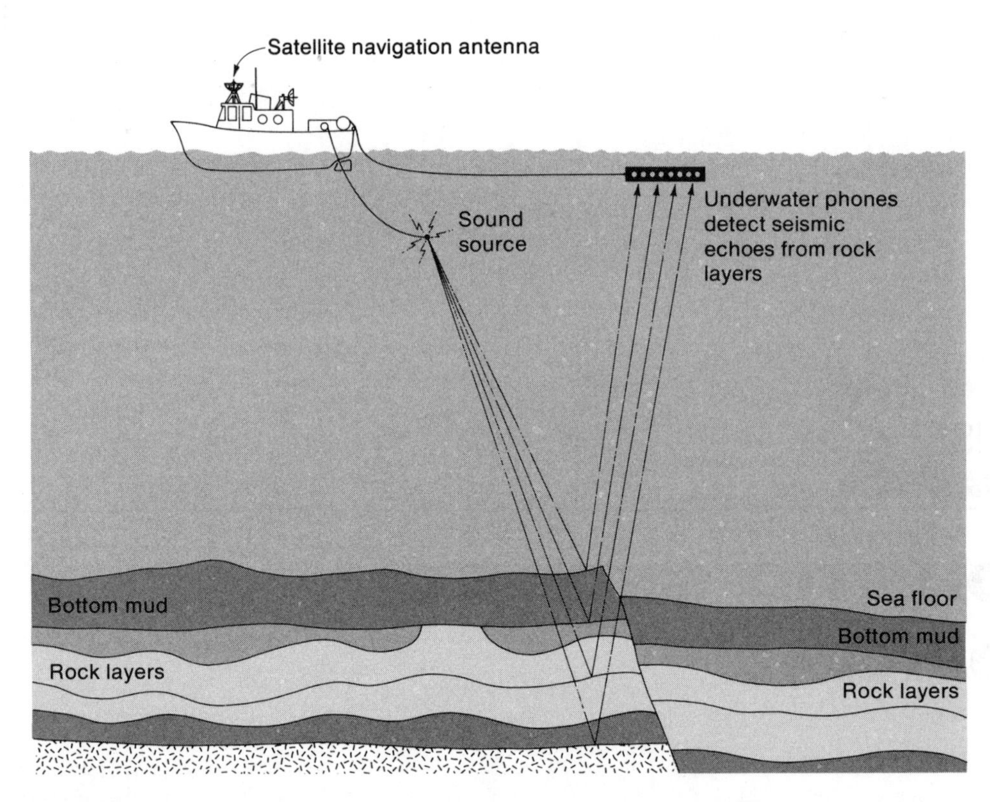

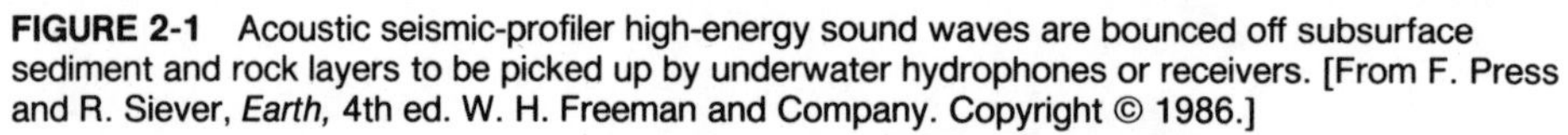
FIGURE 2-1 Acoustic seismic-profiler high-energy sound waves are bounced off subsurface sediment and rock layers to be picked up by underwater hydrophones or receivers. [From F. Press and R. Siever, *Earth,* 4th ed. W. H. Freeman and Company. Copyright © 1986.]

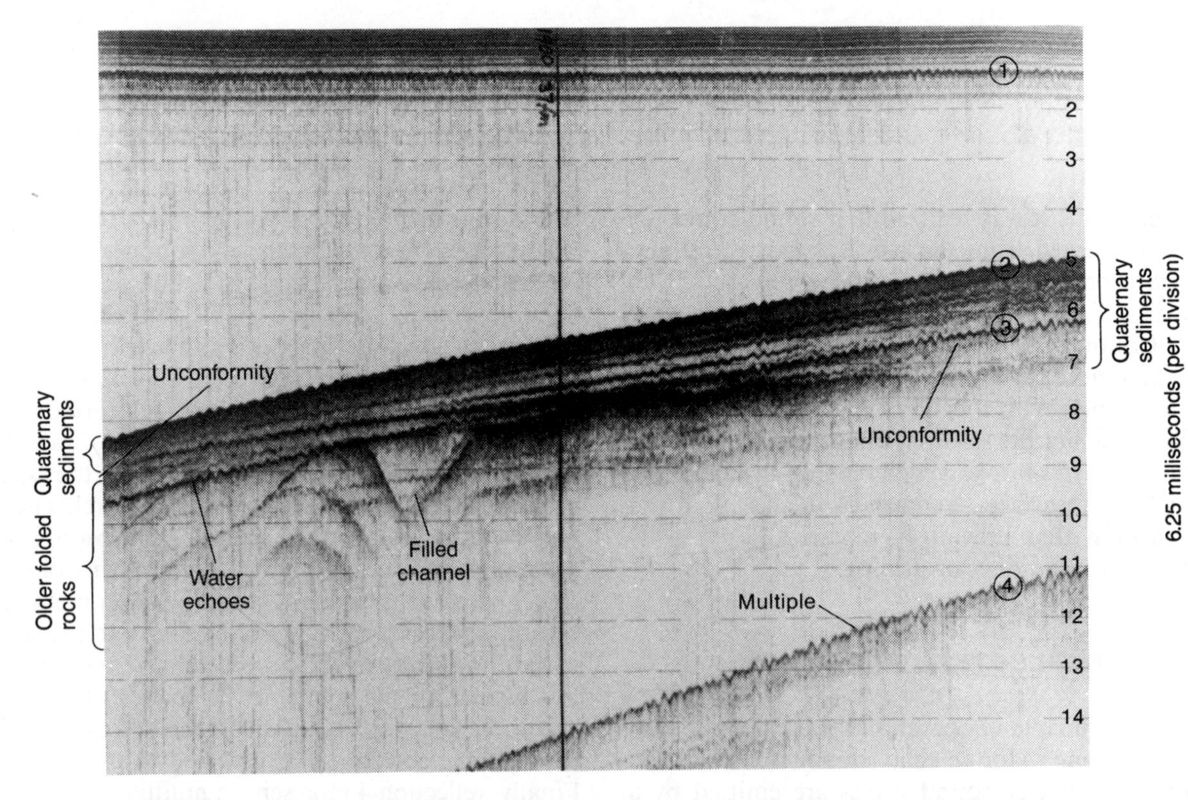

FIGURE 2-2 High-resolution seismic profile of shelf area off San Pedro Bay, California. The circled numerals 1–4 label echoes for the sound-ray paths down schematically in Figure 2-3. [Courtesy Tom Nardin, University of Southern California.]

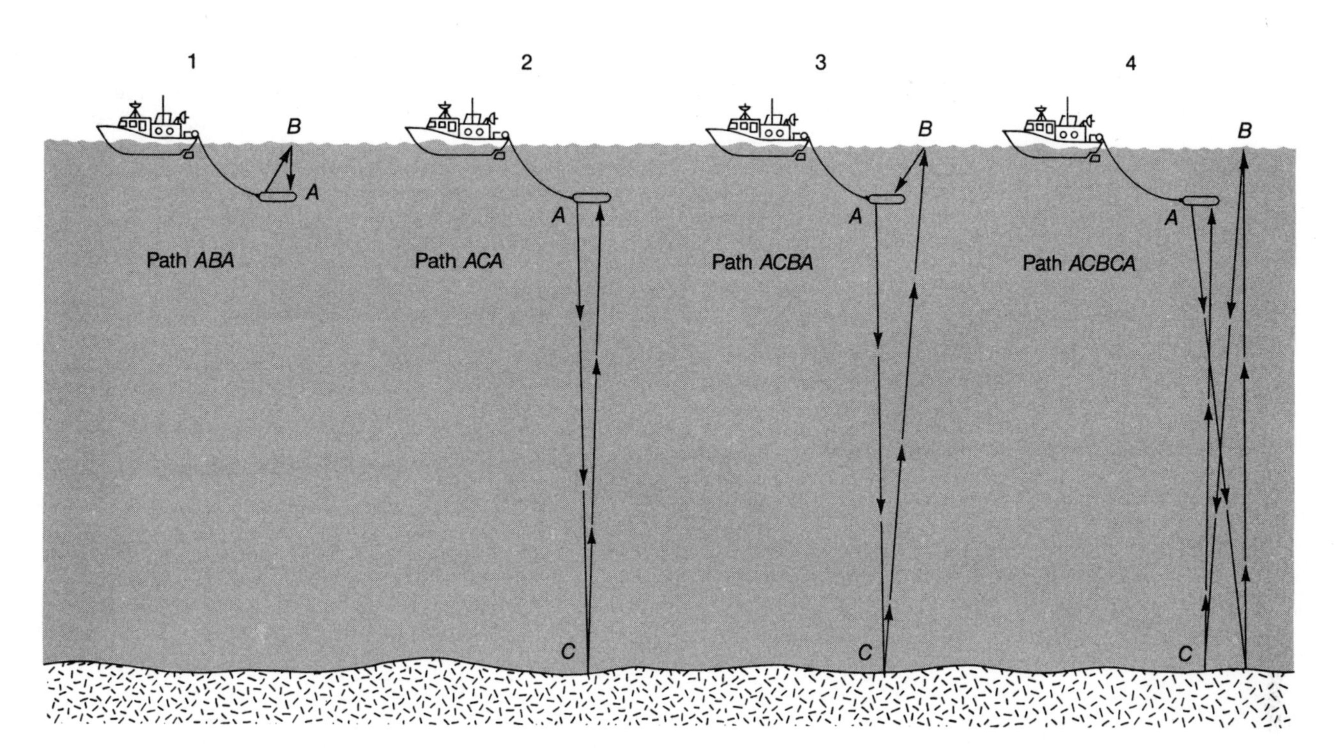

FIGURE 2-3 Schematic representation of a research vessel with an acoustic seismic profiling system in tow below the sea surface. The sound-ray paths shown are for the reflections recorded in Figure 2-2. The letter *A* refers to the sound source and receiver, *B* to the sea surface, and *C* to the surface of the ocean floor. Although the reflections shown are real, the angular relationships are distorted for purposes of clarity.

the pulse travels from the source to the bottom, then to the surface to be reflected back to the bottom, and then to the receiver. Here the sound pulse makes two round trips to the bottom before being intercepted by the receiver: thus the term *multiple.* The true depth of the water is found by subtracting the apparent depth (case 2) from the first multiple (case 4). The most accurate method of determining the depth of the sled carrying the equipment is to subtract the apparent depth (case 2) from the strong reflection represented by 3 (case 3).

HIGH RESOLUTION VERSUS DEEP PENETRATION

The **resolution,** the degree to which details are defined on a record, depends on the **frequency** of the sound waves. It is impossible to obtain high resolution (that is, to discriminate thin sedimentary layers from one another) and deep penetration simultaneously. One reason is that the high-frequency waves needed to obtain high resolution lose their energy much faster as they travel through water and rock than do low-frequency waves. Some seismic-profiler systems are high-frequency systems, which can record layers 15 centimeters thick to depths of 30 meters. Others, among the low-frequency systems, can "see" to depths of 5000 meters but can resolve only layers thicker than 30 meters. Figure 2-4 is a low-frequency seismic-reflection profile across the Aleutian Trench east of Amchitka Island. Note the great depth of penetration and the good reflections from the sedimentary layers that fill the trench and the Kanaga Basin. The entire section is about 120 kilometers across from the Aleutian Ridge to the trench.

An interesting but quite accidental discovery made with high-resolution, or high-frequency, equipment was the recording of rising oil and gas bubbles over oil seeps. Figure 2-5 shows the record of such seeps that occur on the sea floor off Santa Barbara, California. Note that the seep occurs along the axis of an anticline (upwarp), which is the type of geologic structure in which hydrocarbons are trapped in sedimentary rocks. The apparent slopes, or dips, of the sedimentary layers are shown inclined away from the anticlinal axis of the fold: thus the term *anticline.* This structure is also seen on the left-hand side of Figure 2-2.

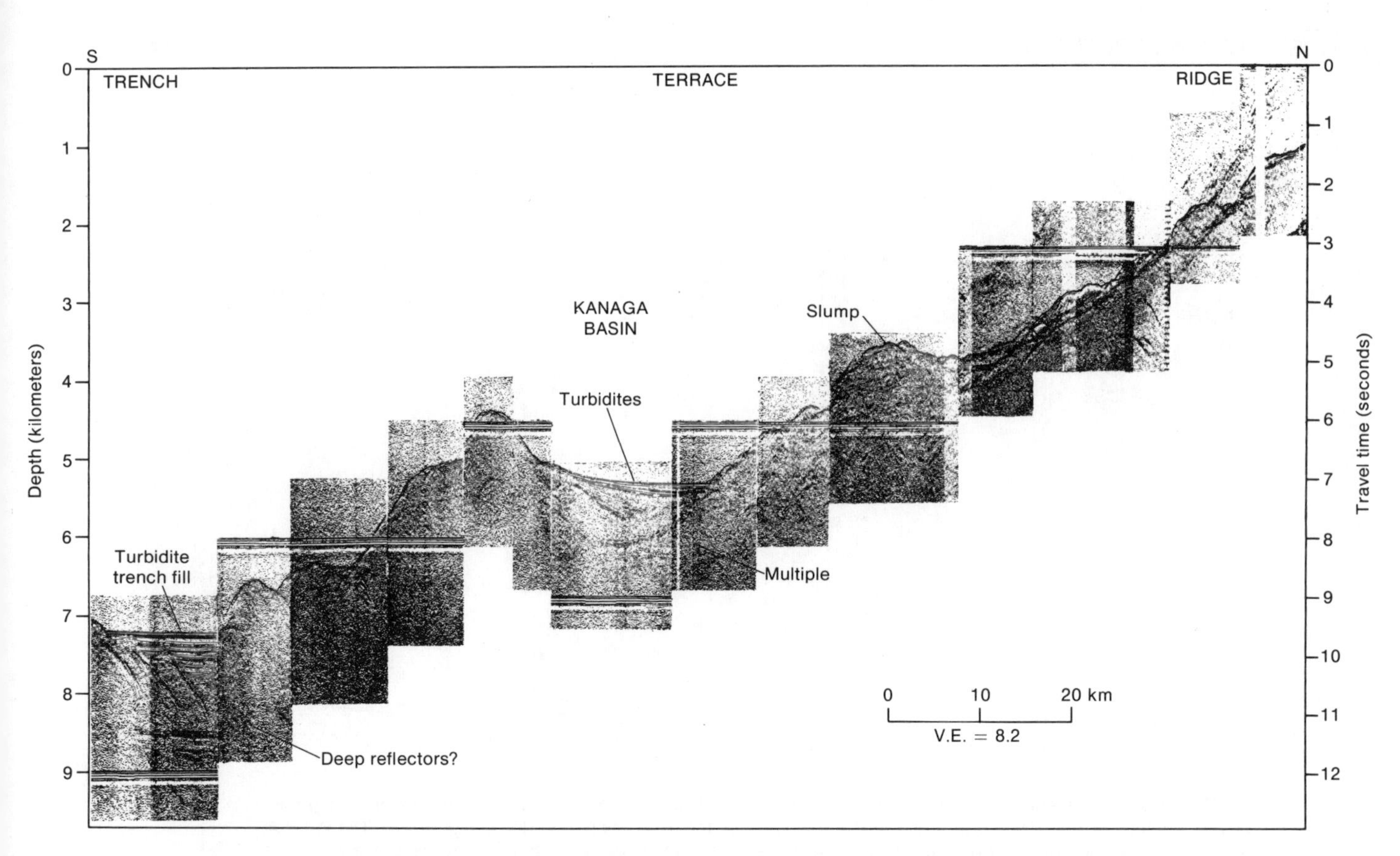

FIGURE 2-4 A low-frequency seismic-reflection profile of deep ocean-basin sediments in the Aleutian Trench. [From Michael S. Marlow et al., "Tectonic History of the Central Aleutian Arc," *Geological Society of American Bulletin,* Vol. 84, No. 5, 1973, pp. 1555–1574.]

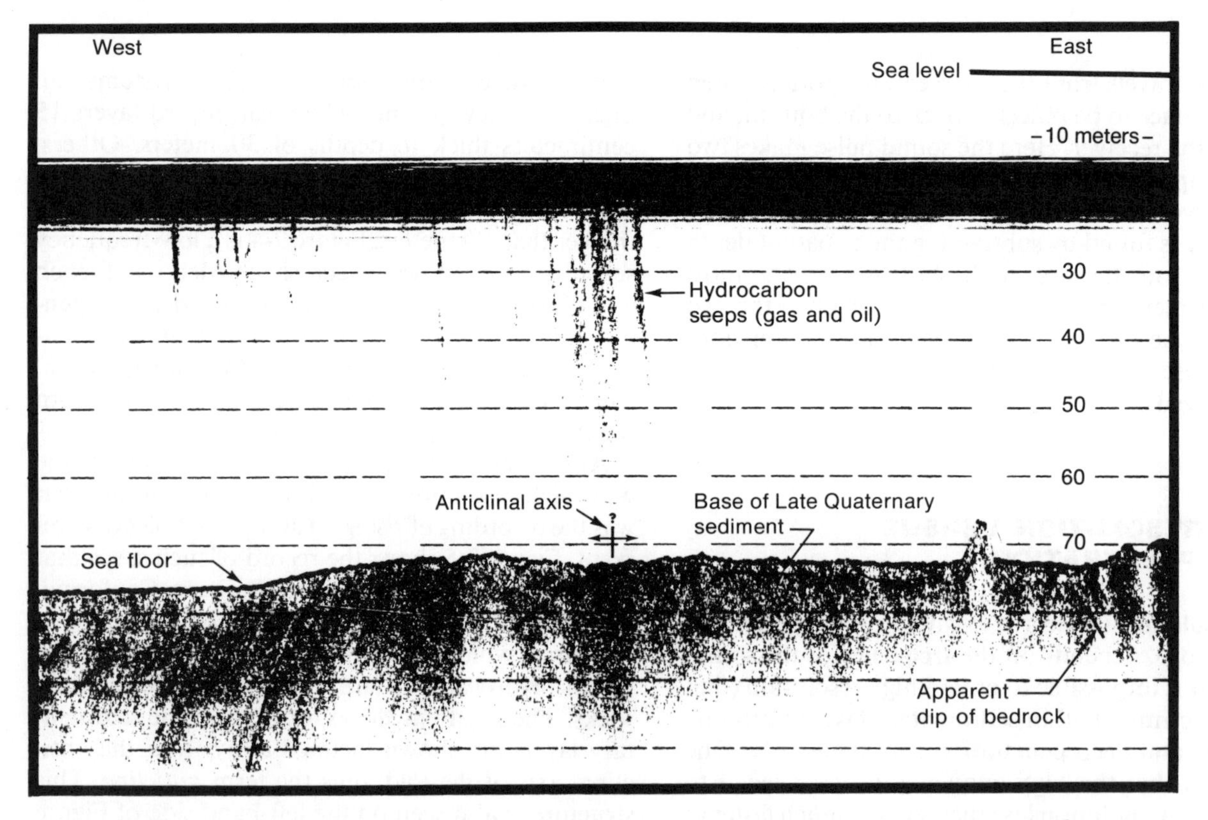

FIGURE 2-5 Profile from high-resolution subbottom record showing natural hydrocarbon seeps in the Coal Oil Point area, southern California. [Courtesy Peter Fischer.]

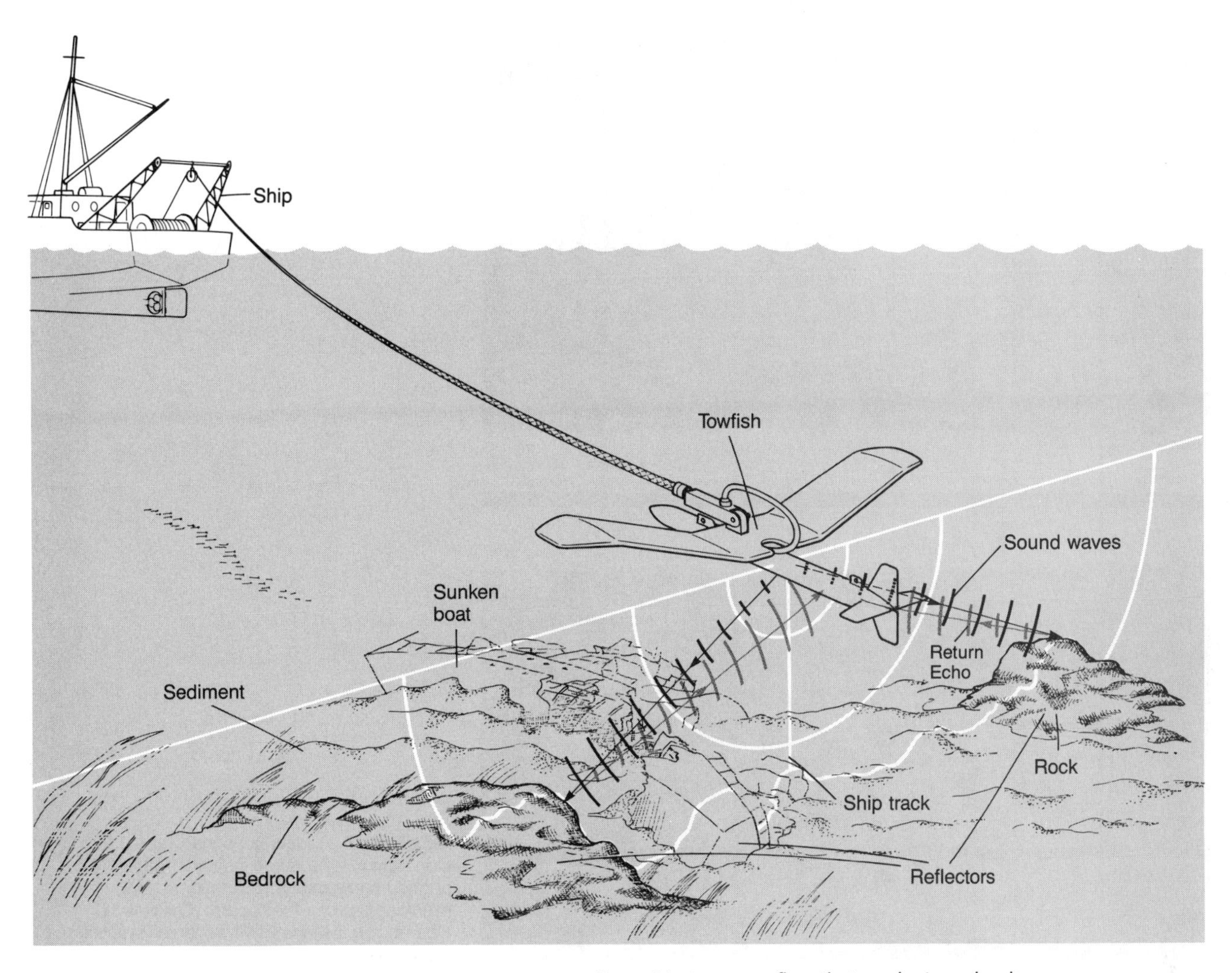

FIGURE 2-6 Typical side-scan sonar system. Note objects on sea floor that send return signals, or "echoes," back to the towfish.

SIDE-SCAN SONAR

Side-looking sonar devices contain electronic circuitry that creates high-intensity, high-frequency bursts of acoustic energy in fan-shaped beams on both sides of the instrument (towfish) that is towed behind the ship. The beams are narrow in the horizontal plane and wide in the vertical plane and thus project onto a wide area of the sea floor (Figure 2-6). Objects on the sea floor, such as rock outcrops, pipelines, or sunken vessels, produce reflections that are received by transducers in the towfish. These "echoes" are amplified and sent to the shipboard recorder where they are processed and displayed on a special multichannel writing mechanism. The echoes are placed side by side on the chart so that a coherent "picture" of the sea floor is obtained (Figure 2-7). This device has proved most useful in engineering site studies for submarine pipelines, piers, offshore oil rigs, and the like.

Figure 2-7, a side-scan sonar record of part of the continental shelf between Point Conception and Point Arguello, California, shows the upturned edges of sedimentary rocks, some of which appear to be folded, trending across the ship's track. The straight lines across the chart represent 150-meter horizontal distances. The slant range of the instrument was 200 meters, which means we are looking at slightly less than 200 meters of the sea floor on each side of the ship's track (blank middle path on the record).

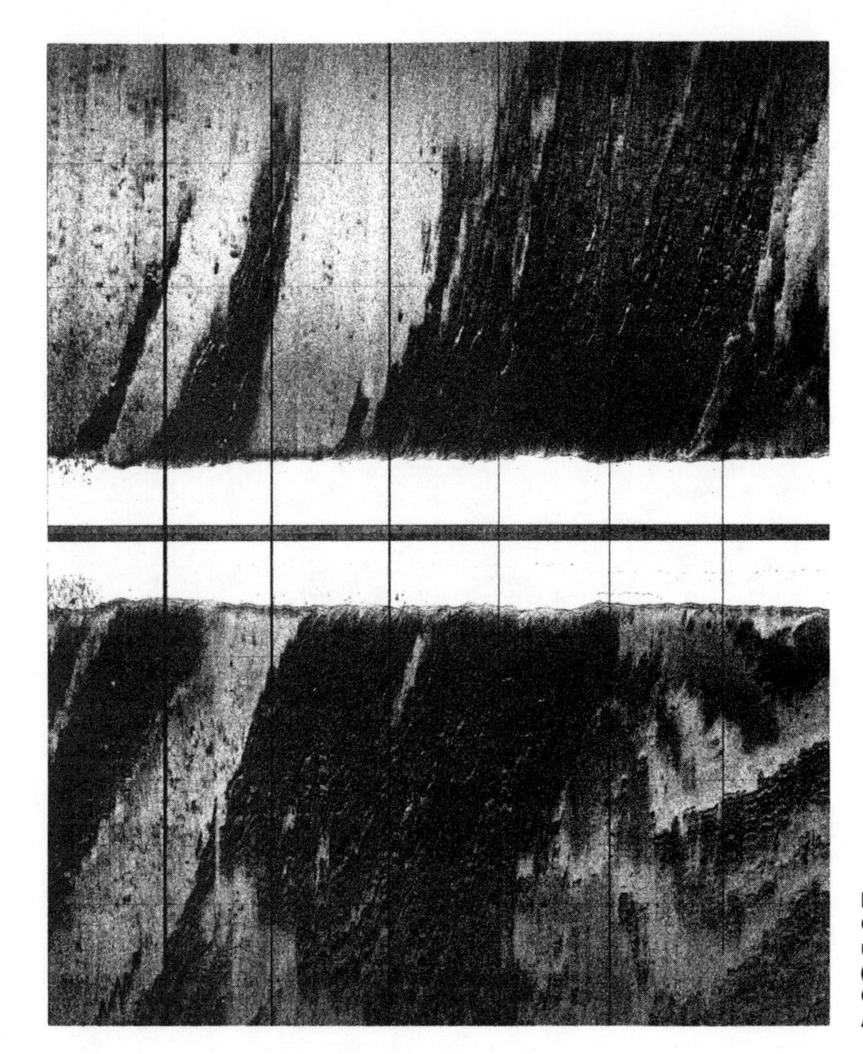

FIGURE 2-7 Side-scan sonargram of Point Conception area taken in July 1983. Note uptilted sedimentary rock layer striking (trending) across the record. [Courtesy C. Chamberlin, Marine Environmental Science Associates.]

DEFINITIONS

Frequency. The number of complete waves that pass a given point in a second, or the number of complete vibrations per second. Short sound waves (high-frequency waves) yield rapid vibrations, whereas the vibration of long sound waves (low-frequency waves) is much slower. High-frequency waves give high-resolution records. Low-frequency waves give poorer resolution but much deeper penetration of the sea floor.

Multiple. A sound wave or pulse that follows a path from the source to the bottom, then to the sea surface, to the bottom again, and back to the receiver. The second round trip from the surface to the bottom and back is called the *first multiple* or *second echo.* As many as three or four multiples may be recorded if the energy source is large enough and the bottom acts as a strong reflector. Multiples may be recognized on the strip-chart record by the fact that each succeeding multiple or echo is displaced downward a distance equal to the water depth (see Figure 1-4).

Reflection. The return or bouncing off (the echo) of a wave from a surface back into its original medium. An example would be the reflection of sound waves that have traveled through seawater from a ship to the ocean bottom and back to the surface again. The same phenomenon occurs when sound waves travel through rock or sediment layers below the sea floor. Reflections are produced primarily by density differences between layers.

Reflector. A surface, usually a rock or sediment layer, that strongly reflects seismic (sound) waves.

Resolution. The degree to which details are defined on a record. High resolution means that fine detail is clearly shown. Resolution is dependent on the frequency of the sound.

Velocity of sound. Ranges from 1450 to 1570 meters per second in seawater. Velocity increases with depth at a rate of 1.7 meters per second per 100 meters; it increases with temperature at a rate of about 4.5 meters per second per degree Celsius; and it increases with salinity at a rate of 1.3 meters per second per part-per-thousand.

EXERCISE 2

REPORT

ACOUSTIC SEISMIC PROFILING

NAME ______________________

DATE ______________________

INSTRUCTOR ______________________

1. Figure 2-8 is a duplicate of Figure 2-2. Each dashed horizontal line on the chart represents a one-way travel time of sound energy of 6.25 milliseconds (0.00625 second). To determine the distance represented by one division, simply multiply the velocity of sound in the medium—seawater or sediment—by 6.25 milliseconds. For seawater each division is 0.00625 second × 4800 feet pet second = 30 feet. Water depths are commonly displayed in fathoms (1 fathom = 6 feet); therefore each division is 5 fathoms of water depth. The same reasoning holds for sediment thickness.

   Fill in the water depths and depths of penetration into sediments on the lines provided at the right-hand side of Figure 2-8. Use 1463 meters per second (4800 feet per second) for sound velocity in seawater, and 1830 meters per second (6000 feet per second) for velocity in sediments and rocks. Depth, D, is equal to velocity, V, multiplied by travel time, T:

$$D = V \times T$$

2. What is the true water depth at the mark near the center of the record in Figure 2-8?

   ______________ fathoms. ______________ feet.

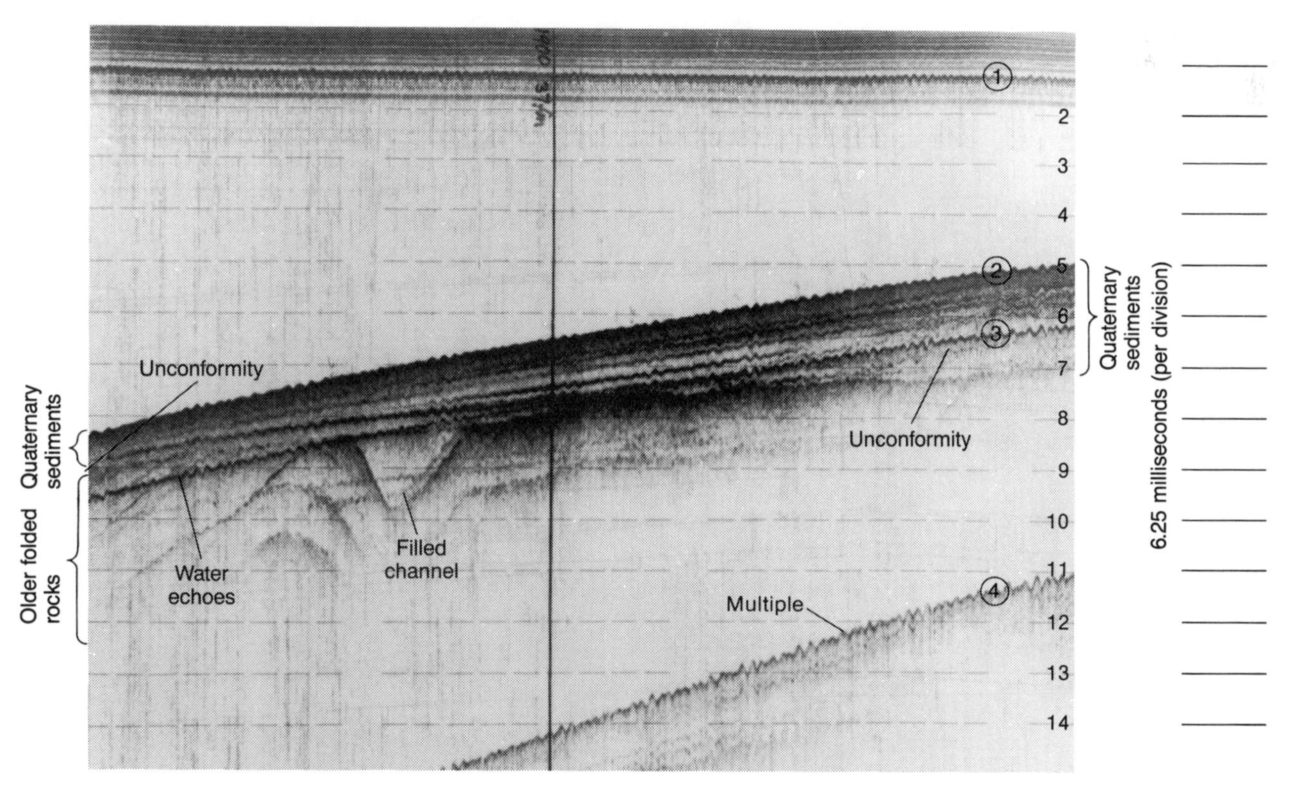

FIGURE 2-8 High-resolution seismic profile of shelf area.

3. At what depth were the energy source and receiver being towed (configuration 1 of Figure 2-3)?

_______________ fathoms.

4. The seismic profile in Figure 2-8 shows older folded sedimentary rocks covered by much younger, flat-lying sediments. An erosion surface (unconformity) truncates the folded beds and represents a time gap in the geologic record during which the older rocks were folded and eroded prior to deposition of the younger deposits. What is the average thickness of the younger deposits overlying bedrock?

_______________ meters.

Assuming a rate of deposition of the young sediments of 40 centimeters per 10,000 years (a very fast rate), how many years did it take to lay down the young, flat-lying layers?

_______________ years.

During what epoch of geologic time were these sediments deposited (see Appendix B)?

5. In the side-scan sonar record of Figure 2-7, why are there smooth areas within and around the rocky outcrops?

6. What is the length of the side-scan record?

_______________ meters.

7. If the ship's course is 270 degrees true (due west), what is the approximate trend (strike) of the outcropping sedimentary layers?

EXERCISE 3

MARINE CHARTS AND NAVIGATION

Since oceanographers rely on surface vessels for transportation, they use marine charts as base maps upon which they can locate their position and plot their data. For this reason it is worth our while to learn some of the basic principles of navigation and seamanship. Navigation has evolved from an art into a science in the course of its 6000-year history. Now, as in the past, the basic tools are the chart, the compass, and a method of determining position. The most critical requirement of all marine work may be precise positioning, because data reported from generalized positions are virtually useless to others wishing to follow up on previous work. This precision would be particularly important in the mining of hard minerals off the sea floor or in offshore oil exploration.

THE COORDINATES

The coordinates of **latitude** and **longitude** are essential in navigation. The lines of latitude are also called parallels of latitude, because they are parallel to the equator and to each other. Measured in degrees of arc along a circle, they specify the angular distance north or south of the equator, from 0° at the equator to 90° at either pole. Each degree is divided into 60 minutes of arc (1° = 60′) and each minute into 60 seconds (1′ = 60″). Latitude is recorded with its hemisphere notation, north or south; for example, that of Seattle, Washington, is La 47°36′N (Figure 3-1). Lines of longitude, or the meridians, are also expressed in degrees

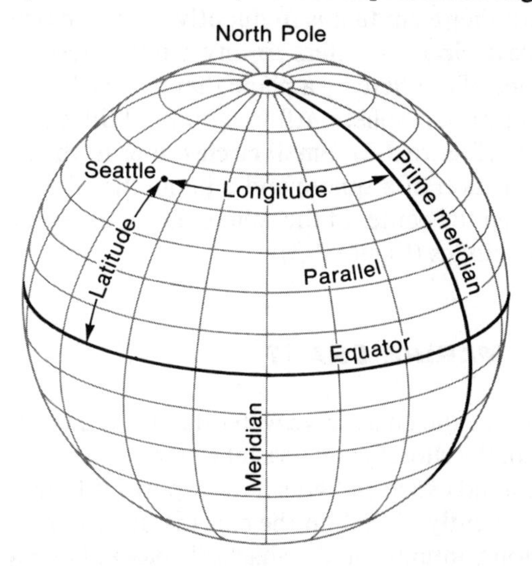

FIGURE 3-1 Meridian and parallel coordinate system for the earth. The prime meridian is 0° longitude, and the equator is 0° latitude. [After Nathaniel Bowditch, *American Practical Navigator*, Hydrographic Office Publication No. 9, U.S. Naval Oceanographic Office, 1966.]

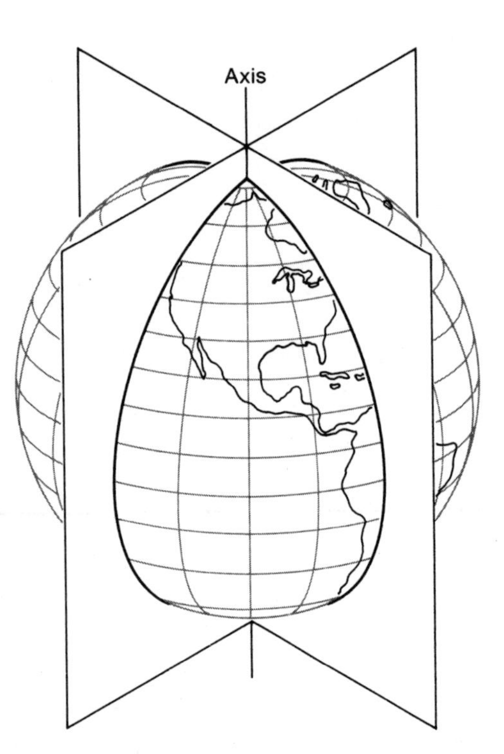

FIGURE 3-2 The planes of meridians (lines of longitude) meet at the polar axis. [After Nathaniel Bowditch, *American Practical Navigator,* Hydrographic Office Publication No. 9, U.S. Naval Oceanographic Office, 1966.]

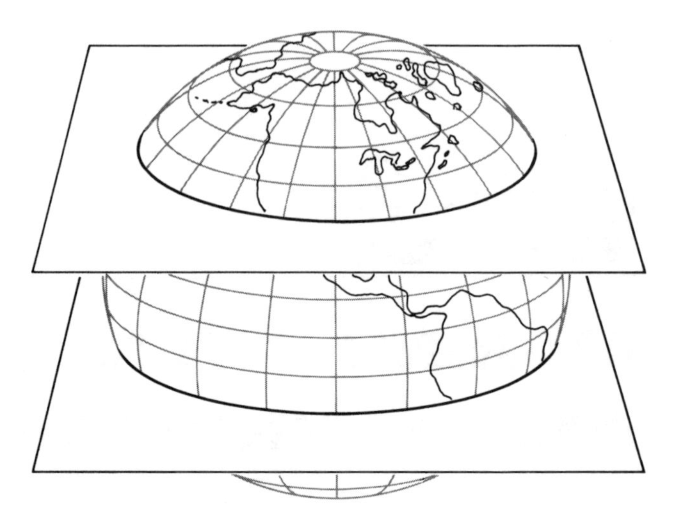

FIGURE 3-3 Parallels of latitude are parallel to the equator and are small circles. The equator is a great circle perpendicular to meridional great circles. [After Nathaniel Bowditch, *American Practical Navigator.* Hydrographic Office Publication No. 9, U.S. Naval Oceanographic Office, 1966.]

and refer to the angular distance on the earth measured from the prime meridian (0°) at Greenwich, England, east or west through 180°. Longitude too should be reported with its hemisphere notations, east or west: for example Seattle, Lo 122°20′W.

Another term that is frequently used in navigation is **great circle.** It refers to any circle traced on the surface of a sphere by a plane that passes through the center of that sphere. All longitudinal lines are great circles (Figure 3-2). **Small circles** refer to the lines of intersection of a sphere and a plane that do *not* pass through the center of the sphere. Lines of latitude are small circles (Figure 3-3).

THE MARINE CHART

Most marine charts give water depths, the configuration of the shoreline in coastal waters, and other navigation aids such as lights and important landmarks. Prominently placed on the chart is the general information contained in the chart title block. Here will be found the name identifying the waters covered by the chart, the units of depth (feet, fathoms, or meters), and the datum plane, or sea-level reference, for the depth measurements, or soundings (Figure 3-4). In most charts, north is the top of the sheet, latitude scales are given on the sides, and longitude scales are at the top and bottom. Meridians and parallels are drawn at given intervals in fine black lines across the chart. The nature of the bottom is specified as hard (hrd), rocky (rky), gravel (g), shells (sh), sand (s), or coral (co).

The coordinates latitude and longitude are used to locate, or "fix," a vessel's position on a marine chart (Figure 3-1). To do so, first find the desired latitude on the scales at either side of the chart and connect the two points. Now locate the desired longitude at the top and bottom of the chart and connect them. Where the lines cross will be the designated position. Note that many charts cover an area of less than 1° of latitude or longitude, so that the scales on the sides of the charts will be in minutes or seconds rather than degrees—an important point to remember when plotting positions.

UNITS OF DISTANCE AND SPEED

On land, distances are expressed in kilometers or **statute miles,** whereas at sea they are in **nautical miles.** The nautical mile is equal to 1′ of latitude. A nautical

FIGURE 3-4 A chart title block.

mile is about 1.15 statute miles. Thus 1° of latitude is 60 nautical miles, or $60 \times 1.15 = 69$ statute miles. Most charts have a bar scale that shows distance in nautical miles or yards. To determine the distance of a given length on the chart, take an ordinary drafting compass or a pair of dividers and spread the points so that they touch the ends of the length to be converted. Transfer the compass points directly to the bar scale to obtain the distance.

The unit of speed used at sea is the **knot** and is defined as 1 nautical mile per hour. It is incorrect to speak of knots per hour because this means nautical miles per hour per hour, a unit of acceleration rather than speed. To convert knots to kilometers per hour multiply knots by 1.85.

PLOTTING A COURSE

In plotting a course at sea we must distinguish between the *magnetic* north (to which the north-seeking pole of a magnetic compass points) and *true,* or geographic, north. For this reason, a *compass rose* appears on all navigational marine charts. This "rose" shows clearly the equivalent magnetic and true north directions. Almost all modern marine compasses are graduated from 0° (north) clockwise through 360°. There are 32 points on the compass, the four cardinal points being north (0°), east (90°), south (180°), and west (270°). The four intercardinal points, midway between the cardinal points, are northeast, southeast, southwest, and northwest. The points between each cardinal and intercardinal point, eight in all, are named for the directions they fall between: for example, north northeast (lying between north and northeast, or at 22.5°), east northeast (67.5°), south southeast (157.5°), and so on (Figure 3-5). The recitation of the full compass circle, known as "boxing the compass," was a remarkable accomplishment by early mariners, but a necessary one. Directions are now expressed largely by degrees rather than by points on a compass, except that cardinal and intercardinal points are used to indicate general directions, as in "northeast wind."

A ship's **course,** expressed in degrees, is the *intended* direction of travel: for example, a course of 180° is due south, and one of 135° is southeast. However, winds, ocean currents, and pilot error may prevent the ship from adhering to a particular course. A ship's **heading** is the direction in which the ship is actually traveling regardless of its prescribed course. All courses and headings are established in reference to true north unless otherwise indicated. A **bearing** is the direction from one point to another and is expressed as an angle from north. The traditional expressions for indicating bearings from a part of the vessel are

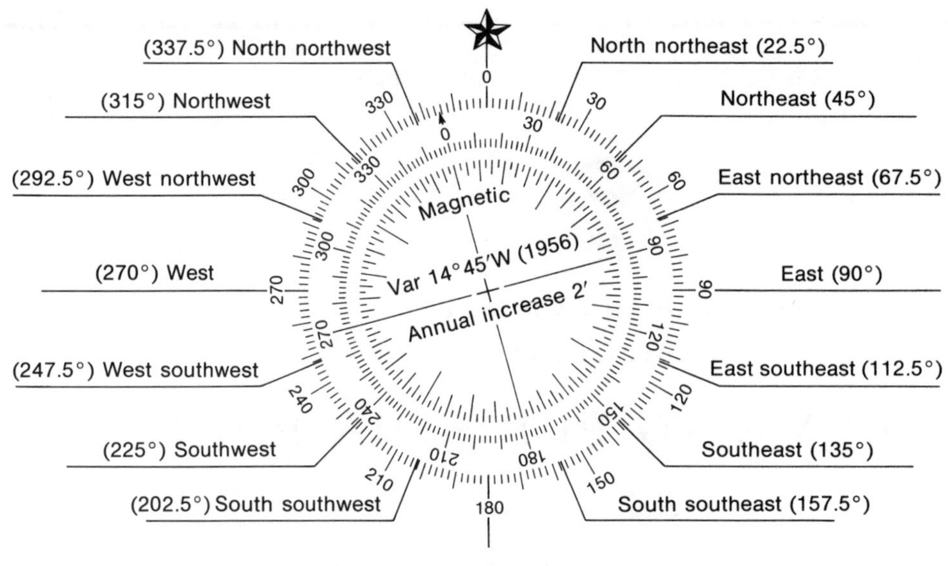

FIGURE 3-5 The compass rose.

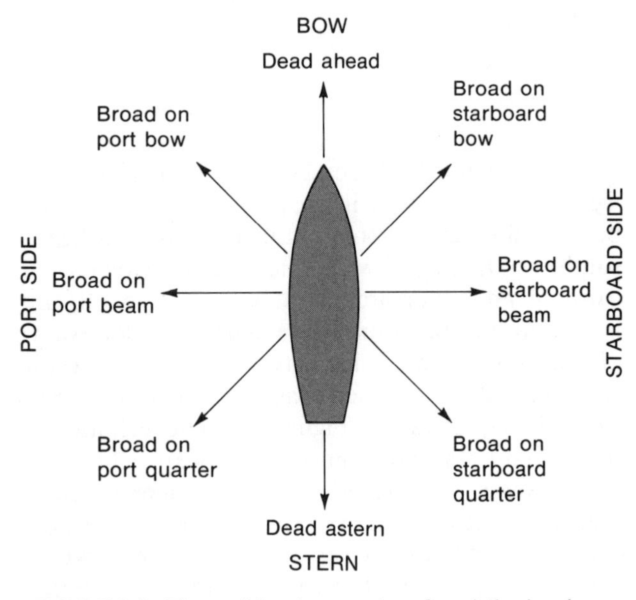

FIGURE 3-6 The traditional expressions for relative bearings.

seldom used today, except for "dead astern" (directly behind), "dead ahead," or "broad on the bow, beam, or quarter" (Figure 3-6).

TIME AND THE EARTH'S ROTATION

Because the earth rotates once on its axis in the course of 24 hours, and a "day" thus is 24 hours long, it can also be said that one 360° rotation is equal to 24 hours. When the sun is directly over Greenwich, England (longitude 0°), the local time is noon, whereas at longitude 180° the local time would be midnight. Thus every 15° of longitude east or west of Greenwich time equals 1 hour (coordinated universal time, sometimes called *Zulu,* or *Z, time*); or 360°/24 = 15° longitude per hour.

DEFINITIONS

Bearing. The horizontal angle between a line connecting an observer and the point being viewed, and a reference direction, usually north. It is recorded as an angle from north (000°) clockwise through 360°.

Course. The direction in which a ship must travel to arrive at a desired destination.

Great circle. The circle traced on the surface of a sphere by any plane that passes through the center of the sphere. All longitudinal lines are great circles.

Heading. The direction of actual travel, regardless of the prescribed course. The heading may be right on course or it may deviate as a result of various influences.

Knot. The unit of speed used at sea. It is equivalent to 1 nautical mile per hour.

Latitude. Angular distance north or south of the equator measured from 0° at the equator to 90° at the poles.

Longitude. Angular distance on the earth measured from the prime meridian (0°) at Greenwich, England, east or west through 180°.

Nautical mile. The basic unit of distance at sea, equivalent to 6080 feet, 1853 meters, or 1.853 kilometers.

Small circle. Any line of intersection of a sphere and a plane that does not pass through the center of the sphere. Lines of latitude are small circles parallel to the equator.

Statute mile. A unit of distance used on land, equivalent to 5280 feet, 1609 meters, or 1.609 kilometers.

EXERCISE 3

REPORT

MARINE CHARTS AND NAVIGATION

NAME ______________________

DATE ______________________

INSTRUCTOR ______________________

1. Answer the questions using the following table:

City	Latitude (La)	Longitude (Lo)
New York, New York	40°38′N	73°50′W
Reno, Nevada	39°30′N	119°46′W
Los Angeles, California	33°42′N	118°15′W
Hana Bay, Hawaii	20°45′N	155°59′W
San Francisco, California	37°19′N	122°25′W

(a) Which city is farthest north?

(b) Is Reno, Nevada, east or west of Los Angeles?

(c) What is the time difference (to the nearest hour) between New York and San Francisco?

(d) What is the time difference (to the nearest minute) between Hana Bay, in the Hawaiian Islands, and Los Angeles?

(e) How far north is San Francisco from Los Angeles?

______________ nautical miles. ______________ statute miles.

2. The distance between two points is 60 statute miles. What is the distance in nautical miles, using the conversion factor 1.15 statute miles per nautical mile.

______________ nautical miles.

3. A speedboat is traveling 40 miles per hour and a sailboat is doing 8 knots. Convert the speedboat's speed to knots and the sailboat's progress to miles per hour.

______________ knots. ______________ miles per hour.

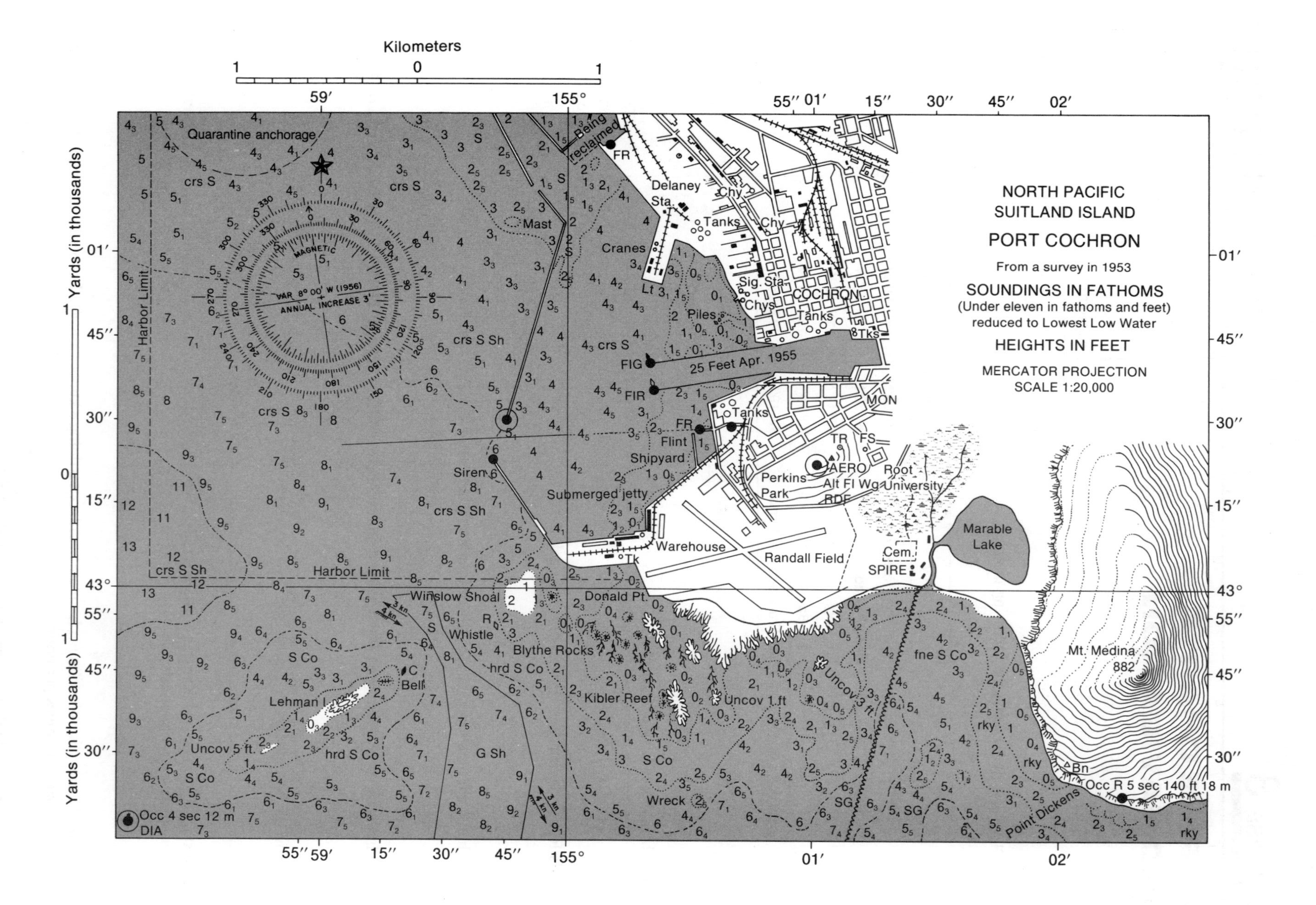
NORTH PACIFIC
SUITLAND ISLAND
PORT COCHRON
From a survey in 1953
SOUNDINGS IN FATHOMS
(Under eleven in fathoms and feet)
reduced to Lowest Low Water
HEIGHTS IN FEET
MERCATOR PROJECTION
SCALE 1:20,000
Kilometers
Yards (in thousands)
Quarantine anchorage
Harbor Limit
VAR 8° 00′ W (1956)
ANNUAL INCREASE 3′
MAGNETIC
Being reclaimed
Delaney Sta.
Tanks
Chy
Cranes
Mast
Sig. Sta.
COCHRON
Chys
Piles
Lt
25 Feet Apr. 1955
FIG
FIR
FR
Flint
Shipyard
Siren
Submerged jetty
Warehouse
Tk
MON
TR
FS
AERO
Perkins Park
Alt Fl Wg
RDF
Root University
Randall Field
Cem.
SPIRE
Marable Lake
Mt. Medina 882
Donald Pt.
Winslow Shoal
Whistle
Blythe Rocks
Kibler Reef
Uncov 1 ft
Uncov 3 ft
Uncov 5 ft.
Lehman I.
Bell
hrd S Co
fne S Co
S Co
G Sh
crs S Sh
crs S
rky
Wreck
SG
Bn
Point Dickens
Occ R 5 sec 140 ft 18 m
Occ 4 sec 12 m
DIA
3 kn
4 kn
155°
43°
59′
01′
02′
55″
45″
30″
15″

4. A ship is on course 90 degrees true. The navigator wants to anchor when a certain lighthouse is "broad on the starboard beam." What is the true compass bearing to the lighthouse at this point?

_______________ degrees true.

5. The following questions relate to the fictitious chart of Port Cochron in Figure 3-7. To answer them, you will need parallel rulers or two triangles and an inexpensive drafting compass.

(a) Locate a ship station at 43°00′45″N, 154°59′45″E and mark it with numeral 1. What depth is the water at this station?

_______________ fathoms.

What is the bottom sediment here?

(b) From Station 1, which you have just labeled, determine the bearing on Lehman Island. The compass rose on the chart is an aid for measuring direction. Place a straightedge on the two points and transfer that direction to the rose by parallel motion with parallel rulers or two triangles.

The bearing is _______________ degrees true and _______________ degrees magnetic.

Do the same for the AERO (aeronautical) beacon on land.

The bearing from Station 1 is _______________ degrees true and _______________ degrees magnetic.

(c) From Station 1 you set off on your course to Lehman Island. The vessel is suddenly engulfed in fog. The navigator fixes your position by determining the radar distance to the south light (0.7 kilometer) and the north light (0.8 kilometer) of the harbor entrance. With dividers or a compass locate your position west of the harbor entrance and determine the actual course made good and the ship's distance from Lehman island.

Course made good _______________ true; _______________ magnetic.

Distance is _______________ kilometers.

6. Answer the following questions by referring to the chart of Newport Beach, California, in Figure 3-8 (on page 28).

(a) What is the course to Lasuen Knoll from the entrance jetties to Newport Bay?

_______________ true.

(b) What is the distance in nautical miles?

(c) How long would it take you to cruise to Lasuen Knoll at 9 knots?

(d) Locate by latitude and longitude a station in Newport Submarine Canyon where you would find mud?

_______________ (La); _______________ (Lo).

(e) What is the depth at this station in the canyon?

_______________ fathoms.

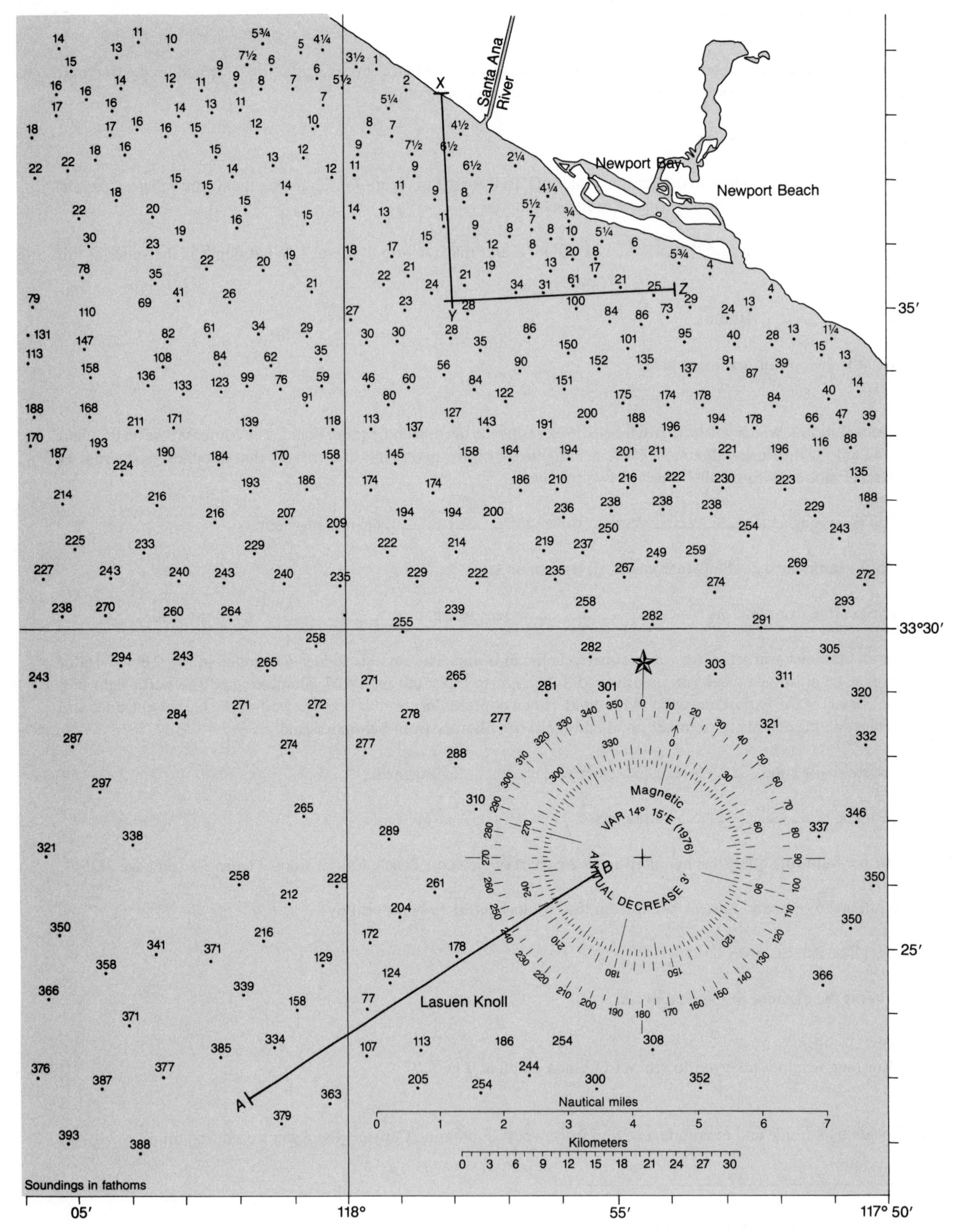
Santa Ana River
Newport Bay
Newport Beach
X
Y
Z
A
B
Lasuen Knoll
Magnetic
VAR 14° 15'E (1976)
ANNUAL DECREASE 3'
Nautical miles
0 1 2 3 4 5 6 7
Kilometers
0 3 6 9 12 15 18 21 24 27 30
Soundings in fathoms
35'
33°30'
25'
05'
118°
55'
117° 50'

FIGURE 3-8

EXERCISE 4

SEA-FLOOR SPREADING AND PLATE TECTONICS

It has long been recognized that such features as mountains, earthquakes, and volcanoes are not randomly distributed upon the surface of the earth. In 1912 the German meteorologist Alfred Wegener attempted to explain this distribution by arguing that if continents could move vertically, then horizontal motion, or *continental drift,* was also possible. The theory of *sea-floor spreading*—a process whereby new sea floor is created as adjacent crust is moved apart to make room—was first proposed in 1960 as a viable alternative to Wegener's theory. These two theories have been joined recently into a unifying model of crustal development called *plate tectonics.* Geophysical evidence collected in the course of the past two decades has given the theory firm scientific support.

THE THEORY OF PLATE TECTONICS

In essence, it has been proposed that the earth's outer shell (called the **lithosphere,** a zone about 100 kilometers thick) is composed of a number of rigid plates. Each of these **crustal plates** moves in a different direction, and thus plate boundaries are sites of tectonic activity where earthquakes, volcanism, and mountain building occur. Six large plates and a number of smaller ones have been identified (Figure 4-1). The source of energy for driving plate motions is heat produced by the decay of radioactive atoms in the earth's interior. Although these atoms exist in very small concentrations, the combined total heat production is large enough to keep the earth's interior quite hot and provide a constant flow of heat to the earth's surface. Because of this heat production, thermal gradients develop in the earth's interior, causing hot, low-density material to rise toward the surface while cold, higher-density material sinks. The convective motion thus generated in the earth's interior moves the lithospheric plates around on the surface of the earth. In a few areas, strong plumes may rise through the earth's mantle and penetrate the lithosphere to produce a hot spot with magmatic activity, which may or may not lie on a plate boundary. Examples of hot spots include Hawaii, Iceland, Easter Island, the Galápagos Islands, and Yellowstone (Figure 4-2).

There are three types of plate boundaries: constructive, destructive, and conservative (Figure 4-3). Where plates are separating, **magma** rises and cools to form new ocean crust. This boundary also called a

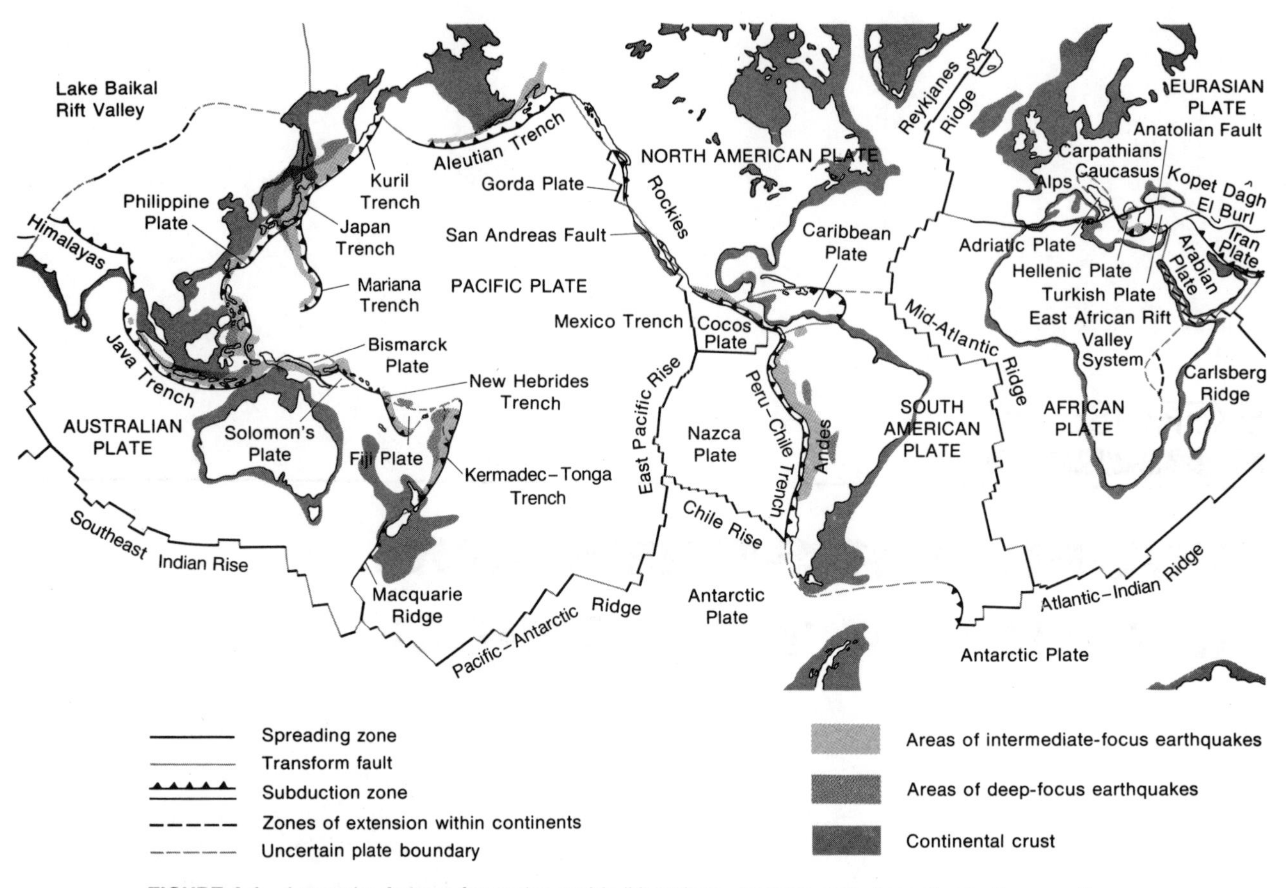

FIGURE 4-1 A mosaic of plates forms the earth's lithosphere, or outer shell. According to the recently developed theory of plate tectonics, the plates are not only rigid but are also in constant relative motion. The boundaries of the plates are of three types: spreading centers, transform faults, and subduction zones. Triangles indicate the leading edge of a plate. [After John F. Dewey, "Plate Tectonics." Copyright © 1972 by Scientific American, Inc. All rights reserved.]

spreading center and is expressed topographically as a mid-ocean ridge. Examples of ridges include the Mid-Atlantic Ridge and East Pacific Rise. Of course, if new sea floor is continuously created at spreading centers, old sea floor must be destroyed. This occurs at destructive boundaries, also known as **subduction zones,** which are expressed topographically by the presence of trenches. As old oceanic crust is subducted, it will be heated, causing partial melting. Magma produced in this fashion will rise, causing volcanism on the overriding plate near the trench. Examples of subduction zones include the Aleutian Trench and Aleutian Islands, and the Peru–Chile Trench and Andes Mountains. The third type of boundary lies along numerous offsets on mid-ocean ridges. These form conservative plate boundaries, where plates slide past one another with no crustal creation or destruction. These are also known as **transform faults** and may be expressed topographically as fracture zones. Examples of these are the Romanche, Oceanographer, and Mendocino fracture zones.

The boundaries between plates are seldom smooth, and thus plates "catch" on each other, deforming elastically on the edges until rock failure occurs to produce the abrupt motion we know as earthquakes. The largest and most devastating of these events often occur in subduction zones, but transform faults such as the San Andreas (in California) may also produce very large earthquakes. Occasionally two continental plates will collide in a subduction zone, but because continental crust has too low a density to sink into the earth's interior, the result is the pushing up of large mountain ranges. Collisions such as these created the Himalayas and the Alps.

PALEOMAGNETISM AS A RECORDER OF SEA-FLOOR SPREADING

The critical evidence for sea-floor spreading is based on earth magnetism (the force exerted by the earth's magnetic field). In the past (approximately every half

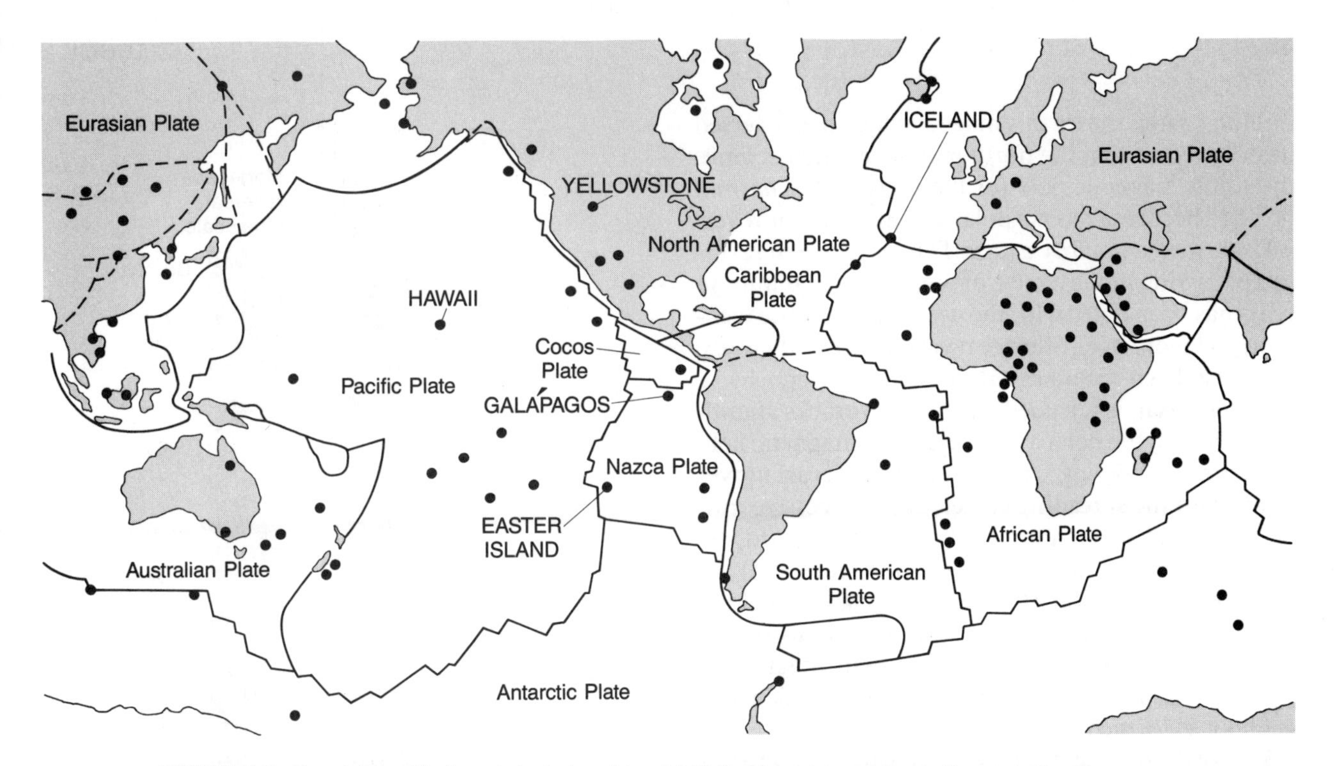

FIGURE 4-2 Population of hot spots includes at least 122 that have been active in the past 10 million years. They are found on all the major plates and on both oceanic and continental crust, but their distribution is decidedly nonuniform. There is a concentration along mid-ocean ridges, and in particular along the Mid-Atlantic Ridge; what is even more conspicuous is that of the 122 hot spots 43 are on the African Plate. Together with other evidence, this abundance of hot spots suggests that the African Plate is stationary over the mantle. If the African Plate is adopted as a frame of reference, other areas that have many hot-spot volcanoes, such as Antarctica and Southeast Asia, are found to be moving only slowly; on fast-moving plates hot-spot volcanism is rare. [After K. Burke and J. T. Wilson, "Hot Spots on the Earth's Surface." Copyright © 1976 by Scientific American, Inc. All rights reserved.]

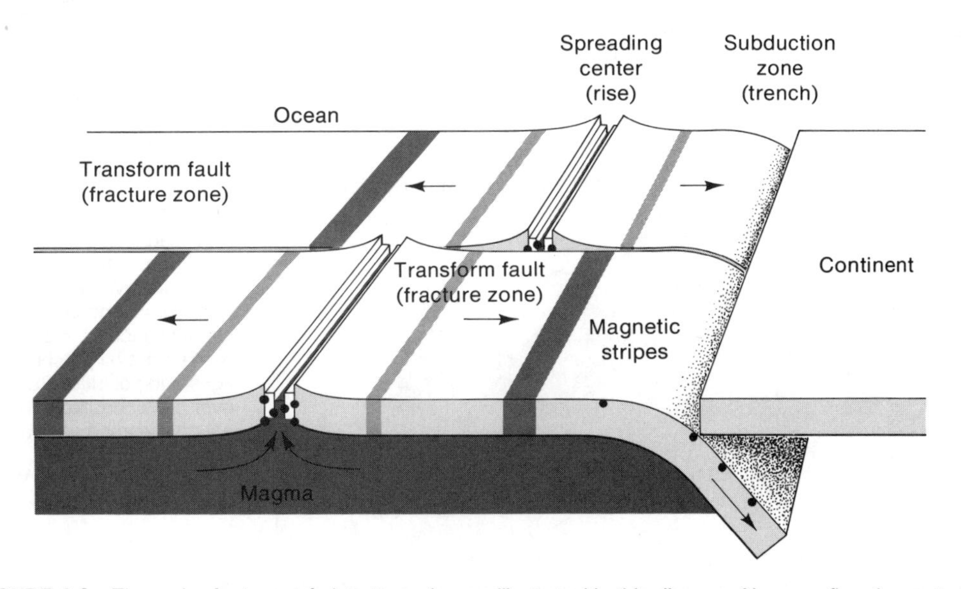

FIGURE 4-3 The major features of plate tectonics are illustrated in this diagram. New sea floor is created at the spreading center, or oceanic ridge, as magma rises from the mantle below. The magma solidifies as lava in cracks and flows, thereby preserving the magnetic polarity of the earth's magnetic field at the time the eruption took place. The subduction zone, or trench, forms where an ocean plate collides with a continental plate. A transform fault forms where two parts of a plate slip past one another, producing earthquakes (the foci, or points of rupture, are indicated by the heavy dots). Note that shallow-focus earthquakes (occurring close to the earth's surface) take place along the transform faults and under rises, and that deep-focus earthquakes occur at subduction zones where one plate plunges beneath another. [After Don L. Anderson, "The San Andreas Fault." Copyright © 1971 by Scientific American, Inc. All rights reserved.]

million years), the earth's magnetic field has reversed its polarity, so that the north magnetic pole becomes the south magnetic pole (or the present-day *normal* polarity becomes *reverse* polarity). Each major reversal is termed a *polarity epoch.* The sequence of reversals occurring in the course of the past several million years has been dated with the use of radiometric techniques (Figure 4-4). Shorter reversals, termed *events,* have also been recorded within the longer epochs.

When basalt is intruded and cools in cracks at mid-ocean ridges, the polarity of the earth's magnetic field at the time of cooling is preserved. As the crust moves away from the spreading center, each successive unit of the cooled magma gradually moves outward, revealing zones of *fossil magnetism* (Figure 4-5), in which the orientation of the earth's magnetic field at the time of formation is preserved. This fortunate preservation allowing us to detect fossil magnetism gives us a method for measuring the rate at which new sea floor is formed.

Where the rocks have the same magnetic polarity as the present-day normal magnetic field we find stronger than average magnetisms and we have a positive anomaly; where the rocks preserve reverse polarity we measure weaker than average magnetic field

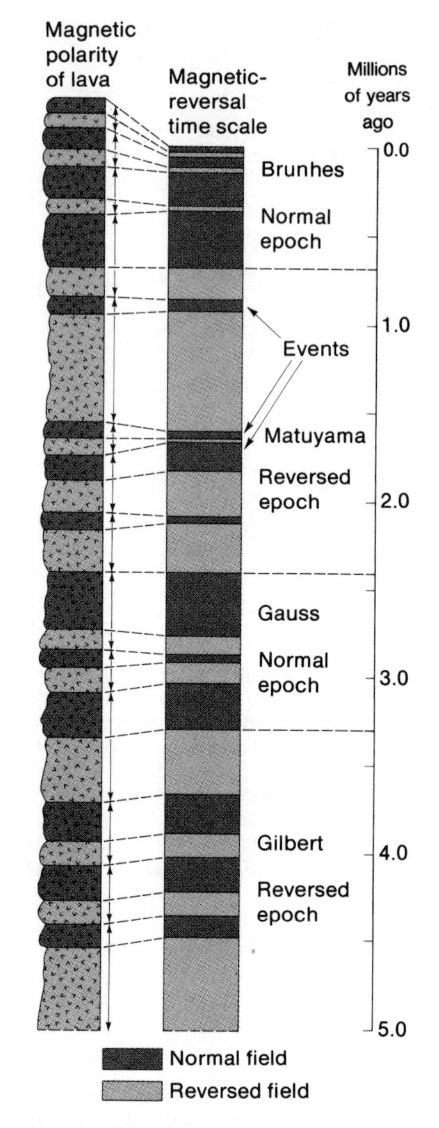

FIGURE 4-4 Schematic illustration of how magnetic polarities of lava flows are used to construct the time scales of magnetic reversals over the past 5 million years. In no one place is the entire sequence found; the sequence is worked out by patching together the ages and polarities from lava beds all over the world. Note that each magnetic epoch is named after a famous magnetician. [From F. Press and R. Siever, *Earth,* 4th ed. W. H. Freeman and Company. Copyright © 1986.]

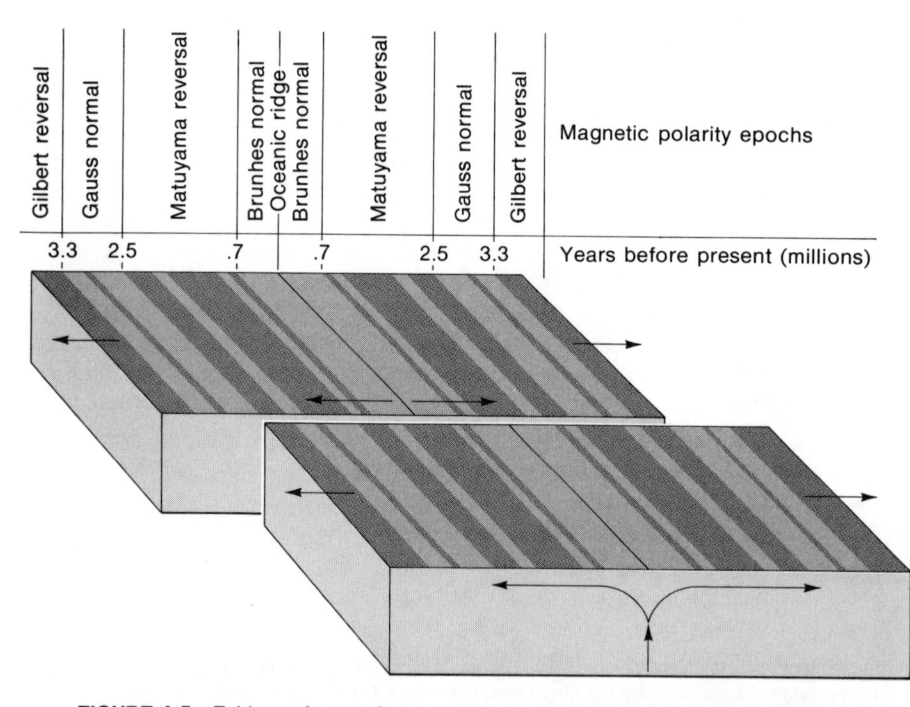

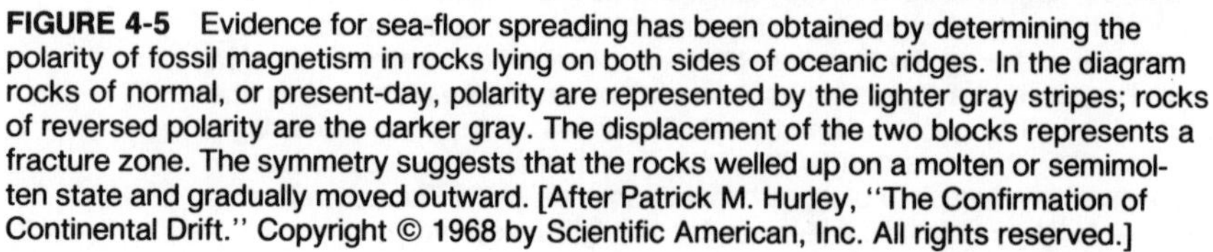
FIGURE 4-5 Evidence for sea-floor spreading has been obtained by determining the polarity of fossil magnetism in rocks lying on both sides of oceanic ridges. In the diagram rocks of normal, or present-day, polarity are represented by the lighter gray stripes; rocks of reversed polarity are the darker gray. The displacement of the two blocks represents a fracture zone. The symmetry suggests that the rocks welled up on a molten or semimolten state and gradually moved outward. [After Patrick M. Hurley, "The Confirmation of Continental Drift." Copyright © 1968 by Scientific American, Inc. All rights reserved.]

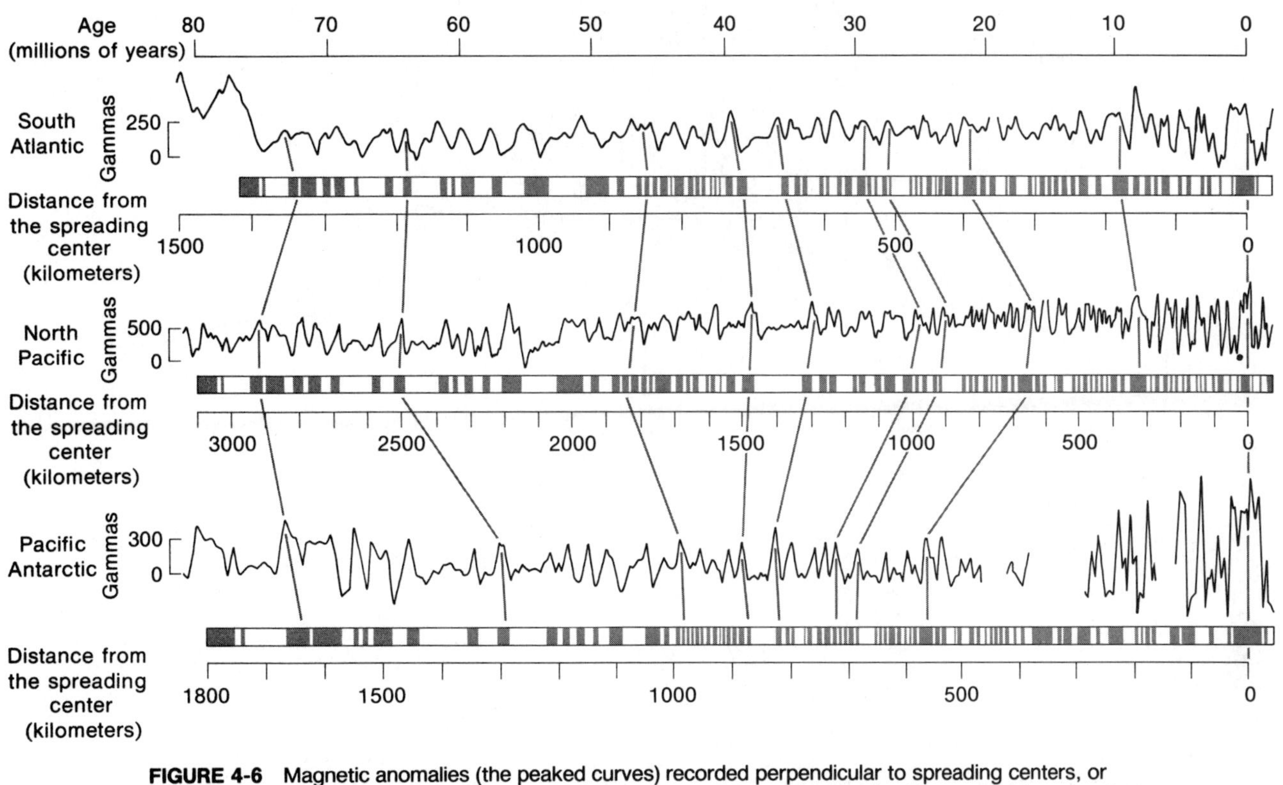

FIGURE 4-6 Magnetic anomalies (the peaked curves) recorded perpendicular to spreading centers, or ridges, in the major ocean basins reveal a similar sequence of magnetized rocks. The vertical lines passing through the three curves correlate equivalent magnetic anomalies from different ocean basins. Anomalies are plotted in gammas, which is a measure of magnetic field strength. To put these anomalies in perspective, note that they constitute only 1–2 percent of the earth's average magnetic field. They are spaced differently because the spreading rates are different at each ridge. However, they all show the same sequence of magnetization, just as different-sized trees from the same region would show the same sequence of rings regardless of the absolute width of each ring. [After J. R. Heirtzler, "Sea-Floor Spreading." Copyright © 1968 by Scientific American, Inc. All rights reserved.]

and we have a negative anomaly. We can determine the rate at which new sea floor is formed by measuring the distance from the ridge crest to a magnetic anomaly of known age. To calculate the spreading rate, simply divide the distance traveled by the age of the oldest reliably dated anomaly. With the aid of this technique, plate velocities of from 1 to 10 centimeters per year have been calculated (Figure 4-6).

HYDROTHERMAL CIRCULATION

Although mid-ocean ridges are areas in which relatively large temperature increases should be observed in sediments beneath the sea floor, measurement of these gradients has shown that they are only slightly larger than those observed at locations far away from ridge crests. At first, these observations were difficult to reconcile with the predictions of the plate tectonic model. Then, suggestions were made that perhaps seawater enters fractures in the oceanic crust and flows toward hot, recently formed rock. Where it encounters hot rock, water would be heated, causing it to rise and exit near ridge crests (Figure 4-7). This suggestion seemed reasonable because hot springs and geysers are frequently observed on continents near magma sources. This hydrothermal flow would provide a mechanism for cooling the newly formed crust and thus reducing the temperature gradients.

In 1977, an expedition went to the Galápagos spreading center to search for vents of hot hydrothermal waters. Using the submersible *Alvin,* they explored the sea floor for evidence of hot springs. Their search was successful, and resulted in some remarkable additional discoveries. Plumes of hot water were found exiting from fissures in the crust. These plumes carry sulfide in solution, which bacteria can combine with oxygen from seawater in order to derive energy for growth. The bacteria can then be consumed by

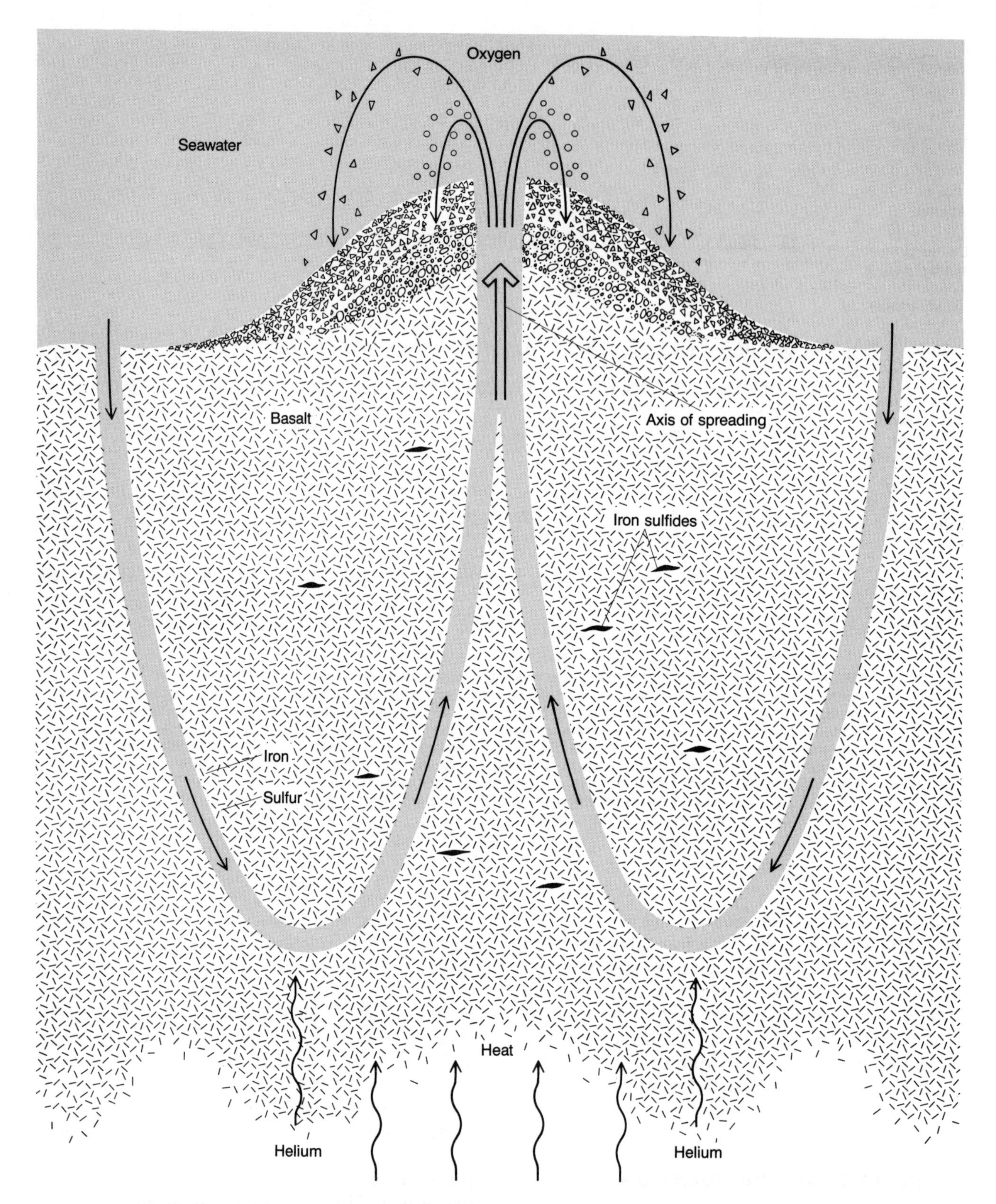

FIGURE 4-7 Hydrothermal process is modeled schematically. Seawater enters the fractured basalt of the crust near a spreading center and penetrates to a depth of several kilometers. Because of the high temperatures in the zone of magma injection along a spreading center, the water is heated to a temperature of a few hundred degrees Celsius. The heated water extracts a number of elements, including metals, from the basalt and also loses a few. Thermal convection drives the now metal-rich water back up to the sea floor, where it is discharged through hot springs. Some of the metals are deposited as sulfides within the crust. The metals remaining in solution are deposited on the ocean bottom. The discovery of mantle-derived helium along oceanic spreading centers suggests that hydrothermal systems also bring gases up. [After E. Bonatti, "The Origin of Metal Deposits in the Oceanic Lithosphere." Copyright © 1978 by Scientific American, Inc. All rights reserved.]

higher organisms, including large clams and tube worms, allowing a unique ecosystem to develop, fueled by chemical energy. Similar vents have since been found at a number of other locations on ridge crests, some of which may vent water nearly 300 °C in temperature.

Although each vent may have only a modest flow and be active for only a few thousands of years, there may be large numbers of them. It has been estimated that the entire volume of the ocean is circulated through ridge crests in 10–20 million years. Examination of the composition of these hot waters has answered some long-standing questions regarding ocean chemistry (see Exercise 9), because reactions between seawater and hot rocks help regulate the abundance of some elements in seawater. Metals are leached from hot rocks and carried upward by hydrothermal flow. As the water rises and cools, some of these metals may precipitate in the fractures or fissures that permit flow, forming ore deposits. Other metals are carried into bottom water and provide a significant fraction of the dissolved metal flux to the sea.

DEFINITIONS

Crustal plate. A portion of the lithosphere bounded by one or more of the three types of boundaries: fracture zones (transform faults); mid-ocean ridges or rises (spreading centers); and trenches (subduction zones).

Lithosphere. The coherent rigid outer shell of the earth that includes both the crust and upper mantle. This layer is about 150 kilometers thick under the continents and 50–70 kilometers thick under the ocean basins.

Magma. Molten rock material within the earth. Magma that has reached the surface of the earth is called *lava.*

Spreading center. A mid-ocean rise or ridge where molten material (basalt) rises to create new sea floor.

Subduction zone. An elongate region along which a crustal plate descends below another one. For example, one such zone is the Peru–Chile Trench, along which the Nazca Plate descends beneath the South American Plate. During subduction, downward motion produces earthquakes, and eventually the sinking rocks remelt to create local volcanic activity.

Transform fault. A fault that crosses a mid-ocean ridge and connects the offset ends of the ridge. As new material is added at the ridge crests it produces the motion shown in Figure 4-1.

EXERCISE 4

REPORT

SEA-FLOOR SPREADING AND PLATE TECTONICS

NAME ____________________

DATE ____________________

INSTRUCTOR ____________________

1. Examine Figure 4-3 and state where you would expect to find the highest temperature gradients in sediments.

   Why are large temperature gradients not found?

   Where would deep-focus earthquakes occur? (The focus is the location inside the earth where an earthquake starts).

   Why would you not expect to find deep-focus earthquakes under spreading centers?

2. What types of plate boundaries occur between the following?

   (a) The North American and Pacific plates

   (b) The Nazca and South American plates

   (c) The South American and African plates

3. What stresses (tension, compression, or shearing) characterize the following types of junctions?

   (a) Spreading centers

   (b) Subduction zones

   (c) Transform faults

4. (a) A transform fault (single line) is shown offsetting mid-ocean ridges (double lines) in the sketch below. Use XXX to show the possible locations of earthquake epicenters (the point on the earth's surface directly above an earthquake's focus) along its strike. Use arrows to show the directions of plate motions.

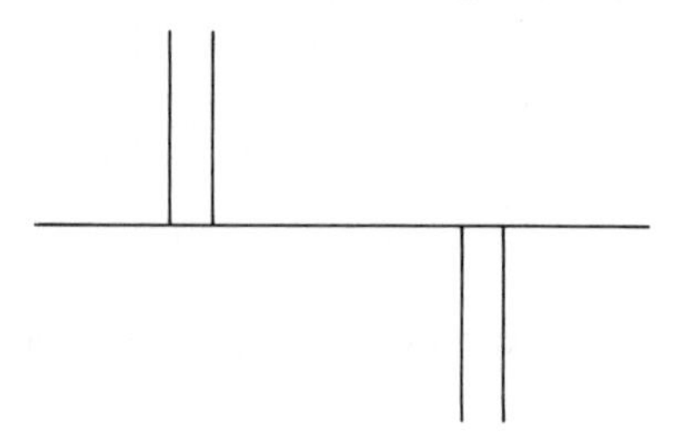

(b) Explain the distribution of earthquake epicenters on the sea floor.

5. (a) Refer to Figure 4-6 and calculate the half-spreading rates in centimeters per year for the three areas shown in the figure.

South Atlantic

North Pacific

Pacific Antarctic

(b) Which of the ocean basins exhibits the slowest spreading rate?

(c) Which one is spreading the fastest?

(d) What is the collision rate of the North American Plate and the Pacific Plate? (Assume that the North American Plate is moving at the same rate as the opening of the South Atlantic and that the two plates move in opposite directions.)

(e) About how long did it take for the Atlantic Ocean to open to its present width at the equator?

(f) Using the base of the continental slope as the original junction of the continents, estimate how much time would be required to open the Atlantic Ocean.

During what geologic time period did these events take place?

6. The approximate duration of the short polarity events shown on the time scale in Figure 4-4 is 10,000–50,000 years. What is the significance of the 3 percent uncertainty in dating when applied to the duration of the very short events or to polarity epochs that are many millions, or tens of millions, of years old?

7. The Hawaiian Islands have been created by the passage of the Pacific Plate over a hot spot. On which Hawaiian Island would you expect to find active volcanism and the youngest rocks? (Indicate east or west, or identify the islands on a good map of the sea floor.)

OPTIONAL QUESTIONS

8. On a good map of the Pacific sea floor, find the following hot-spot ridges in the Pacific: the Emperor Seamount Chain hot-spot ridge, the Tuamotu Archipelago–Line Island hot-spot ridge, and the Austral–Gilbert–Marshall Island hot-spot ridge. If these oceanic island chains are produced by hot spots, explain the change in the lineation of each of them (mainly westerly for the Hawaiian, Tuamotu, and Austral parts; and mainly northerly for the Emperor, Line, and Gilbert–Marshall parts).

9. If the Pacific Plate is moving at 10 centimeters per year, and this speed has not changed, estimate the time (years before present) when the change described in Question 8 took place.

10. On a good map showing detailed bottom topography, locate the Walvis Ridge and Rio Grande Rise in the South Atlantic. It is believed that these features were produced as volcanic ridges from a hot spot currently at about 40° S on the Mid-Atlantic Ridge. If this is true, why do the lineations of the hot-spot ridges and the lineation of the fracture zones (both produced by plate tectonics) show a different sense of movement (lineations are in different directions)?

EXERCISE 5

GEOGRAPHY OF THE MARINE ENVIRONMENT

In order to understand more fully some of the concepts presented in other exercises, it is worthwhile to familiarize yourself with the geography of the sea floor and oceans. Knowing the location of the major geographic features will enhance your appreciation of sea-flooring spreading, current patterns, and the distribution of marine sediments.

PHYSIOGRAPHIC RELATIONSHIPS

Marine scientists always look for associations in their data. For example, it has become evident that the largest features of the ocean basins have a regular and predictable pattern of interrelationships. For example, the principal submarine mountain chains, or ridges, traverse the major basins near the middle. They are flanked on each side by deep plains or plains with low hills. These abyssal plains and hills extend to the continents and meet the terrace-like continental edges, which terminate either in deep trenches or in gently sloping aprons of sediments. Such an apron, or blanket, of sediment is called the *continental rise* and consists of deposits that have accumulated at the base of the *continental slope.* Further examination has shown that where an apron of sediment occurs at the base of a continental slope, it adjoins a deep flat plain of sediment, but if a trench is at the base of the slope, the neighboring deeper part of the ocean basin is hilly and has little sediment cover.

It has also been recognized that the mid-ocean ranges are broken by large fracture zones, or faults. These various features are shown in Figures 5-1 through 5-3. We will study them and their origins in more detail in this and other exercises.

PRINCIPAL MORPHOLOGIC FEATURES

The major relief features of the ocean floor are listed in Table 5-1. The *first-order* morphologic features of the earth's crust are the continents and ocean basins. In the ocean basins, the *second-order* features are of five types: **continental margins; deep-ocean floor; mid-ocean ridges** and **rises; fracture zones;** and **island arcs** and **trenches.** (The first three types and their subdivisions are shown in Figure 5-1.) The five major second-order features and their smaller-scale, or third-order, features as well as certain special forms are described in the list of definitions at the end of this exercise.

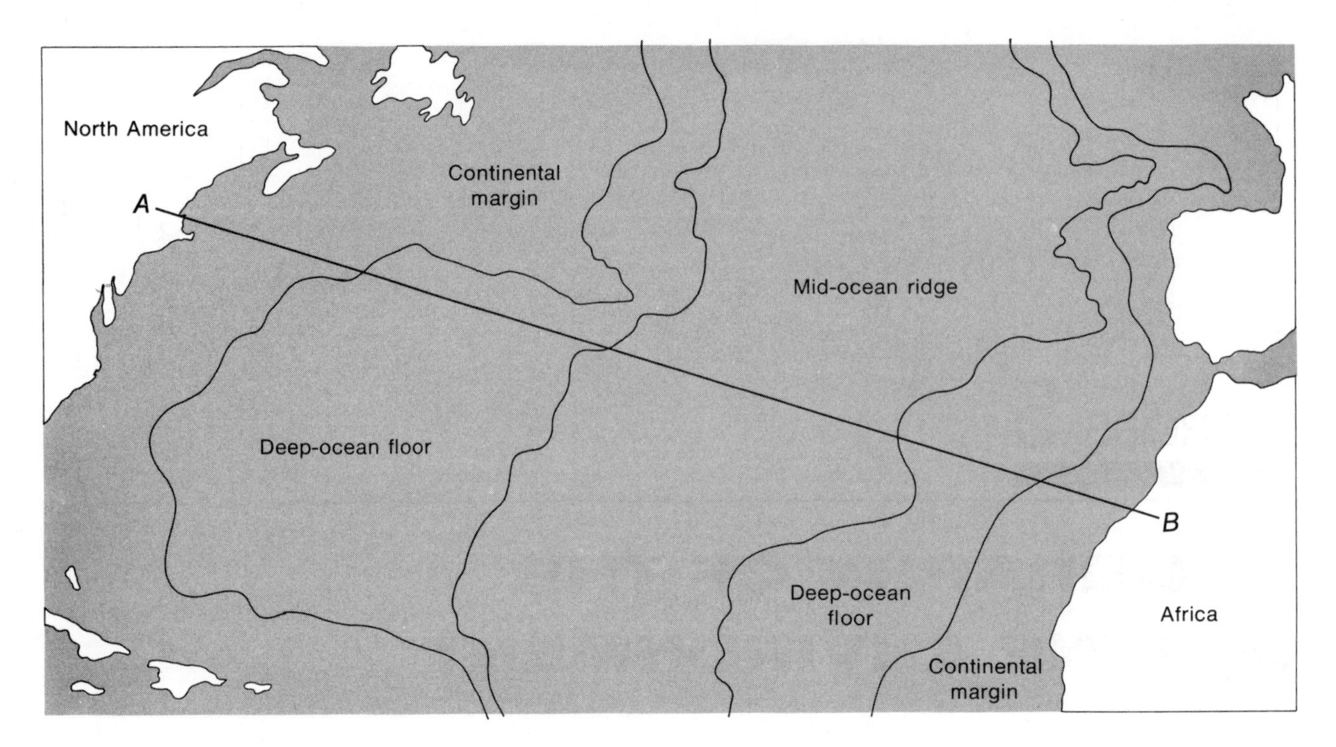

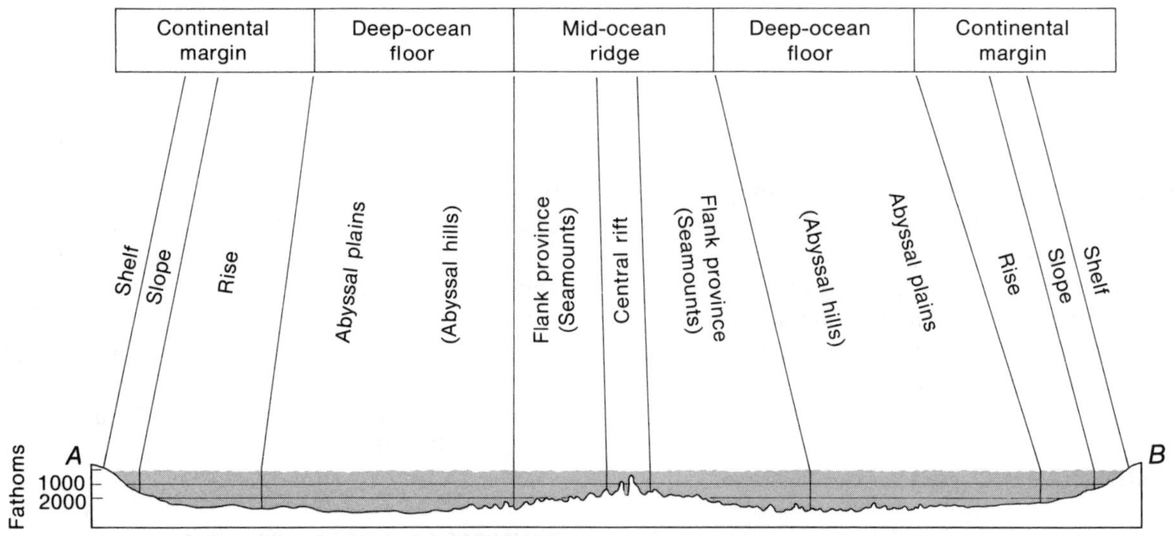

FIGURE 5-1 Some second-order morphologic features of the ocean basin. The top part is a simplified map of the Atlantic Ocean basin showing the continental margins, the deep-ocean floor, and the mid-ocean ridge. The bottom is a profile of the sea floor along the line *AB* showing the second- and third-order features of the North Atlantic Ocean basin. [After B. C. Heezen, M. Tharp, and M. Ewing, Geological Society of America Special Paper 65, 1959.]

RELATIONSHIP BETWEEN GEOGRAPHIC FEATURES AND MAJOR PROCESSES

Another useful application of the study of morphologic features of the ocean floors is that when we visualize the relationships in charts or diagrams such as those shown in the figures, we are immediately led to questions. Why are forms of the continental margins different from ocean to ocean? Why do some mid-ocean ridges occur in the centers of ocean basins? Why are rises, channels, and abyssal plains always associated? These association problems were answered for some of the major features when we discussed plate formation and motions in Exercise 4 on sea-floor spreading.

For a beginning, given the general acceptance of the theory of sea-floor spreading and plate motion, we can see in Figure 5-3 that margins around the Pacific Ocean are narrow and steep and commonly have trenches at the base of the slope. Margins around the

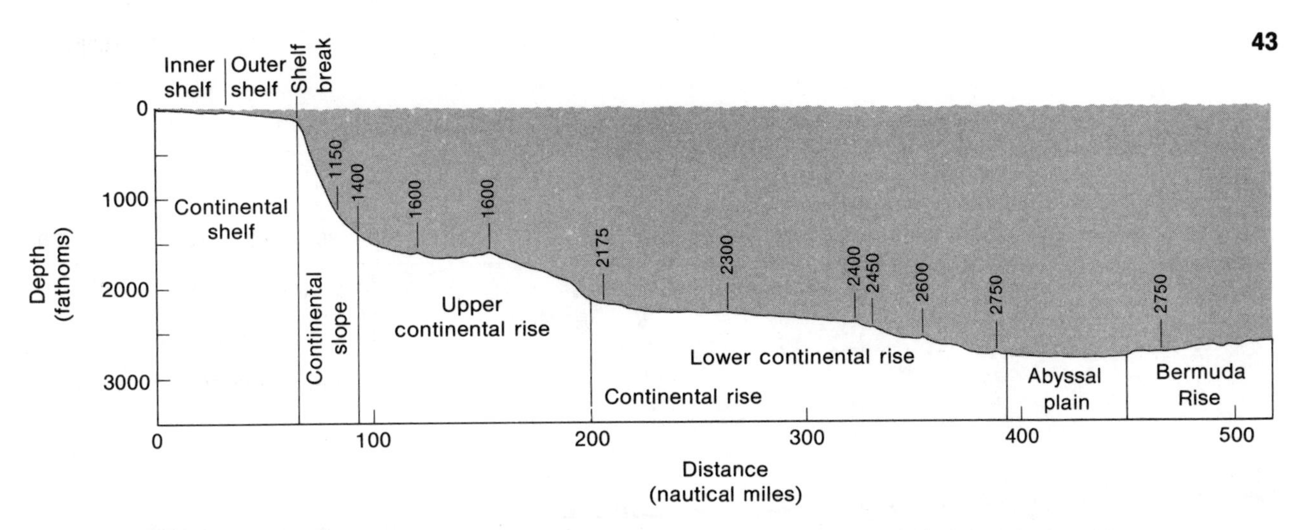

FIGURE 5-2 Profile of an Atlantic type of continental margin and its major morphologic subdivisions. The topography shown extends from the northeastern United States to the rise surrounding the island of Bermuda. [After B. C. Heezen, M. Tharp, and M. Ewing, Geological Society of America Special Paper 65, 1959.]

Atlantic are wide and are typically bounded at the base of the slope by broad rises that merge with the deep abyssal plains. If we further recall that trenches are evidence of plate collisions and that wide margins with rises are moving on broad crustal plates, we can begin to understand the reasons for the differences in ocean-floor features in different oceans. We will discuss the processes that modify or form these features in other exercises.

THE GEOGRAPHY OF COASTS

Where ocean and continental plates collide, we usually see that the coasts are high and mountainous, as along the Chile–Peru coasts. Where a margin lies on a plate, the coasts are typically wide plains that merge with broad river valleys, as on the Atlantic coast of South America. It has been observed that the great majority of the world's largest rivers empty into wide margins in marginal seas or in Atlantic-type ocean basins. In Exercise 6 we will examine and analyze the results of these relationships. If we visualize a continental plate ramming into an ocean plate with the colliding edges crumpled up and intruded by volcanic rock, it is to see that the drainage divide to that continent will be in the high volcanic and faulted mountains behind the colliding coast. From those high mountain ranges the surface will slope gently all the way back to the trailing-edge coast. This is roughly what we would see in a cross section from west to east across South America. The huge Amazon River basin is formed by the broad, gently sloping surface of the continent behind the western Andes barrier along the Pacific coast. Thus, the geography of coasts is strongly influenced by larger earth processes, and can tell us much about those processes.

TABLE 5-1
Major relief features of the ocean floor

First-order features of the earth's crust

- Continents
- Ocean basins

Second-order features of the ocean-basin floor

- Continental margins
- Deep-ocean floors
- Mid-ocean ridges
- Fracture zones
- Island arcs and trenches

Third-order features

- Continental margins
  - Typical Atlantic forms
    - Continental shelf
    - Continental slope
    - Continental rise
  - Typical Pacific forms
    - Continental shelf
    - Continental slope
    - Marginal trough
  - Special forms
    - Continental borderland
    - Marginal plateaus
  - Deep-ocean floors
    - Abyssal plains
    - Abyssal hills
    - Seamount trends
    - Deep-ocean channels
    - Gaps and local rises
  - Mid-ocean ridges
    - Crest provinces
    - Flank provinces
    - Rift zones
  - Fracture zones
    - Scarps
    - Depressions
  - Island arcs and trenches
    - Volcanic island arcs
    - Insular slopes
    - Trenches

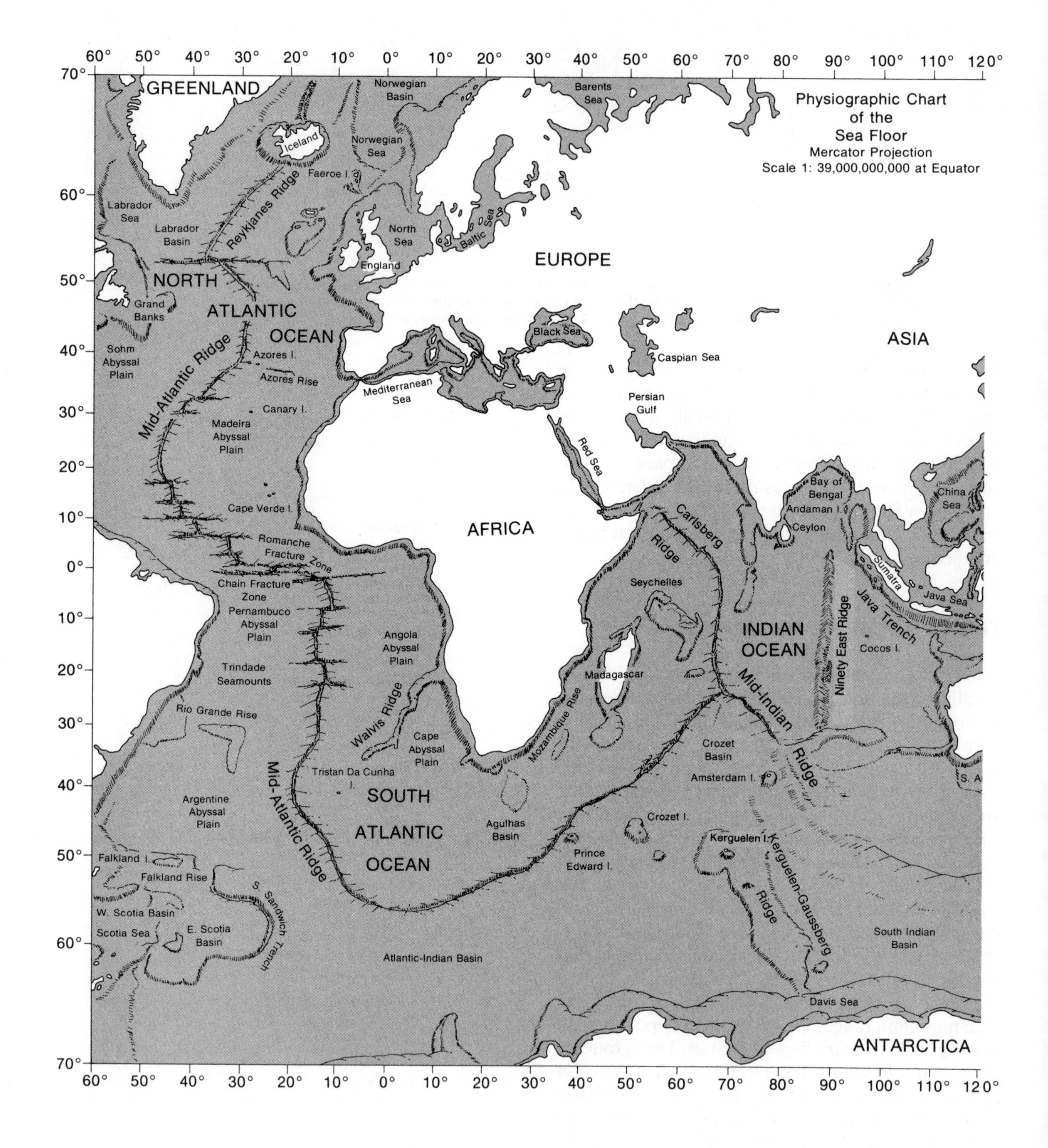

FIGURE 5-3 The ocean basins and their major features. [Courtesy Hubbard Scientific Company.]

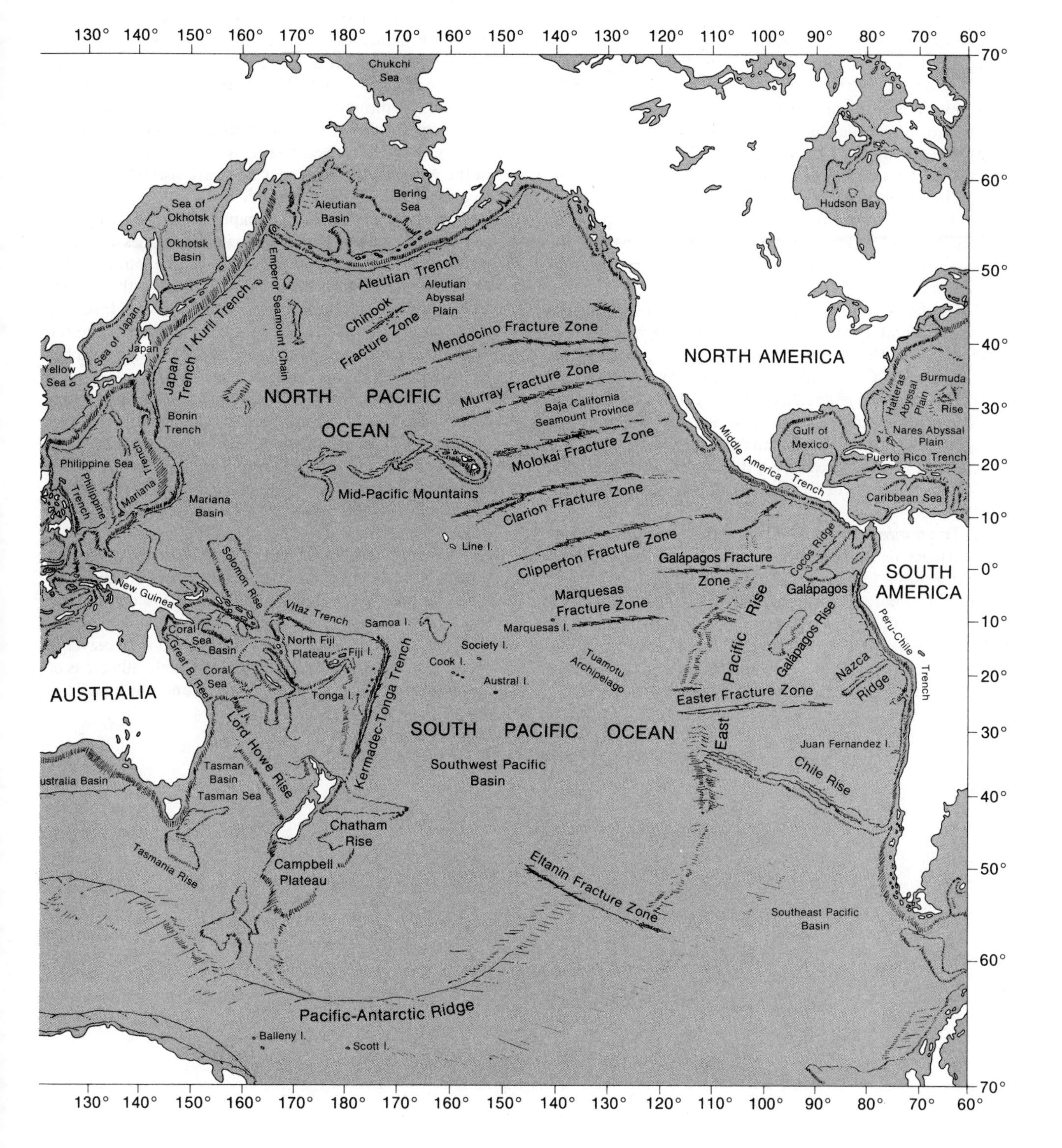

Chukchi Sea
Sea of Okhotsk
Okhotsk Basin
Aleutian Basin
Bering Sea
Hudson Bay
Aleutian Trench
Aleutian Abyssal Plain
Emperor Seamount Chain
Kuril Trench
Sea of Japan
Japan
Yellow Sea
Japan Trench
Chinook Fracture Zone
Mendocino Fracture Zone
NORTH AMERICA
NORTH PACIFIC OCEAN
Murray Fracture Zone
Baja California Seamount Province
Molokai Fracture Zone
Bonin Trench
Philippine Sea
Philippine Trench
Mariana Trench
Mariana Basin
Mid-Pacific Mountains
Clarion Fracture Zone
Middle America Trench
Gulf of Mexico
Hatteras Abyssal Plain
Burmuda Rise
Nares Abyssal Plain
Puerto Rico Trench
Caribbean Sea
Line I.
Clipperton Fracture Zone
Galápagos Fracture Zone
Cocos Ridge
Galápagos
SOUTH AMERICA
Solomon Rise
New Guinea
Vitaz Trench
Samoa I.
Marquesas Fracture Zone
Marquesas I.
Society I.
Cook I.
Austral I.
Tuamotu Archipelago
East Pacific Rise
Galápagos Rise
Peru-Chile Trench
Nazca Ridge
Coral Sea Basin
Great B. Reef
North Fiji Plateau
Fiji I.
Coral Sea
Tonga I.
Kermadec-Tonga Trench
AUSTRALIA
Easter Fracture Zone
SOUTH PACIFIC OCEAN
Lord Howe Rise
Juan Fernandez I.
Australia Basin
Tasman Basin
Tasman Sea
Southwest Pacific Basin
Chile Rise
Chatham Rise
Tasmania Rise
Campbell Plateau
Eltanin Fracture Zone
Southeast Pacific Basin
Pacific-Antarctic Ridge
Balleny I.
Scott I.
130° 140° 150° 160° 170° 180° 170° 160° 150° 140° 130° 120° 110° 100° 90° 80° 70° 60°
70° 60° 50° 40° 30° 20° 10° 0° 10° 20° 30° 40° 50° 60° 70°

DEFINITIONS

Active margin (within plates). Where ocean and continental plates collide, a deep trench is typically formed at the base of steep slopes. Shelves are narrow and the shore is high and characteristically backed by a coast-parallel line of volcanoes. Earthquakes are common to great depths below the continental margin and volcanic activity is common along these active margins.

Continental margins. Consisting of continental shelf, slope, and rise. Several types of margins are recognized.

(1) The *Atlantic type* is characterized by a wide continental shelf, a steeper continental slope descending to the deep sea, and a flatter continental rise at the base of the slope formed by accumulation of sedimentary materials (Figure 5-2).

(2) The *Pacific type* is characterized by a narrow shelf and slope descending into a deep marginal trough, or trench, generally parallel to the continental margin. An example is the area off Chile and Peru in South America.

(3) The *marginal plateau* has a narrow shelf and incipient slope leading down to a shelflike feature similar to the continental shelf. The plateau is similar to the shelf but occurs at a much greater depth. The Blake Plateau off Florida, at a depth of about 1000 meters, is a good example.

(4) The *borderland* consists of a series of offshore basins and ridges or islands, such as those found off southern California.

Deep-ocean floor. Characterized by abyssal plains of depositional (sedimentation) origin, abyssal hills, seamounts (volcanoes), guyots (flat-topped seamounts or tablemounts), channels, and gaps and local rises. See Figures 5-1 and 5-2.

Fracture zone. A linear zone of irregular topography on the sea floor, averaging 100 miles wide and more than 1000 miles long. These are, in fact, great faults along which sea-floor spreading has taken place, an example being the Eltanin Fracture Zone. Fracture zones are characterized by escarpments (scarps) several thousand feet high that separate regions of different depths. The zones are conspicuous for numerous seamounts, irregular topography, and local depressions. Unfortunately, some depressions have been called trenches, such as the Romanche Trench in the Atlantic, but they bear no relationship to deep-sea trenches in the Pacific Ocean.

Island arcs and trenches. Arc-shaped island chains generally curving seaward, with steep insular slopes and deep trenches on the seaward sides. They are found in the northern and eastern Pacific Ocean, and in the eastern Indian Ocean. Examples are Japan and the Japanese Trench, the Aleutians and the Aleutian Trench, and the Marianas Islands and Mariana Trench. In these trenches are the deepest places in the oceans.

Mid-ocean ridges and rises. Ridges are elongate, steep-sided elevations of the sea floor having a central rift (valley) and rough marginal topography that flanks the crest; an example is the Mid-Atlantic Ridge. Rises are broad, elongate, and smooth elevations of the sea floor, such as the East Pacific Rise. Both ridges and rises have a similar origin (see Exercise 4).

Passive margin (within plates). Continental margins with broad shelves, gentle wide slopes, and rises are typically found in locations where the margin of the continental block is part of a larger ocean–continental crustal plate, as in the Atlantic Ocean. Thus, the margin is "passively" moving on the plate. Such margins are generally areas of few earthquakes, no volcanic activity, and low heat flow.

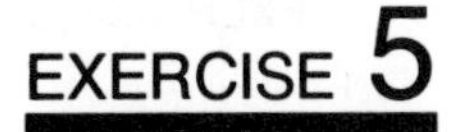

REPORT

GEOGRAPHY OF THE MARINE ENVIRONMENT

NAME ______________________

DATE ______________________

INSTRUCTOR ______________________

1. Locate on the map in Figure 5-3 the features or areas named below. Refer also to the physiographic diagrams of the major oceans by B. C. Heezen and M. Tharp (published by *National Geographic*), if available. On the blank map of the world in Figure 5-4 place the number for the feature at the appropriate location.

Pacific Ocean

1. East Pacific Rise
2. Mendocino and Murray fracture zones
3. Eltanin Fracture Zone
4. Mariana Trench
5. Peru–Chile Trench
6. Sea of Okhotsk
7. Bering Sea
8. Nazca Ridge
9. Kermadec–Tonga Trench
10. Emperor Seamount Chain

Atlantic Ocean

1. Argentine Abyssal Plain
2. Mid-Atlantic Ridge
3. Hatteras and Nares abyssal plains
4. Grand Banks
5. Reykjanes Ridge
6. Walvis Ridge
7. Puerto Rico Trench
8. Scotia Sea or Ridge
9. Falkland Islands
10. Rio Grande Rise

Indian Ocean

1. Carlsberg Ridge
2. Seychelles Islands
3. Arabian Sea
4. Kerguelen Islands
5. Ninety East Ridge
6. Madagascar
7. Red Sea
8. Persian Gulf

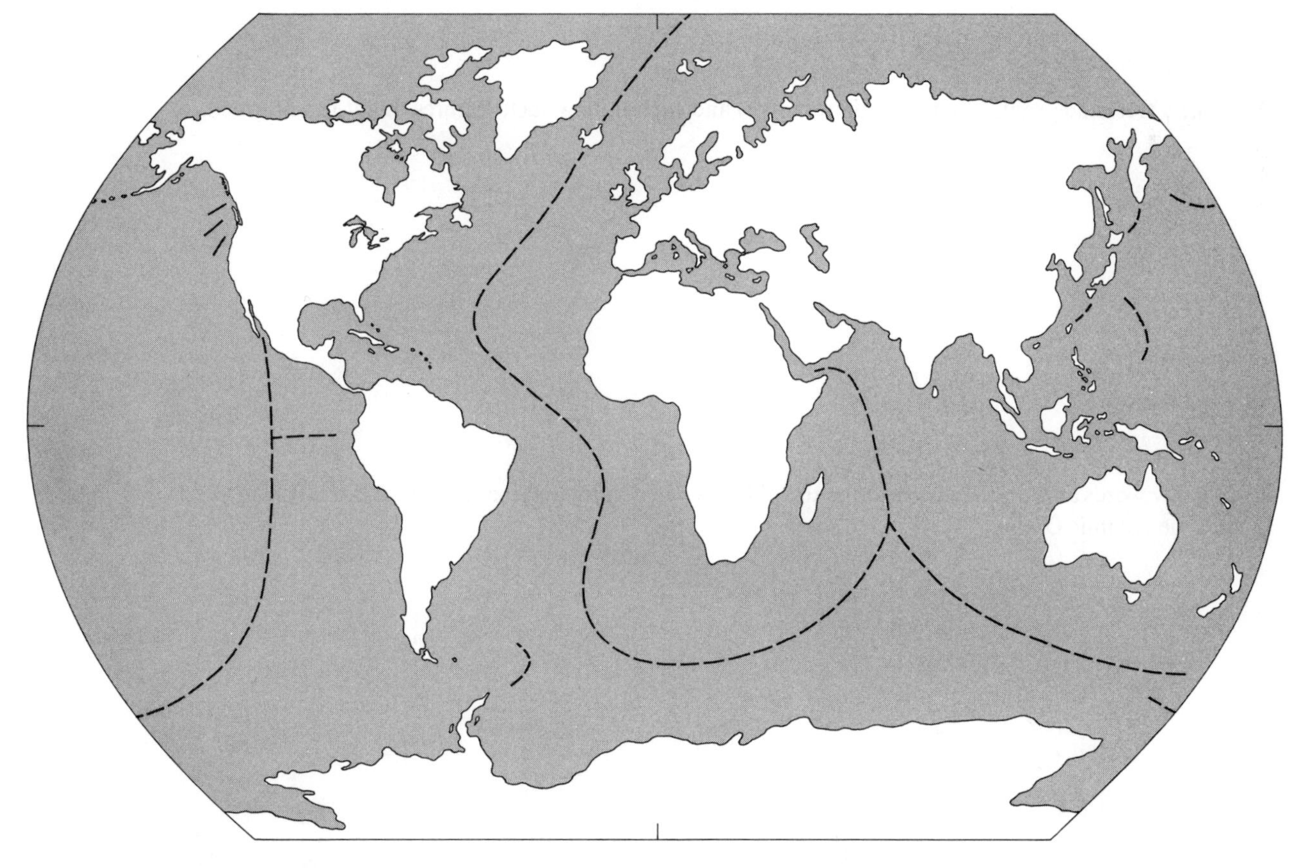

FIGURE 5-4 Worksheet for location of geographic features of the sea floor. The mid-ocean ridge system is indicated by the dashed lines.

2. What kind or class of fault is represented by the Romanche and Eltanin fracture zones? Make a rough sketch of the relative movement and location of earthquake epicenters on this type of fault (see Exercise 4).

3. Why are abyssal plains larger and more numerous in the Atlantic Ocean than in the Pacific?

4. What is the geographic relationship between the Mid-Atlantic Ridge, Mid-Indian Ocean Ridge, and the East Pacific Rise?

5. Make a freehand profile of a tablemount (guyot) and explain how such features might have formed.

6. In a few words and/or sketches describe the difference in continental margins around the Pacific Ocean and the Atlantic Ocean.

7. Where is the deepest spot in the oceans to be found?

8. What appears to be the geographic relationship between the Carlsberg Ridge and the Red Sea trough?

9. What geographic relationship does Iceland have with the Reykjanes Ridge?

10. Name the following from the map in Figure 5-3.

(a) An active subduction zone

(b) A spreading center

(c) A transform fault

(d) A constructive plate boundary

(e) A destructive plate boundary

(f) An active continental margin

(g) A passive continental margin

EXERCISE 6

MATERIALS OF THE SEA FLOOR

The variety of sedimentary materials on the sea floor is rich indeed. As you might expect, marine deposits contain an abundance of **rock** and **mineral** fragments eroded from the land, both on the continental shelves and on the deep-sea floor. However, much of the ocean floor is covered by the microscopic shells, or "tests," of small marine animals and plants. These organic **oozes** (see Figure 6-1) may be either calcium carbonate or silica, but they are the product of biological activity rather than of weathering and erosion of the land. In addition, marine **sediment** is also derived from dust and larger fragments resulting from explosive volcanic action. Some older material forms new minerals by chemical reactions with seawater, and nodules of potentially valuable elements may form atom by atom, around a suitable nucleus. Finally, a very small contribution comes from outer space in the form of tiny meteorites.

FIGURE 6-1 Foraminiferal ooze dredge from a depth of 450 meters off the coast of Central America. [Photograph by Patsy J. Smith, U.S. Geological Survey.]

THE CLASSIFICATION OF MARINE SEDIMENTS

Marine sediments may be classified into three groups according to origin or source regardless of how they were transported to their final resting place.

1. *Terrigenous deposits.* These are derived from the land and include both organic (shell) and inorganic (rock) components. These deposits are generally found in shallow water and contain some coarse material. They are classified as gravel, sand, silt, or mud according to their texture.
2. *Pelagic deposits.* These are formed far from land and settle grain by grain in the deeper parts of the ocean. They are predominantly fine grained or colloidal (less than 10 micrometers in diameter) in size, and cover about 74 percent of the sea floor.
3. *Hemipelagic deposits.* These are mixtures of components of the pelagic and terrigenous sediment groups. They are typically found closer to the continents within the limits of distribution of terrigenous particles but where the rate of contribution of these particles is still slow as compared to that of the pelagic contribution. As you might guess, hemiplagic deposits combine the characteristics of the sediments in the other two groups.

The first group, terrigenous deposits, may be classified according to texture without regard to origin (see Table 6-1). Thus a very coarse-grained deposit would be termed a gravel regardless of its origin, for example, whether it was composed of organic or inorganic components. Terrigenous deposits are difficult to characterize in a general statement because they accumulate in shallow or marginal areas subject to variability in supply and intensity of transporting agents. Deposits off mountainous coasts differ from those off coastal plains, and sediments derived from glacial terrains, such as Antarctica or Greenland, have distinct properties that indicate their origin. Thus each area of terrigenous sedimentation must be considered a unique environment, and only after intensive investigation can conclusions be made about the origin of the deposit.

In comparison, pelagic deposits are deposited in a reasonably uniform deep-water environment and are much less variable than terrigenous deposits. They may be classified as follows, according to the dominant constituent making up the sediment:

1. *Biogenous.* Organic sediments composed of the skeletal (shell) remains of marine plants and animals.
2. *Lithogenous.* Inorganic sediment composed of rock and mineral fragments.
3. *Hydrogenous.* Formed in place on the sea floor by chemical reactions with seawater. Commonly called authigenic (self-forming) by marine geologists and oceanographers.

Pelagic deposits were first classified by Sir John Murray and A. F. Renard in 1981 based upon samples collected during the voyage of H.M.S. *Challenger* (1872–1876). Since that time a great deal of work has been done on pelagic sediments, mostly on the identification and origin of their components.

However, most classification systems follow the general scheme shown below:

Pelagic deposits	Components
Biogenous (organic)	Siliceous ooze Calcareous ooze
Lithogenous (inorganic)	Red clay Brown mud Volcanic ash Turbidites
Hydrogenous	Manganese nodules Zeolites

Now let us examine the various types of pelagic deposits.

TABLE 6-1
Sediments and their grain size*

Name	Grain diameter (millimeters)	Grain diameter (micrometers)†
Gravel		
Boulder	>256**	256×10^3
Cobble	64–256	$64–256 \times 10^3$
Pebble	4–64	$4–64 \times 10^3$
Granule	2–4	$2–4 \times 10^3$
Sand		
Very coarse	1.0–2.0	1000–2000
Coarse	0.50–1.0	500–1000
Medium	0.25–0.50	250–500
Fine	0.125–0.25	125–250
Very fine	0.062–0.125	62–125
Silt	0.004–0.062	4–62
Clay	<0.004**	<4**

* Size scale developed by C. K. Wentworth in 1922.
† One micrometer equals one one-thousandth (0.001) of a millimeter. It is the metric unit commonly used to measure fine-grained sediments and microscopic organisms.
** > means "greater than"; < means "less than."

THE BIOGENOUS SEDIMENTS

Biogenous deposits, known as *oozes,* consist of more than 30 percent skeletal debris and may be divided into the taxonomic and chemical groups in the following list:

Siliceous ooze ($SiO_2 \cdot nH_2O$, opal)

Radiolarian ooze. Small (50–100 micrometers) marine protozoa found in abundance in the equatorial waters of the oceans (Figure 6-2).

Diatom ooze. Small (10–500 micrometers) single-celled plants most abundant in polar regions with nutrient-rich waters (Figure 6-3)

Calcareous ooze ($CaCO_3$, limestone and chalk)

Foraminiferal *(Globigerina)* ooze. Skeletons of microscopic marine protozoan from about 1 to 300 micrometers in size (Figure 6-4).

Coccolith ooze. Plates and fragments of very tiny plants on the order of 1–20 micrometers (Figure 6-5).

Pteropod ooze. Small (1–2 millimeters) shells of a pelagic gastropod; covers a small area of the sea floor compared to other oozes.

Whether a biogenic ooze forms on the sea floor depends upon the amount of biological production in surface waters, upon dilution on the bottom with other sediments, and upon destruction by scavengers or chemical solution.

Carbonate ooze is not known to form on the present sea floor below a depth of 4500 meters, even though there may be high production rates of calcareous organisms in the overlying waters. The reason is that below this depth, which is known as the "snow line," or **carbonate compensation depth,** (CCD), the shells are dissolved.

The exact process and rate of carbonate dissolution are known only generally, but the top of the zone of complete dissolution of calcium carbonate ($CaCO_3$) usually coincides with the top of the **Antarctic bottom water** (see Exercise 10). This water mass is a cold, high-pressure (deep) water that forms in Antarctic surface waters as pack ice forms. It is saturated with carbon dioxide at the surface, but as it sinks toward the deep-ocean floor it becomes undersaturated because of the increased pressure. (Gas saturation is determined by temperature and pressure; solubility increases at low temperatures and at high pressures.) The increased dissolving capacity causes solution of the carbonate particles that fall through this water. Another factor related to dissolution is the amount of production of calcareous organisms in the surface waters. Where production is high, bacterial destruction of organic matter falling through the deeper waters uses dissolved oxygen and increases the carbon dioxide content. These factors also increase the capacity of the deeper waters to dissolve carbonate. Therefore, increased production is compensated by in-

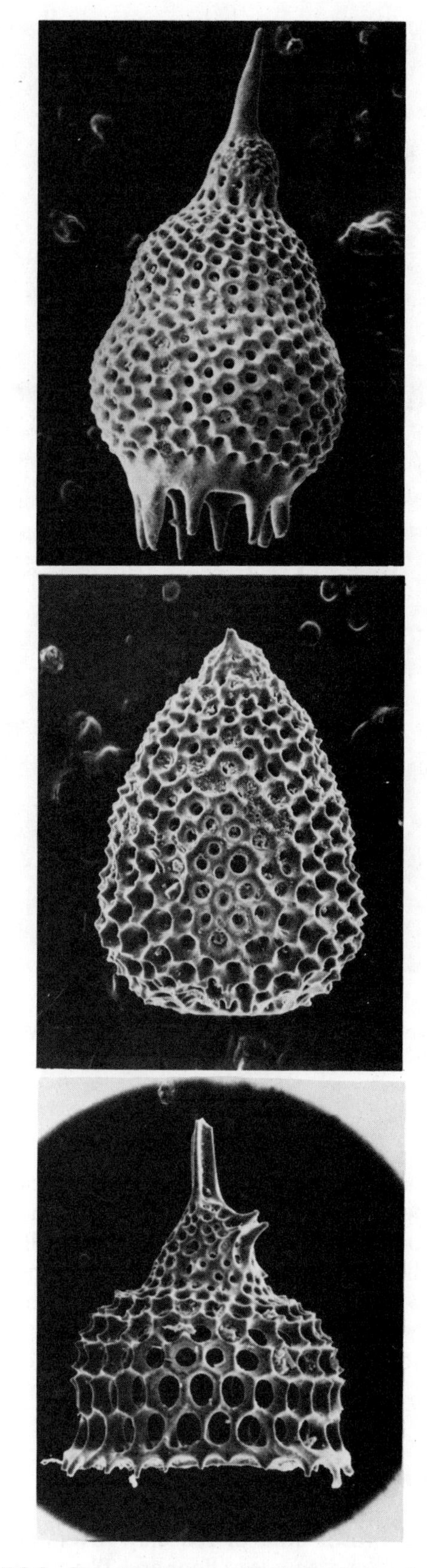

FIGURE 6-2 Scanning electron microscope photos of modern radiolarians magnified many times. [Courtesy Fritz Theyer.]

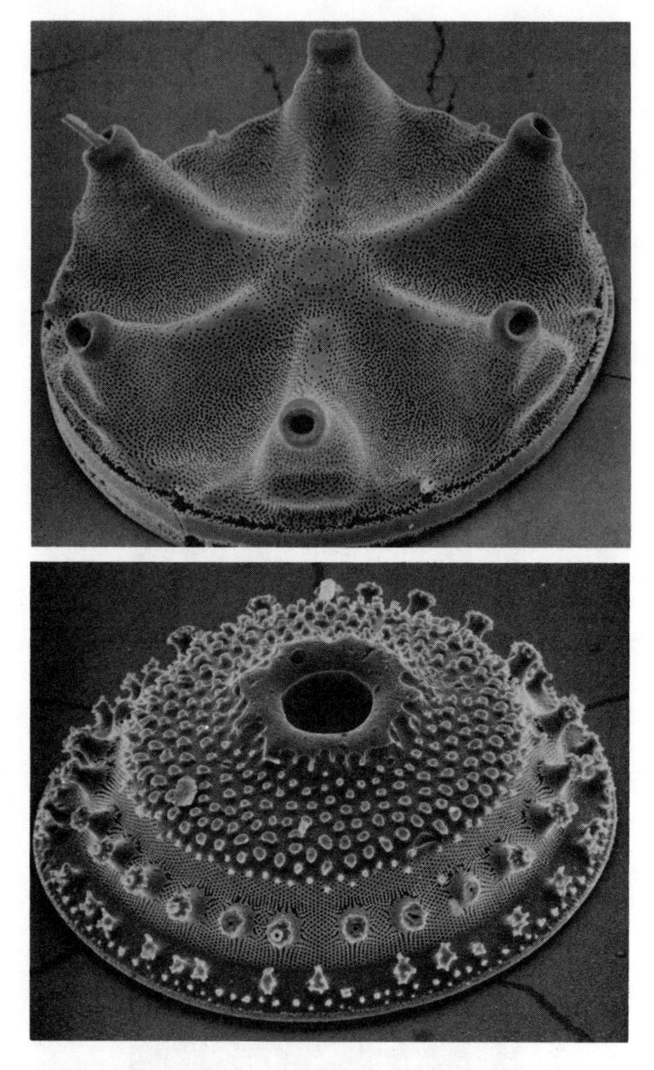

FIGURE 6-3 Diatom tests magnified many times with the scanning electron microscope. [Courtesy Anderson, Wilcoxon and Associates, Inc., San Diego, California.]

creased solution (the zone of solution is called the **lysocline**), although the net result is carbonate addition to oceanic sediments on the sea floor. All carbonate is dissolved below the carbonate compensation depth, which has varied over geologic time in relation to changes in ocean productivity. Its present level, at about 4500 meters, varies somewhat in each ocean basin.

THE LITHOGENOUS SEDIMENTS

So-called brown mud or red clay covers about 38 percent of the deep-ocean floor and 28 percent of the total sea floor. Because of the great length of time it takes for this fine-grained material to settle through the water column, there is ample time for iron in the sediment or in the water to react with dissolved oxygen to create a brown or reddish coating on a sediment grain. These are among the most slowly accumulating deposits known, the average rate being about 1 millimeter per 1000 years. The origin of these sediments is atmospheric dust, fine-grained rock debris from land, micrometeorites, volcanic ash and dust, and the insoluble residues from dissolved carbonate oozes. Red clay is found almost everywhere on the sea floor below 4500 meters, except for the deep trenches, where a mixture of coarser terrigenous material is to be found.

Terrigenous sediments are also mostly lithogenous, since they are derived from rock weathering. These sediments are the major contribution to the ocean sedimentary budget, although most terrigenous sediment is deposited on or near the margins of the continents. The major contribution of terrigenous sediment comes from a few very large rivers. The three largest Asian rivers, the Ganges, Yellow, and Yangtze rivers, contribute about one-quarter of the world total of terrigenous sediment. Approximately 75 percent is contributed by the 20 largest rivers, almost all of which drain into the Atlantic or into marginal seas. Because of the plate collision and coastal ranges around much of the open Pacific Ocean margin, few large rivers enter directly into that ocean, although several enter marginal seas (the Yellow Sea or China Sea, for instance). Where these large rivers dump sediment into the ocean, their contribution dominates the sea-floor sediments for thousands of kilometers away from the river mouth or major delta. The coast-parallel mountain ranges limit the drainage area available to streams and thus limit the sediment discharge to active margins.

THE HYDROGENOUS SEDIMENTS

Where the process of sedimentation is very slow, as in the red clay areas, manganese nodules may form (Figure 6-6). The nodules grow slowly, particle by particle, on a nucleus of skeletal material or mineral matter. Therefore, a slow rate of sedimentation of other components is required or else they would be buried and removed from interaction with seawater. It is estimated that manganese nodules cover about 10 percent of the deep Pacific Ocean floor, and a somewhat smaller area of the Atlantic and Indian ocean basins. They are composed mostly of iron and manganese oxides, but are of economic interest for the small amounts of copper, nickel, and cobalt they contain.

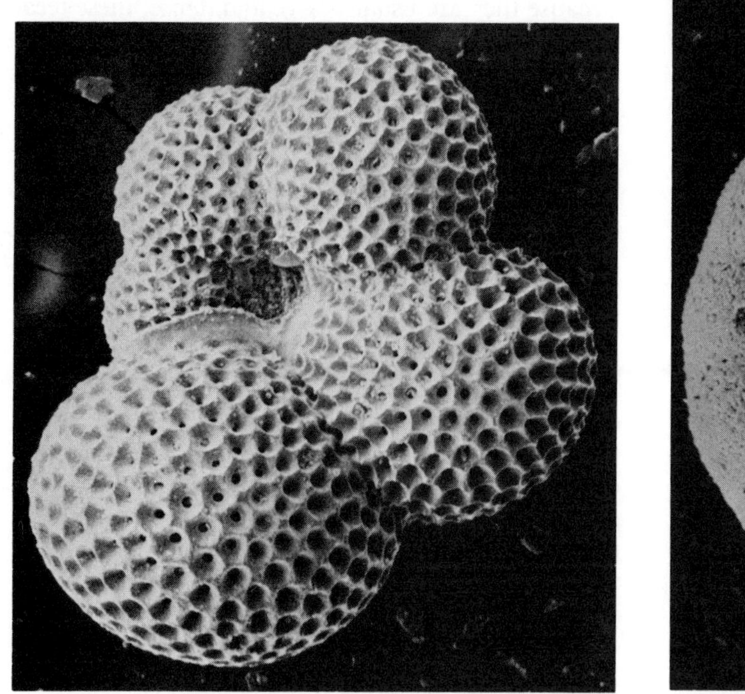

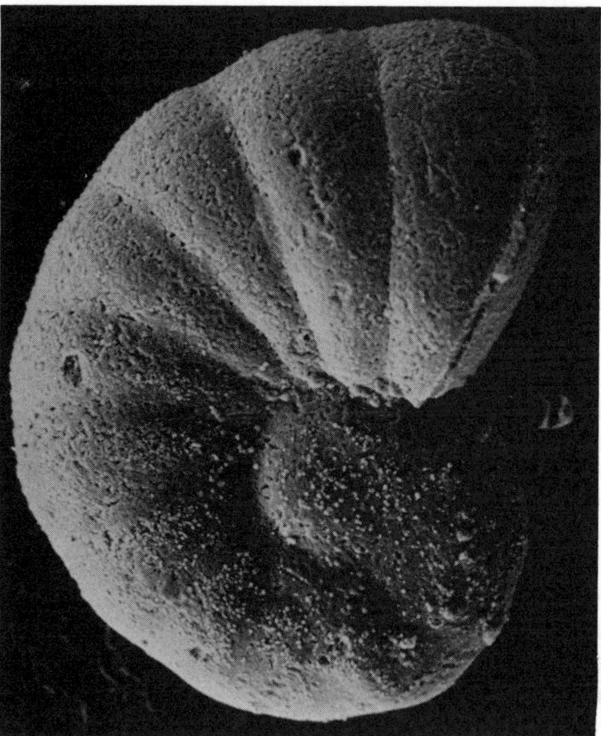

FIGURE 6-4 Tests of foraminifera enlarged many times. Note the fine structure and ornamentation of the shell. [Courtesy Anderson, Wilcoxon and Associates, Inc., San Diego, California.]

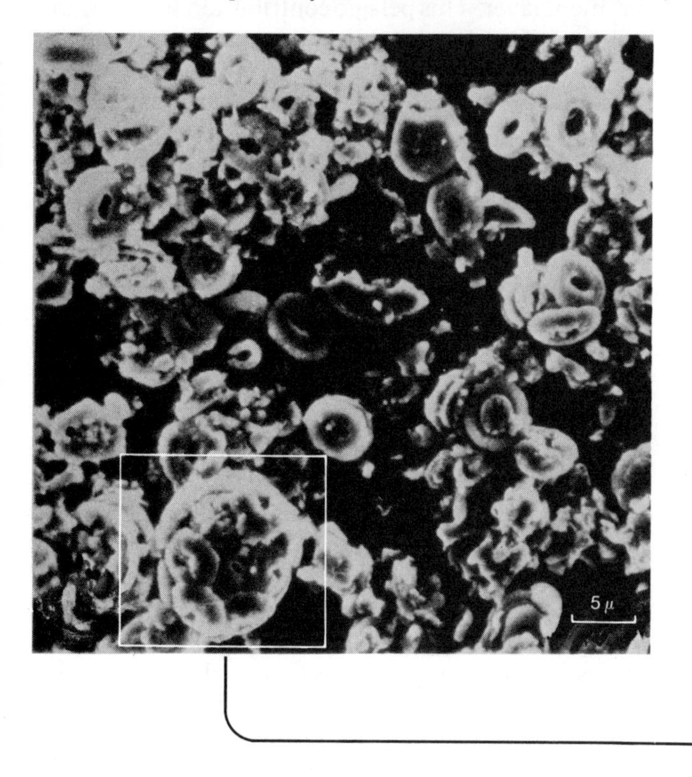

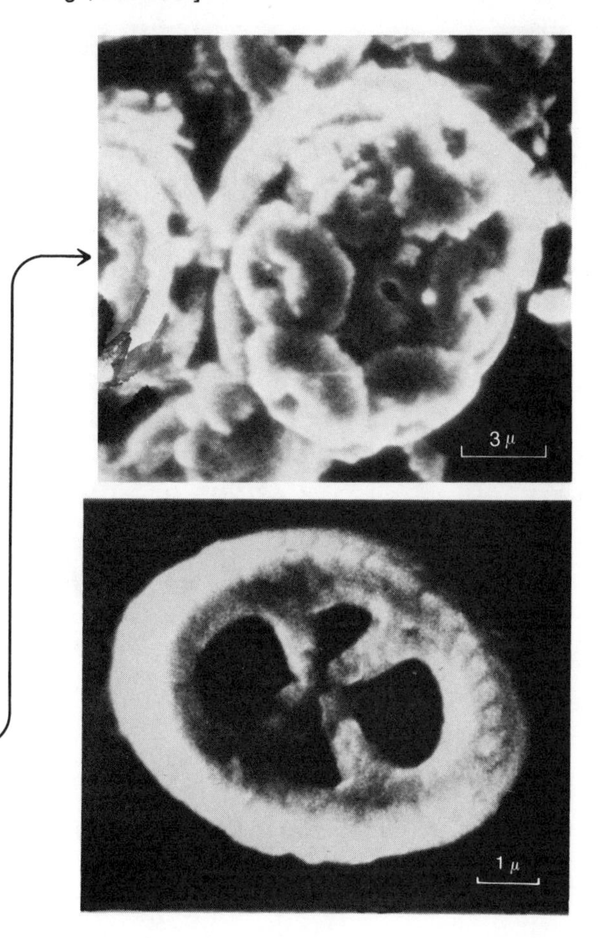

FIGURE 6-5 Calcareous plants (coccoliths) showing occurrence in ooze and enlarged details of skeleton. [From H. W. Menard, *Geology, Resources, and Society*. W. H. Freeman and Company. Copyright © 1974.]

We will probably see nodules mined from the sea floor in large quantities within the next decade. Nodules have precipitated faster during some parts of the geologic history of the oceans than during others. Also, they tend to precipitate on ridge flanks faster than on older, deeper sea floors away from the ridges.

SUSPENDED SEDIMENT TRANSPORTATION

Much fine silt and clay remains in suspension after it has been introduced to the ocean from a river. The rate at which these small particles (usually less than 0.03 millimeter, or 30 micrometers, in diameter) settle is very slow, and so very slow currents can keep them suspended. However, coastal waters are also biologically very productive, and therefore the surface waters are populated by many microscopic and larger organisms that constantly filter or sweep the water for food particles. As they capture these particles they also capture the lithogenous fine mineral particles, which pass through the gut and are excreted as aggregates of particles with some binding organic matter. Because they are usually large and dense, these fecal pellets fall rapidly. Thus the organisms "vacuum clean" the waters, stripping out particles and sending them to the bottom as pellets. This explains why almost 90 percent of the terrigenous sediments contributed to the oceans settle out on the margins within a few hundred kilometers of their sources. This also explains why so little fine sediment escapes to the deep-ocean floor and why the pelagic clays are mainly windborne dust or oceanic volcanic dust.

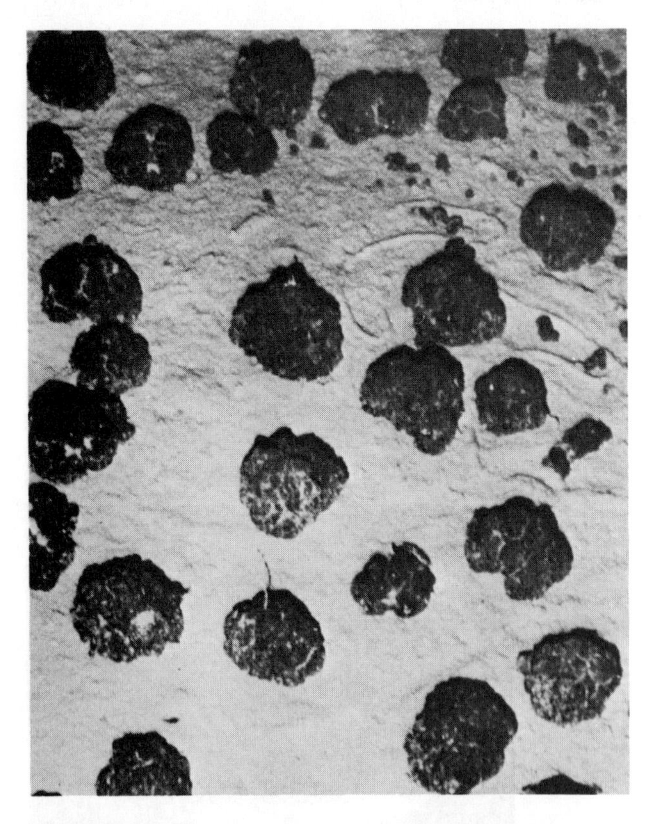

FIGURE 6-6 A field of grapelike manganese nodules in the abyssal Pacific at a depth of 5145 meters. [From *The Face of the Deep* by Bruce C. Heezen and Charles D. Hollister. Copyright © 1971 by Oxford University Press, Inc. Reprinted by permission.]

The suspended particles sink out of the surface waters, or pellets that are smaller and less dense are broken up as they sink and contribute to deeper layers of fine particle concentration. These zones of higher particle content typically are found at or near the ocean floor, at changes in water density, and at the surface near the continents. They have been called **nepheloid layers,** a designation derived from the Greek word for cloud. Although the concentrations are very low and would be difficult to see with the unaided eye, the deepest major nepheloid layer in the Atlantic contains about 100,000,000 tons of material at any given time. These layers move with the currents, and material settles out to add to the deep-sea sediment layer. This pelagic contribution settles more or less uniformly over the deep-ocean topography.

BOTTOM-TRANSPORTED PELAGIC SEDIMENTS

In addition to particles that settle grain by grain on the deep-sea floor, there are more coarsely textured bottom-transported components that reach abyssal depths. These materials are transported by dense sediment-laden turbidity currents that slide from continental slopes or submarine canyons and travel along the bottom to the deepest ocean floor. Such currents have actually been observed and are known to attain velocities greater than 20 kilometers per hour. Transport by turbidity currents explains sand and silt layers, some containing land plants and shallow-water shells, that are found in abyssal-plain and trench sediments. As the sediment load in the current settles out, distinctive deposits, called *turbidites,* in which individual grains grade upward from coarse to fine, are formed. These currents suggest a transporting mechanism for part of the thick blanket of sediment found on the abyssal plains of the Atlantic Ocean and the great wedge of terrigenous material trapped in deep trenches around the Pacific Ocean.

Because the largest amount of terrigenous sediment

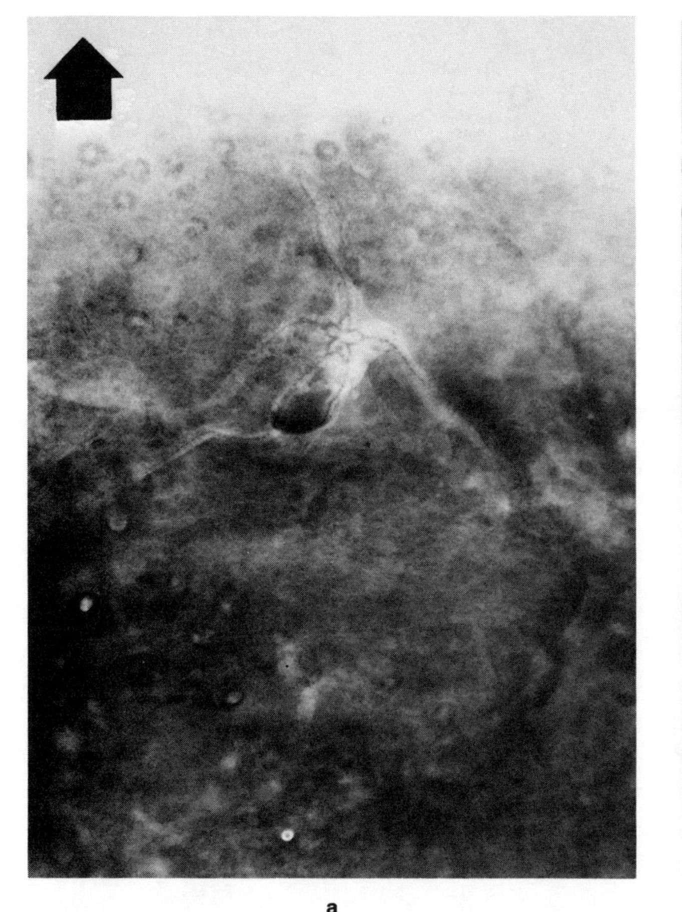

a

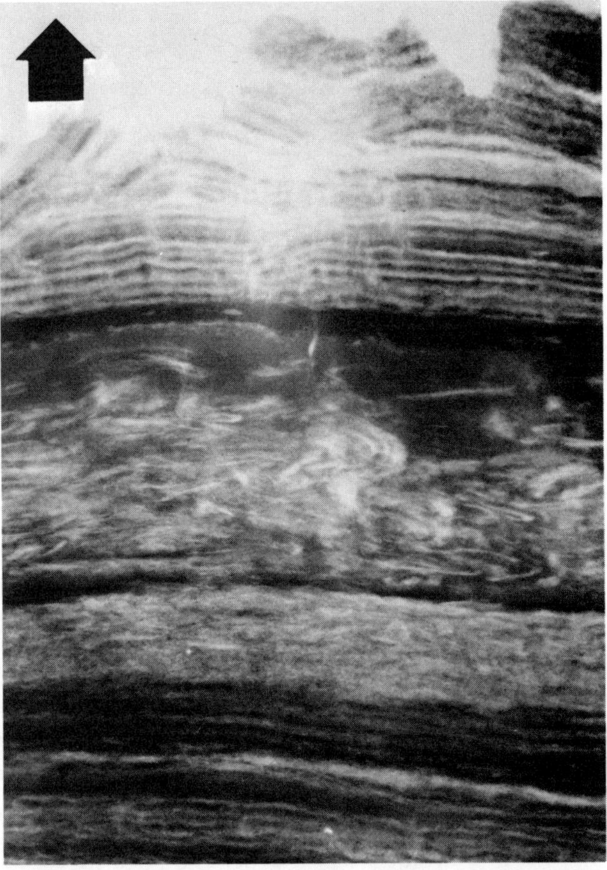

b

FIGURE 6-7 (a) Excellent slab photograph example of the activity of burrowing organisms as they mix bottom sediments. This sample was deposited in depths of about 100 meters off southern California. There was enough oxygen in the bottom waters to support animal populations. The x-ray shows a starfish in its burrow and a small clam. The sediments have been thoroughly stirred and mixed many times. [Courtesy D. S. Gorgline, University of Southern Californa.] (b) Print of an x-ray photograph of a slab from a core of modern deep-water sediments in Santa Barbara Basin off southern California. Water depth is 600 meters. Because there is no oxygen in the bottom water, no animals live on or in these sediments, and so the annual additions of clays are preserved as thin layers. The thickness of layers is a function of the amount of rainfall each year and resulting changes in river delivery of sediment to the sea floor. At a depth in the core of a few centimeters there is a disturbed zone that indicates that the surface sediment has moved relative to the lower sediment. This slippage also explains the irregular broken surface of the core.

first accumulates on the slopes of the continental margins, as discussed earlier, these deposits eventually slide to the base of the continental slope. These large **slides** are sometimes triggered by local earthquake shocks. As the masses of sediment slide they generate turbidity currents as well. Where sedimentation rate is fast these masses can be very large, sometimes many kilometers on a side and hundreds of meters thick. Along the base of the Atlantic margin they form irregular features of large dimensions. Much smaller slides, or **mass movements,** are common wherever sediment accumulates on slopes. This process may be the most important means by which sediment is carried to the deep-sea floor.

INTERPRETATION OF MARINE SEDIMENTS

The ocean basins are a major receptacle for weathering products of **rocks** and **minerals** on land. These products are transported as discrete particles or as dissolved matter in water and may be deposited along riverbanks or in lakes, lagoons, or the ocean basins. In time the deposits become sedimentary rocks, and by studying the mineral composition, structures, and textures of ancient rocks we are able to determine the physical environment in which the sediments were deposited. This is of practical significance because we now know that large oil accumulations may exist in sand bodies in ancient deltas that were deposited

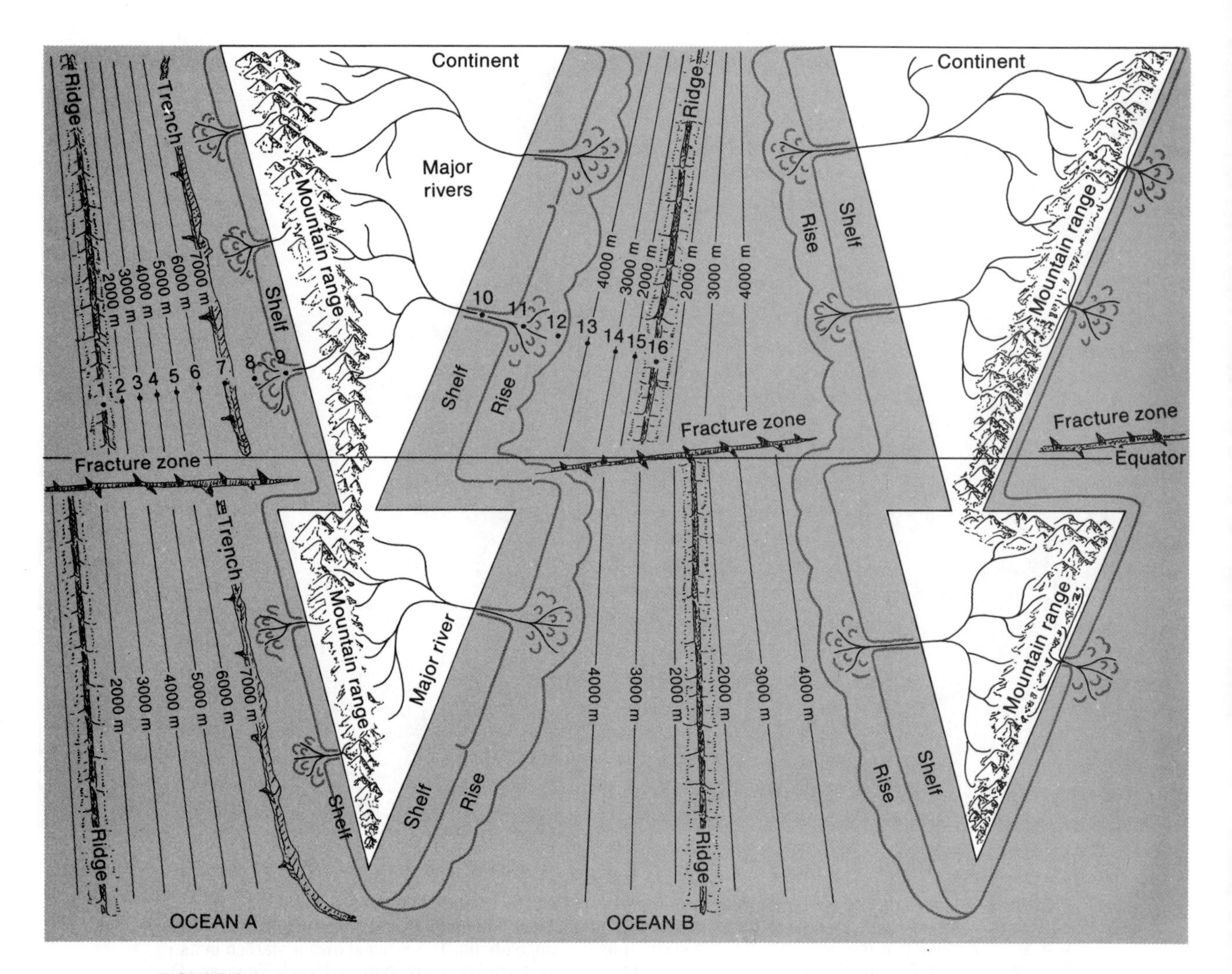

FIGURE 6-8 Map of hypothetical ocean basins and continents showing location of core holes on sea floor. Contours indicate depths of the sea floor in meters.

where a river entered the sea. These coarse-grained layers make excellent reservoirs for oil and gas, and finding them is facilitated by knowing something of the sediments and sedimentary patterns in the region. Similarly, deep-sea deposits are quite different from deposits formed on the continental shelves, and these dissimilarities aid in reconstructing ancient geography. In other words, by studying modern sediments we may be able to interpret the depositional environment of ancient rocks. Was the rock deposited in deep or shallow water, fresh or salt, near shore or far from land? These and other questions may be answered by geologic research and a better understanding of modern sedimentary processes.

Sediments deposited on the sea floor are not yet at the end of their journey. Deep currents can resuspend the fine muds much as windstorms on land move sediment as dust storms or sandstorms. Animals living in and on the bottom sediments constantly dig and burrow and stir the deposits in their search for food; therefore, the original layering of the settling particles is usually destroyed by this constant mixing. Only where the bottom water is devoid of oxygen because of rapid utilization of the oxygen do we see the original layering preserved. Examples of both of these conditions are shown in Figure 6-7. In the oxygen-free **(anoxic)** areas the organic matter deposited in the sediments also tends to be preserved, and these types of deposits may be the major eventual oil sources, or **source rocks.**

Material carried to the sea cannot accumulate indefinitely. If dissolved substances were not extracted from seawater, the chemistry of the oceans would change in a few million years, and if sediments were not converted into sedimentary rocks to become part of the land, the ocean basins would fill completely in a few hundred million years. Thus a cycle recurs whereby rocks are weathered to form sediment that is

TABLE 6-2
Core descriptions

Core number	Description	Percentage of calcium carbonate in top 10 centimeters	Water depth (meters)
1	Fragments of volcanic rock, some ooze	—	1500
2	10 meters of *Globigerina* ooze, some volcanic fragments and volcanic ash in top meter	90	2000
3	10 meters of ooze, manganese nodules at top	80	3000
4	9 meters of ooze, manganese nodules at top	75	4000
5	3 meters of reddish clay and ooze over 3 meters of *Globigerina* ooze	15	5000
6	8 meters of red clay	3	6000
7	9 meters of red clay	5	7000
8	1 meter of sand grading from fine at the top to coarse at the bottom, over 50 centimeters of muddy clay, over 60 centimeters of graded sand, over 3 meters of red clay	15	4500
9	6 meters of alternating gray clays and brown silts with some fine to coarse sand layers, one sand layer at base with gravel at bottom	—	1000
10	4 meters of coarse sands and gravels	—	500
11	7 meters of silts with sand and gravel layers averaging about 40 centimeters thick	—	1500
12	9 meters of gray muds with fine sand and silt layers averaging about 20 centimeters thick	—	3000
13	10 meters of gray and brown clays and muds, several fine sand and silt layers 10 centimeters thick	—	4000
14	10 meters of brown mud with a few silt layers about 5 centimeters thick	50	3000
15	10 meters of brownish ooze	70	2000
16	Rock fragments and volcanic ash, rock is basalt	—	1500

transported to a depositional site to become rock again. Certainly this "recycling" of earth materials has occurred many times throughout earth history.

SAMPLE COLLECTION

Samples of the sea bottom can be collected by a number of different methods. Dredges of many designs, consisting of a simple frame with a net or bag attached, can be dragged along the bottom to collect materials. Weighted tubes can be dropped into the bottom and a core of the materials collected—a procedure that is much like pushing a cookie cutter into dough. Samplers with spring-loaded jaws can be used to grab a bite of the bottom materials.

The best samplers are those that collect an undisturbed core from unconsolidated bottom sediments, because cores display the sedimentary history of a given area. Layers of sediment represent a finite period of time, and the kinds of sediment tell us what the conditions were at the time of deposition. We can use the biogenous material to tell us about water temperature and, if radioactive materials are present, we can use various chemical methods to determine ages.

SAMPLE LOCATION AND ANALYSIS

A list of hypothetical core samples, and their description, is given in Table 6-2. Figure 6-8 is a hypothetical chart of oceans and continents in which we see the locations and approximate depths of these core samples, designated by sample number. Study both the table and the figure carefully, and refer to them as necessary in order to answer Questions 1–8 in the report form.

DEFINITIONS

Anoxic. Without oxygen; denotes water or sediment from which free oxygen has been removed by oxidation of organic matter or by iron or sulfide in oxidation processes.

Antarctic bottom water. Forms around the margin of Antarctica where formation of sea ice produces a cold, saline, dense water that sinks to the deep-ocean floor.

Carbonate compensation depth (CCD). Depth in the ocean where the solution of carbonate equals the supply of carbonate; it changes with time in response to changes in ocean biological production and ocean circulation driven by world climate changes.

Lysocline. Depth zone in the oceans in which solution of carbonate occurs; lower boundary is the CCD.

Minerals. Naturally occurring, inorganic, crystalline substances with a definite range of chemical composition and reasonably definite physical properties.

Nepheloid layer. Zone of increased suspended sediment content (turbidity) in the ocean water column. Such zones typically occur at the surface in coastal waters, at mid-depth associated with density changes, and at the bottom.

Ooze. A deep-sea sediment consisting of at least 30 percent of skeletal remains of microscopic floating organisms. The remainder is mostly fine-grained clay minerals.

Reservoir rocks. After burial and lithification, coarse deposits of sand or gravel have a high porosity and if oil and gas are generated nearby, these rocks can become saturated with hydrocarbons to form "pools," or reservoirs, much as sponges sop up water in their openings to hold a large amount of fluid.

Rocks. An aggregate of one or more minerals rather large in area. The three classes of rocks are the following: (1) *Igneous rock.* Crystalline rocks formed from molten material. Examples are granite and basalt. (2) *Sedimentary rock.* A rock resulting from the consolidation or cementation of loose sediment that has accumulated in layers. Examples are sandstone, shale, and limestone. (3) *Metamorphic rock.* Rock that has formed from preexisting rock as a result of heat, pressure, or chemically active fluids.

Sediment. Loose fragments of rocks, minerals, or organic material that are transported from their source for varying distances and deposited by air, wind, ice, and water. Other sediments are precipitated from the overlying water or form chemically in place. Sediment includes all the unconsolidated materials on the sea floor.

Slide (mass movement). A mass of sediment initially accumulated on a slope that eventually moves downslope as a mass; can be triggered by earthquakes or overloading.

Source rocks. Where conditions permit the accumulation of organic matter, the slow change in this material after burial can produce hydrocarbons. Typically, high organic content is found in fine sediments that accumulated in areas of high biological production and low oxygen content in the receiving waters; these areas are the potential sources of oil and gas.

EXERCISE 6

REPORT

MATERIALS OF THE SEA FLOOR

NAME

DATE

INSTRUCTOR

1. (a) Plot the data for percentage of calcium carbonate (see Table 6-2) versus depth (see Figure 6-8) from cores 1 – 16 on the graph below. Use a dot for each sample.

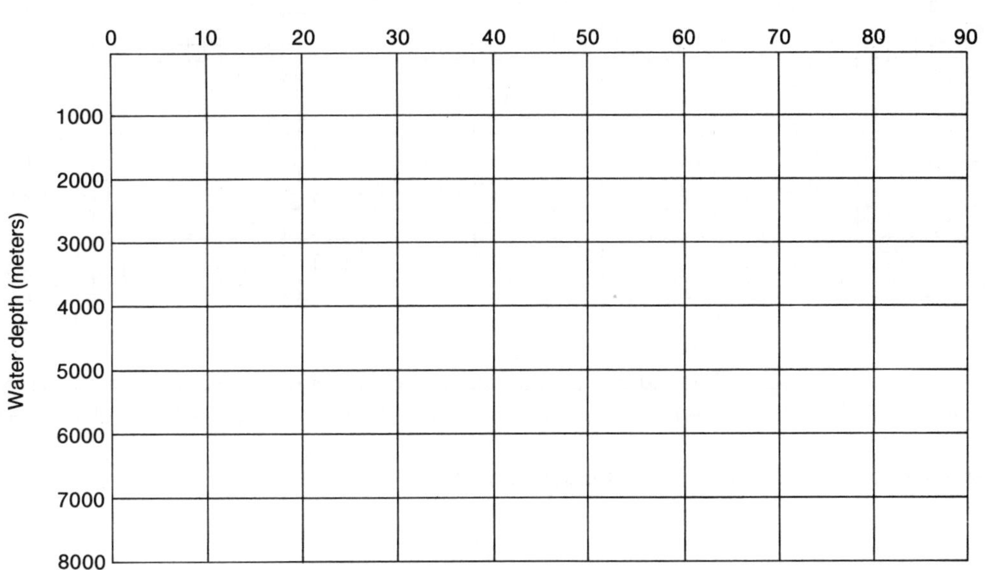

(b) Give an explanation for this sharp reduction in percentage of calcium carbonate below a depth of 4000 meters in both oceans.

2. (a) From Figure 6-8 plot a depth profile of each line of cores (1–9 in Ocean A, 10–16 in Ocean B) on the graph below.

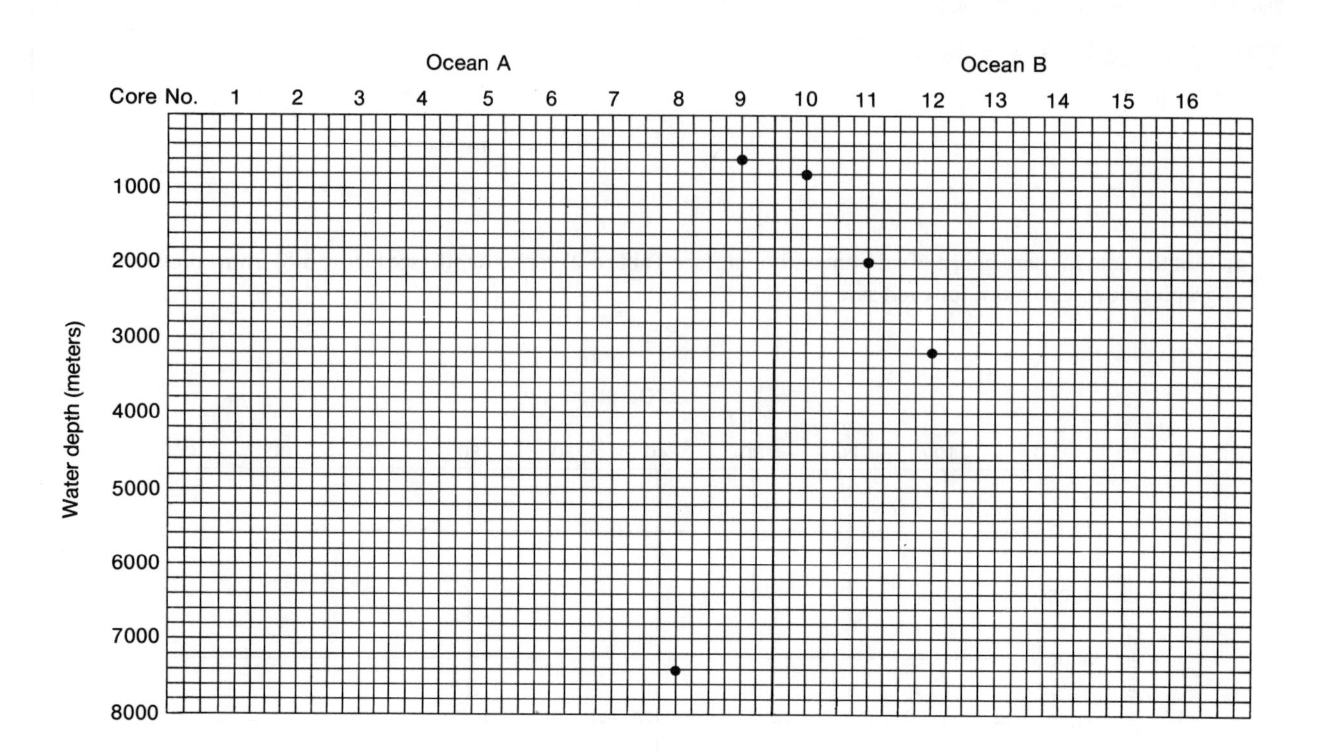

(b) On the profile use colored pencils to indicate the dominant types of sediment present, and label the sediment types. Use red for clays, blue for oozes, yellow for sand, silt, or terrigenous clays, and black for volcanic rock or basalt.

3. Why do the sediments change as the continent is approached?

4. Why are there no red clays in the Ocean B profile?

5. If oozes and red clays are pelagic (oceanic), why are there none in cores 8–13?

6. How do you explain the oozes below the red clay in core 5?

7. There are sands in core 14 in Ocean B at 3000 meters in depth but none in core 3, which is also at 3000 meters, in Ocean A. Suggest the reason for this difference.

8. Red clay on the deep-sea floor accumulates at an average rate of about 1 millimeter per 1000 years. Oozes accumulate at rates that are often ten times faster.

   (a) How much time would be required to deposit 1 inch of red clay?

   (b) How much time would a core 10 meters in length represent if it contained 5 meters of ooze and 5 meters of red clay?

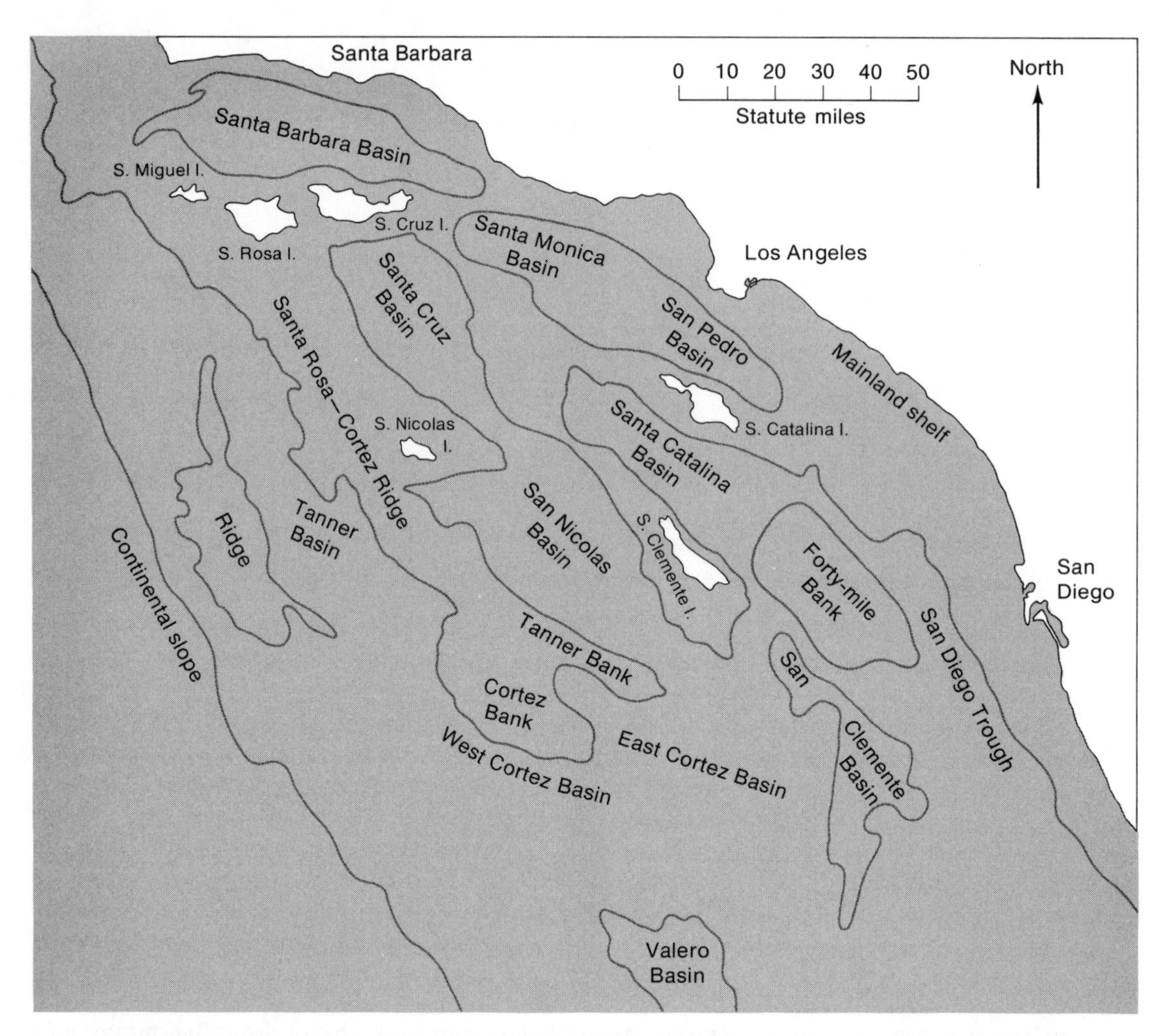

FIGURE 6-9 Simplified diagram of borderland off southern California showing major sedimentary basins and intervening ridges. Note that the channel islands represent portions of ridges that are above sea level.

9. Off the coast of California a special form of sea floor has developed because of the collision of a mid-ocean ridge and a continental plate. The resulting topography, shown in Figure 6-9, is called a *continental borderland* and is a checkerboard of deep basins, shallow banks, and islands. In the deep basins some representative sedimentation rates are as follows: San Pedro Basin, 54 centimeters per 1000 years; Santa Catalina Basin, 29 centimeters per 1000 years; San Nicolas Basin, 12 centimeters per 1000 years.

(a) Study the figure and explain the differences in sedimentation rates.

(b) The Pliocene sediments on land in the Los Angeles Basin are about 2100 meters thick. Assuming the soft sediments compact to about 35 percent of their original thickness (65 percent compaction), how long would it take for a similar thickness of sediment to accumulate in the San Pedro Basin? (Effective rate of sedimentation is 0.35×54 centimeters per 1000 years = 18.9 centimeters per 1000 years, say 19 centimeters per 1000 years.)

Is this a reasonable figure in view of the lengths of the epochs of the Tertiary period?
(See the time scale in Appendix B.)

EXERCISE 7

SOLAR RADIATION AND HEAT BALANCE

The solid earth, the oceans, and the atmosphere constitute a system that receives and exchanges radiant energy from the sun. This energy is not distributed around the earth equally: the equator is so oriented that it receives the most, whereas the polar regions receive the least. Much of this incoming energy is radiated back to space, but there remains a net surplus, or gain, at the equator and a deficit at the poles. In order that the tropics not become unbearably hot and the poles even more frigid, a portion of this surplus must be transferred to cooler regions. Because the solid earth is a poor conductor of heat, most of the transfer required to maintain stable average annual temperatures around the world is accomplished by the circulation of the atmosphere and oceans.

Oceans currents transfer heat poleward in both hemispheres, thereby limiting the extension of ice from these regions. Heat is a *conservative* property, that is, one that remains constant for the earth as a whole for long periods of time. This means that the earth's heat gains will, in the long run, be balanced by heat losses and vice versa. For instance, if the oceans retained all the heat they gained from the sun, they would boil in 300 years! The example of heat balance we will use in this exercise is a lake in Minnesota; however, the means by which the lake gains and loses heat would be just as applicable to the world ocean, which covers 72 percent of the planet.

Source: Adapted from George Rapp, C. Matsch, R. Bartells, and E. Rapp, *Jour. Geol. Educ.*, **15**, 6, 1967, pp. 221–222. (Special Issue sponsored by the Council on Education in the Geological Sciences as CEGS Programs Publication Number 1, Problems in Physical Geology.)

HEAT AS A FORM OF ENERGY

Heat is generally understood to be a form of energy that causes a body to rise in temperature, to melt, to evaporate, or to undergo some other change as a result of its interaction with the source of the energy. The sun is the chief source of heat directed at the earth's surface; however, the sun gives off a broad spectrum of radiant energy, heat—or the energy radiated in the infrared part of the spectrum—being just part of the total (Table 7-1). Two characteristics of all forms of electromagnetic radiant energy are (1) that they travel with the speed of visible light (186,000 miles per second), and (2) that they can be transmitted through a vacuum.

TABLE 7-1
The energy distribution in the sun's spectrum

Type of radiation	Percentage of total energy
Infrared and heat rays	50.0
Visible light rays	40.5
X-rays and gamma rays	9.0
Ultraviolet rays	0.5

This radiation travels through the vacuum of outer space (taking 8.3 minutes to reach the earth from the sun) to interact eventually with the atmosphere, hydrosphere, and solid earth. Heat may also be transmitted by **conduction**—that is, through a medium—as for example when an iron bar warms up as it is heated on one end; or it may be transmitted by **convection,** whereby it is conveyed upward by a moving warm mass, such as one of water or air.

The region of the earth's outer atmosphere that is nearly at right angles to the sun's rays receives about 2 calories* of solar radiation per square centimeter per minute. This value is known as the *solar constant* and is a relatively fixed value over the long run at a given geographic location.

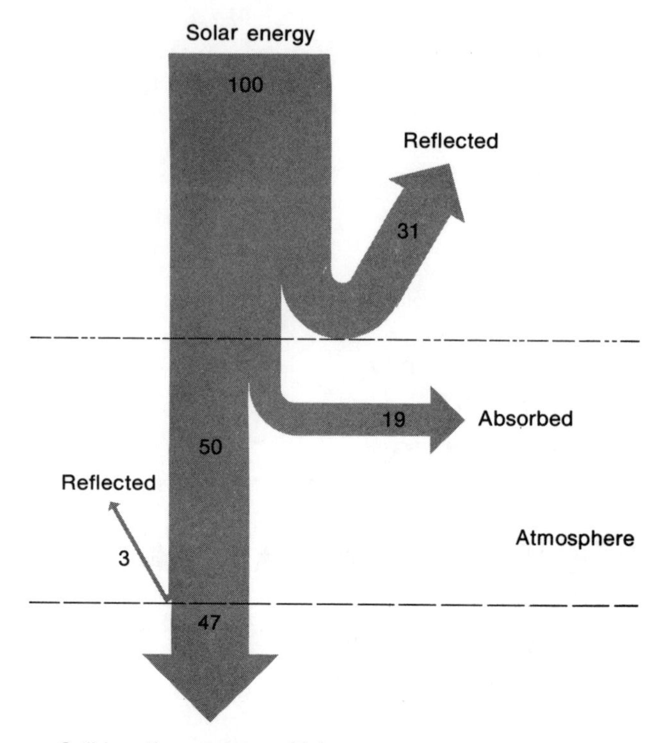

FIGURE 7-1 On a day with average cloud cover, insolation is about half the value of solar energy delivered to the outer atmosphere. The numerical values represent the percentages of solar energy. [After Robert H. Romer, *Energy: An Introduction to Physics.* W. H. Freeman and Company. Copyright © 1976.]

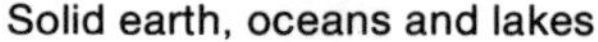

AMOUNTS OF HEAT REACHING THE EARTH'S SURFACE

The amount of heat actually received at a given point on the earth's surface **(insolation)** depends on (1) the angle between the sun's rays and that portion of the surface; (2) the length of exposure; and (3) the extent of cloud cover. Items 1 and 2 depend on latitude and season of the year.

On an average July 1 in Minnesota approximately 950 calories per square centimeter reach the upper atmosphere. As the solar radiation travels through the atmosphere, the gamma rays and x-rays are rapidly absorbed by molecules and atoms of oxygen and nitrogen. **Scattering** and reflection cause about a 10 percent decrease in the energy emitted by the entire spectrum. Water vapor in the atmosphere will absorb as much as another 10 percent of the incoming radiation. Consequently, if there is no cloud cover, no more than 75 percent of the sun's radiant energy will reach the ground. The average number of calories per day, or the amount of insolation, actually reaching each square centimeter of Minnesota's surface during July is about 525. In contrast, a heavy cloud cover may absorb and reflect enough of the effective incoming radiation to reduce the quantity to nearly zero. The diagram in Figure 7-1 shows the amount of absorption and reflection for a day with average cloud cover.

HEAT ABSORPTION AND LOSS

When sunlight strikes a body of water the radiant energy is gradually absorbed through a surface layer many meters deep. About half of the energy, including all of the infrared, is absorbed in the top 1 meter.

When water changes from liquid to the gaseous state the more rapidly moving water molecules (that is, those having the most energy) leave the surface of the water, carrying with them a large quantity of kinetic energy (so that a cooling of the surface results). The amount of heat required to evaporate a gram of water is about 600 calories, and is known as **latent heat.** The rate of evaporation depends not only on the temperature but also on vapor pressure. Wind promotes evaporation because it sweeps the vaporized water molecules away from the surface, thereby re-

* A calorie is defined as the amount of heat necessary to raise the temperature of 1 gram of water 1 Celsius degree at a standard initial temperature.

TABLE 7-2
Data required to calculate lake storage of solar energy*

Month	Incoming radiation, Q_s	Back-radiation, Q_b	Evaporative transfer, Q_e
May	475	125	50
June	525	125	175
July	525	125	250
August	475	125	225
September	350	125	175

* All values are in calories per square centimeter per day.

ducing the vapor pressure and allowing evaporation to continue.

In a mirror-like effect, some of the incoming radiation is reflected from the water surface back into the atmosphere. Also, because the water body contains heat, it **back-radiates** energy to the atmosphere in the infrared portion of the spectrum. It is estimated that the oceans back-radiate an average of 40 percent of the solar energy they receive.

The approximate values for the effective incoming radiation, Q_s, effective back-radiation, Q_b, and evaporative transfer, Q_e, that occur in a typical Minnesota lake in the course of a summer (from May through September) are given in Table 7-2. All values are given in calories per square centimeter per day and are based on the assumption that an average of 55 percent of the sun's energy penetrates the atmosphere. The amount of solar energy that remains stored in the lake for any given month can be determined by subtracting the amounts of back-radiation and evaporative transfer from the total incoming radiation: Storage $= Q_s - (Q_b + Q_e)$. See also Figure 7-2.

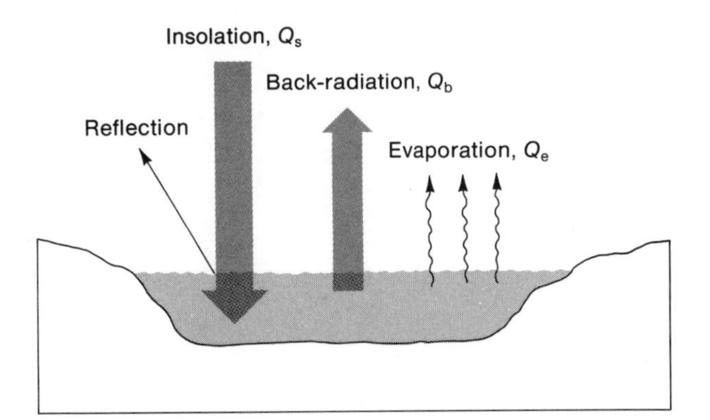

FIGURE 7-2 A simplified representation of the heat budget of a lake. Of the effective incoming radiant energy (insolation, Q_s,) on a body of water, most is lost or returned to the atmosphere by back-radiation, Q_b, in the infrared part of the spectrum and by evaporation, Q_e, as latent heat. If insolation is greater than back-radiation and evaporative transfer, heat is stored and the lake warms up: Q_s is greater than $Q_b + Q_e$. If insolation is less than the heat-loss mechanisms, heat is lost and the lake cools: Q_s is less than $Q_b + Q_e$. For both lakes and the ocean, thermal equilibrium prevails, meaning that over long periods of time, heat absorbed must equal heat lost: $Q_s = Q_b + Q_e$.

DEFINITIONS

Back-radiation. The energy, in the infrared portion of the spectrum that is returned to space from the heated earth's surface. It is a function of surface temperature and increases as temperature increases.

Conduction. Transmission of heat through a conductor such as an iron or copper bar, a rock, a car engine, or other.

Convection. A process of movement of fluid medium as a result of different temperatures in the medium and hence different densities. The process moves both the medium and the heat.

Isolation. The heat entering the atmosphere, ocean, and land from the sun.

Latent heat. The heat required to transform ice to liquid water or liquid water to vapor. That transforming ice to liquid is called *heat of fusion* and is 80 calories per gram of water. That changing liquid water to vapor is called *heat of vaporization* and is 540 calories per gram for fresh water. These reactions are reversible, that is, 540 calories must be added to 1 gram of liquid to change it to vapor, and when the vapor condenses back to liquid, it gives up to 540 calories. Latent heat may be viewed as "stored" heat.

Scattering. The interaction of light with particulate matter or atoms and molecules.

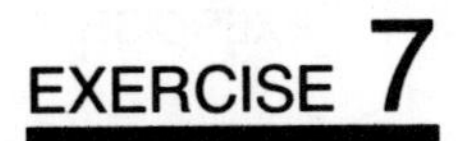

EXERCISE 7

REPORT

SOLAR RADIATION AND HEAT BALANCE

NAME ______________________

DATE ______________________

INSTRUCTOR ______________________

1. Refer to Table 7-2 and then compute the amount of energy storage in the lake for each month.

Month	Storage calories per square centimeter per day
May	______
June	______
July	______
August	______
September	______

2. In the spaces below, complete the histogram (bar graph) to show the average incoming radiation in the lake for each of the five months. As in the box for May, subdivide each graph to indicate the portion that is back-radiated, that which is transferred by evaporation, and that which is stored.

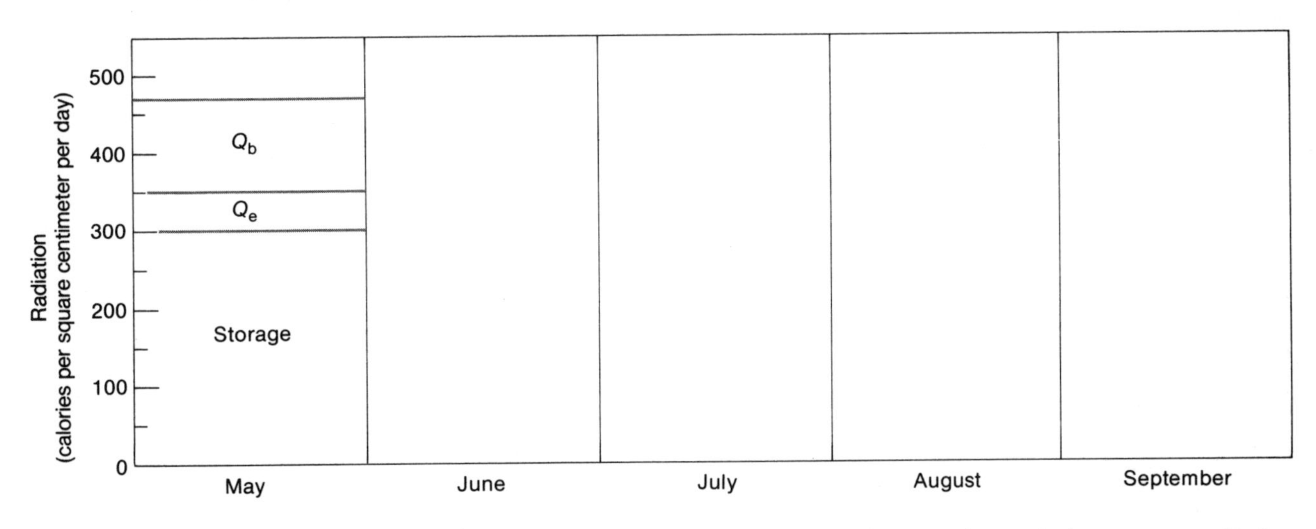

3. Assuming no heat loss to a possibly cooler atmosphere at night, what would be the approximate rise in temperature (to the nearest degree Celsius) of the top 1 meter of water in the lake from June 15 to July 15?

Does your answer seem reasonable? If not, does it seem too high or too low? If it seems too high or low, explain what assumptions were made that might have influenced the numerical result.

4. Extremely heavy cloud cover can reduce the effective incoming radiation to nearly zero. What would be the storage of solar energy in the lake during 15 days in June and 15 days in July if on half of these days very heavy cloud cover occurred and on the other half the surface received the average quantity of incoming radiation given in Table 7-1?

Calculate the change in temperature for the top meter (assuming that any loss of energy would occur only in the top meter).

EXERCISE 8

SEAWATER TEMPERATURES

As energy enters the ocean from the sun, the radiation is absorbed and is stored in the mass of the ocean water as heat. The energy also evaporates seawater and thus increases its salinity, but for the purposes of this exercise we will focus our attention on the portion of incoming radiation that heads the water.

Heating a substance leads to expansion and therefore a decrease in the **density** (mass per unit volume). Conversely, cooling a substance increases its density, and in the case of seawater changes in density brought about by seasonal heating and cooling constitute one of the processes that generate ocean currents. Change in the amount of salts dissolved in seawater also influences the density. The reduction of volume by the pressure of the overlying waters is the third factor that determines the density of seawater at any point. of the three—temperature, salinity, and pressure—temperature is the most important. Therefore, a simple model of oceanic circulation in surface waters in a single region can be constructed on the basis of temperature alone.

Understanding oceanic circulation is important because the water motion influences the distribution of energy and materials on the earth, defines marine biological environments, and has a major impact on climate. Coastal currents strongly influence local climates and produce the differences in temperature and humidity on a given summer day between, say, San Diego, California, and Savannah, Georgia, two points that, although on different coasts, are at about the same latitude and would therefore be expected to have similar climates. The major factor is the temperature of the surface water in each area because it controls or regulates local air temperatures and humidities.

We can see the importance of temperature as an influence on water density if we make a quantitative comparison of the effect on density of changes in salinity and temperature. We will exclude pressure effects from our investigation since they are very slight in water shallower than about 1000 meters. A change in salinity of one part salt per thousand parts of water (1 part per thousand, or 1‰) has more effect on density than a change in temperature of 1 Celsius degree. For instance, the density difference produced by a change in salinity of 1‰ is 0.001 gram per cubic centimeter; and the density difference produced by a temperature change of 1 Celsius degree is, as a rule, between 0.00005 and 0.00035 gram per cubic centimeter. When we consider the surface waters of the oceans as a whole, however, we see that temperature is the more important factor because its variations (ranging from −2 to 35 °C) are much greater than the salinity variations (which range only from 33‰ to 37‰). We will be concerned with temperature in this exercise, and with the second factor, salinity, in Exercise 9.

THE PERMANENT THERMOCLINE

The permanent **thermocline,** the water layer within which temperature decreases rapidly with depth, acts as a density barrier to vertical circulation; that is, we may view this thermocline as the floor of the low-density warm **surface,** or **mixed, layer** and the ceiling for the cold dense bottom waters (Figure 8-1). Large-scale vertical movements of the water between bottom and surface, are inhibited by the strong contrast in density between the two layers. However, in the polar regions, surface waters are much colder and therefore denser, so that little temperature variation exists between surface waters and deeper waters in these areas. Here vertical circulation takes place as surface waters sink to replenish deep waters in the major oceans.

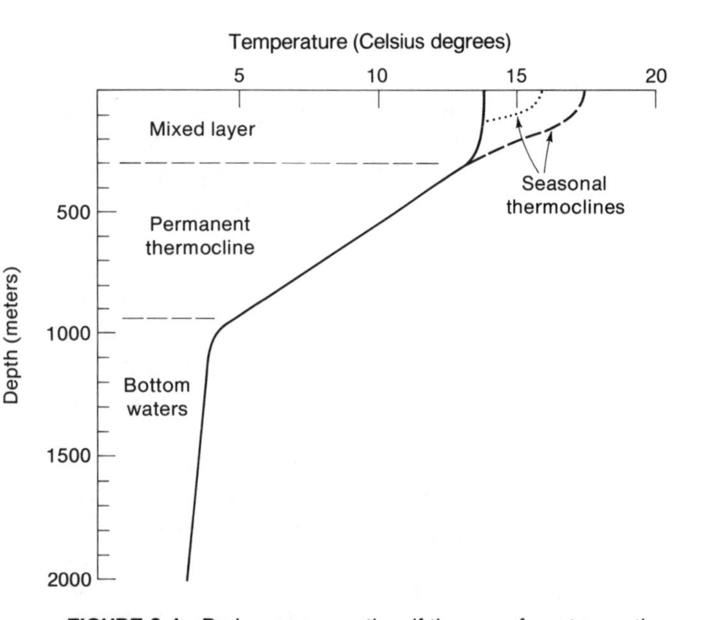

FIGURE 8-1 During warm weather, if there are few storms, the upper 50–100 meters of the ocean warms, resulting in minor shallow thermoclines as shown in the figure. The solid curve shows the winter condition in the mixed layer, and the permanent thermocline below; the dotted curve shows the seasonal thermocline that prevails after spring warming and that is present only in the upper mixed layer; the dashed curve is the seasonal thermocline in extreme summer condition, also present only in the mixed layer.

SURFACE TEMPERATURES

The distribution of surface temperatures in the major oceans is shown in Figure 8-2. Note that the **isotherms** tend to warp toward the equator on the east sides of the oceans, and poleward on the west sides. This is due to the major circulation pattern by which warm water is carried from the equator toward the poles on the west sides, and cooler water from the subarctic regions moves toward the equator on the east sides.

UPWELLING AND COASTAL CLIMATES

Another influence on surface temperatures, in the coastal waters on the eastern sides of ocean basins, is **upwelling,** or the rise of cooler waters from lower depths. First, the **Coriolis effect** causes water that has been set in motion by winds or other forces to be deflected to the right in the Northern Hemisphere, and to the left in the Southern Hemisphere. However, owing to friction, the surface layer drifts at an angle of 45 degrees to the wind direction. Water at successive depths moves increasingly to the right (or left) until at some depth it moves in a direction opposite to that of the wind (Figure 8-3). Meanwhile, velocity decreases with depth throughout the spiral to its base at a depth of about 100 meters. This rotary motion downward is called the **Ekman spiral.** As a consequence of this spiral, the net transport of water is 90 degrees to the right of the wind direction in the Northern Hemisphere (and 90 degrees to the left of it in the Southern Hemisphere).*

This means that—in the waters off the coast of Florida, for example—winds blowing from the south along the Atlantic coast drive surface waters to the right, or offshore, and this water is replaced by cooler nutrient-rich water from depths of 200–300 meters. Although upwelling occurs at many places along both coasts of the United States, it is particularly important along the coasts of California, Oregon, and Florida. It may be detected by temperature measurements, because it produces colder water inshore, and the isotherms warp upward toward the shore. The main

* For further information on the Coriolis effect and the Ekman spiral, see Exercise 11.

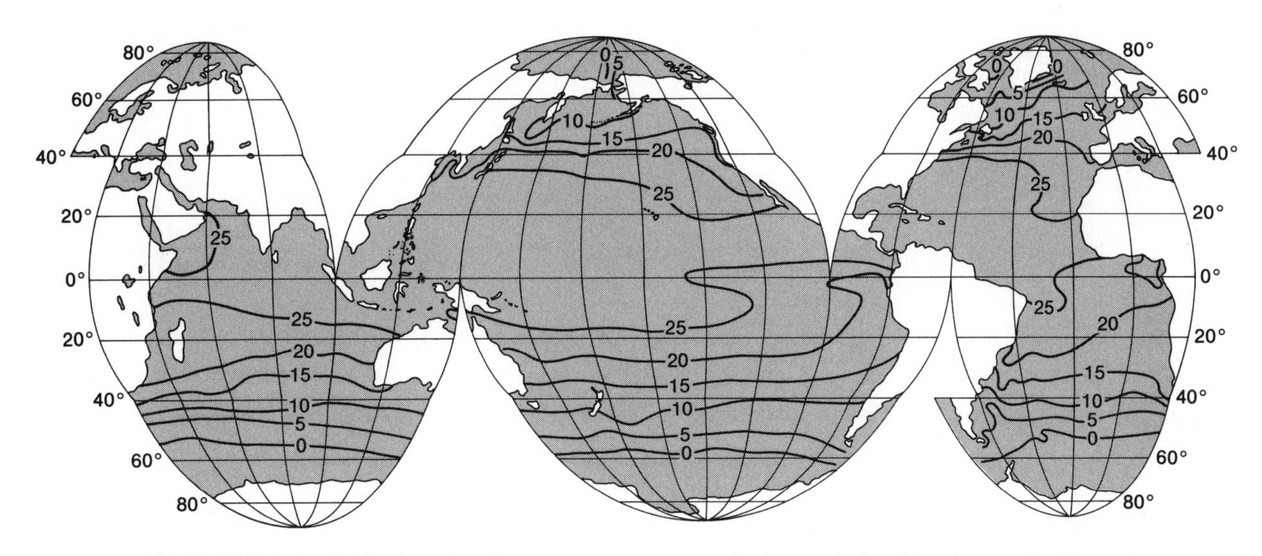

FIGURE 8-2 The distribution of surface ocean temperatures (in degrees Celsius) for the month of August. [Goode Base Map Series. Copyright by the University of Chicago Department of Geology.]

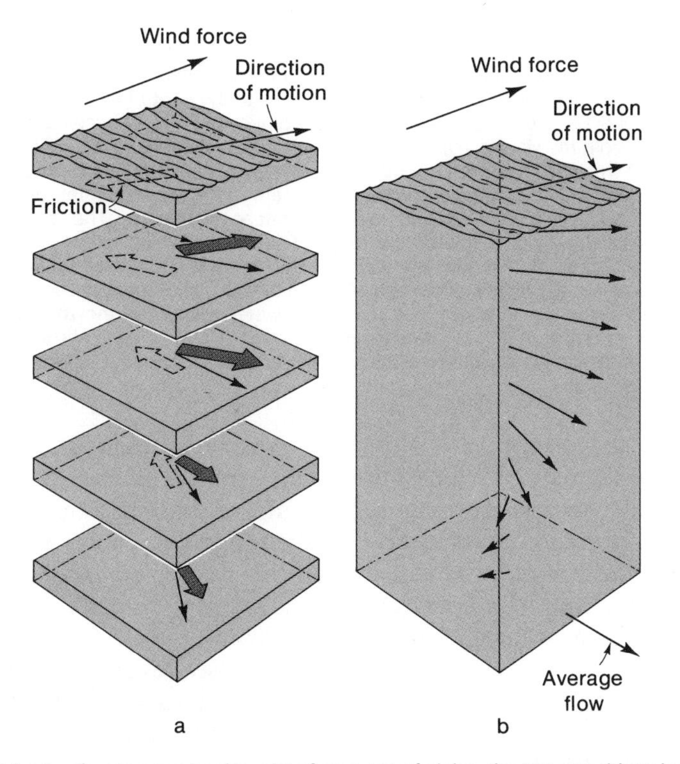

FIGURE 8-3 (a) A body of water can be thought of as a set of slabs, the top one driven by the wind and each one setting the one below it in motion by friction. Each succeeding layer moves with a slower speed and is directed in a sprial motion—more to the right in the Northern Hemisphere, more to the left in the Southern Hemisphere—until friction becomes negligible. (b) Although the direction of movements varies for each layer, the average, or "net," flow of water is 90 degrees to the right of the prevailing wind force. [After R. W. Stewart, "The Atmosphere and the Ocean." Copyright © 1969 by Scientific American, Inc. All rights reserved.]

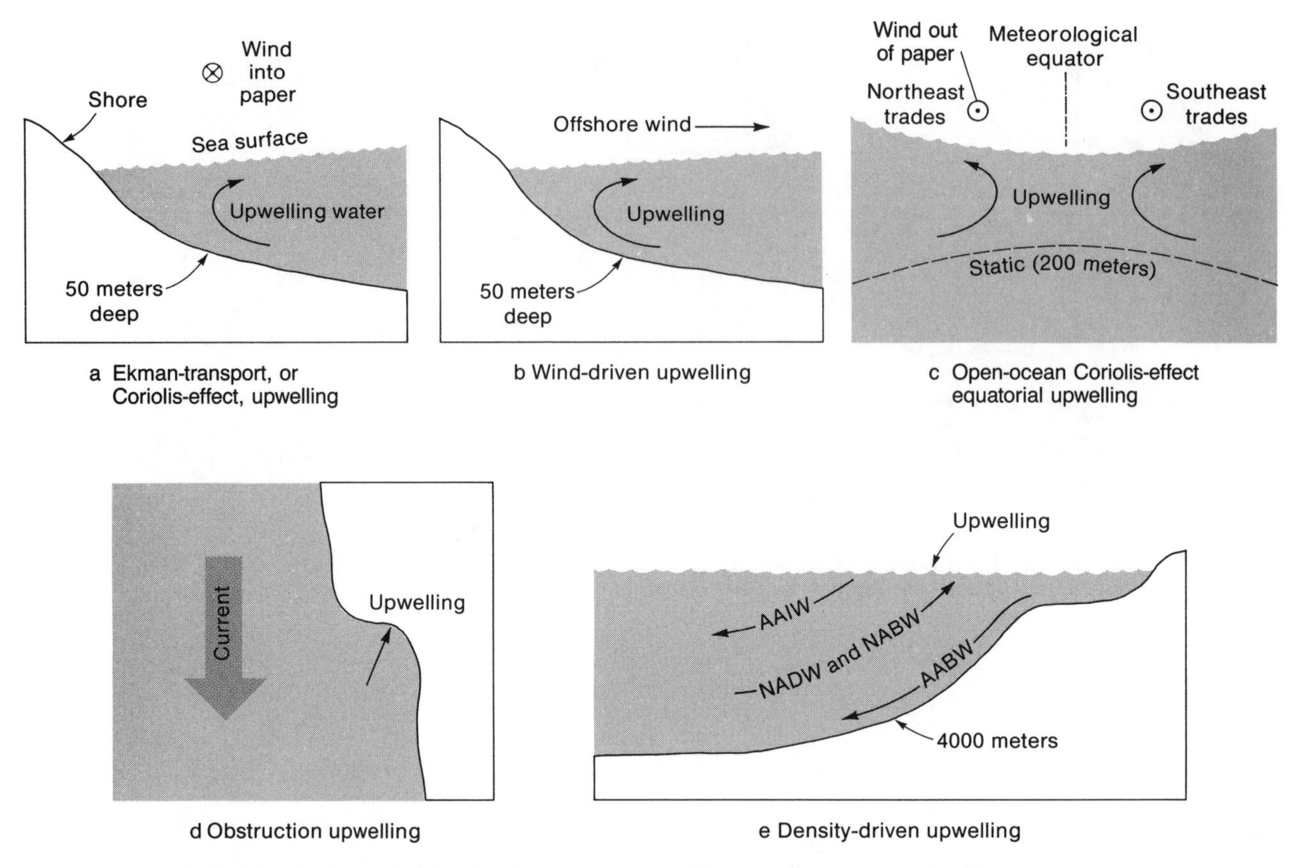

FIGURE 8-4 Diagrams of various kinds of upwelling. (a) Ekman-transport, or Coriolis-effect, upwelling. In the Northern Hemisphere net transport of water is 90 degrees to the right of the wind direction, and in the Southern Hemisphere it is 90 degrees to the left of the wind direction. The encircled cross indicates that the wind direction is away from the viewer. (b) Wind-driven upwelling. Offshore winds will blow water away from the continent, and that water will have to be replaced by upwelled water. (c) Open-ocean Coriolis-effect equatorial upwelling. Wind-induced divergence near the equator will transport water to the right north of the equator and to the left south of it, thus causing upwelling right at the equator. The encircled dots indicate that the direction of the trade winds is toward the viewer. (d) Obstruction upwelling. A current moving past a headland or other obstruction will draw water away from the obstacle and upwelling will occur. This is a map view of a headland or point. (e) Density-driven upwelling. In the course of thermohaline circulation, denser water sinks and replaces less dense water. The less dense water upwells as shown in this example from the South Atlantic Ocean. AABW stands for Antarctic bottom water, AAIW for Antarctic intermediate water, NADW for North Atlantic deep water, and NABW for North Atlantic bottom water. The NADW and NABW are being upwelled in the Antarctic.

kinds of upwelling are illustrated in Figure 8-4. Note that upwelling may be caused not only by winds, but also by entrainment of water by coastal currents as they move past obstructions or promontories, and by density differences. The opposite process is called **downwelling** (sinking) and is disclosed in temperature measurements by depression, or downwarping, of isotherms toward the shoreline.

THE MEASUREMENT OF OCEAN TEMPERATURES

An important instrument for measuring ocean temperatures at varying depths is the **bathythermograph (BT),** which is shown in Figure 8-5. With this device it is possible to obtain a continuous reading of temperature with depth after determining the surface-water temperature with a bucket thermometer. The bathythermograph enables the investigator to determine the thermal structure of the ocean at selected stations and assess the possibility of vertical motions, such as upwelling.

Deep-sea reversing thermometers (DSRT) are designed to measure the temperature of the deep ocean. They are attached in pairs to a water bottle (Figure 8-6) that is lowered to the desired depth and "tripped," or turned upside down, by a sliding weight called a *messenger.* By design, the act of reversing traps mercury in the column proportional to the temperature at the depth of reversal (Figure 8-7). One DSRT in Figure 8-7 is protected from pressure at

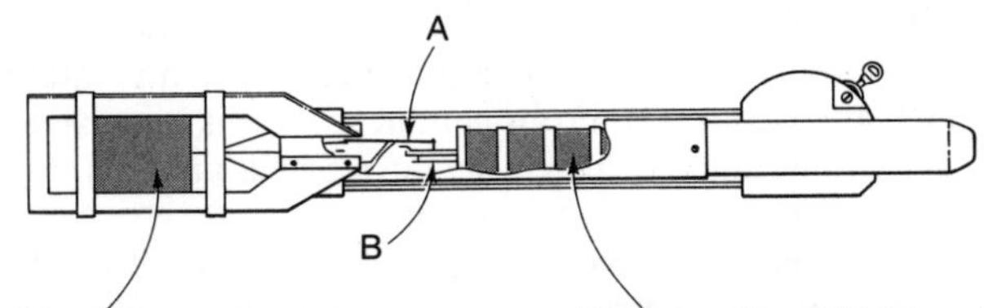

FIGURE 8-5 Cutaway view of a bathythermograph (BT). Temperature is recorded by the pointed stylus, A, on a glass slide, B. The slide moves to the right or left to record depth as the pressure element compresses or expands in response to depth changes.

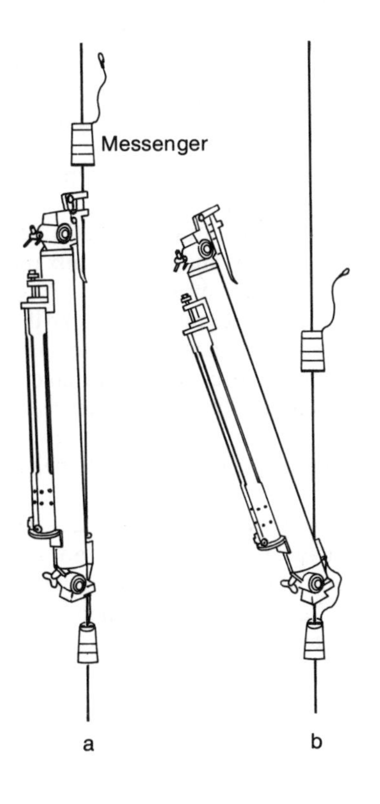

FIGURE 8-6 Nansen bottle on the wire in three positions: (a) before tripping—note the messenger just upper clip; (b) during tripping; (c) after tripping. Observe that the messenger is released and slides down the wire to release the first Nansen bottle in the series on the wire. At (c) the bottle has toppled and is held on the wire only by the bottom clamp, and the messenger at the bottom of that bottle has been released. (U.S. Navy Hydrographi Office, Publication No. 607, 1955.)

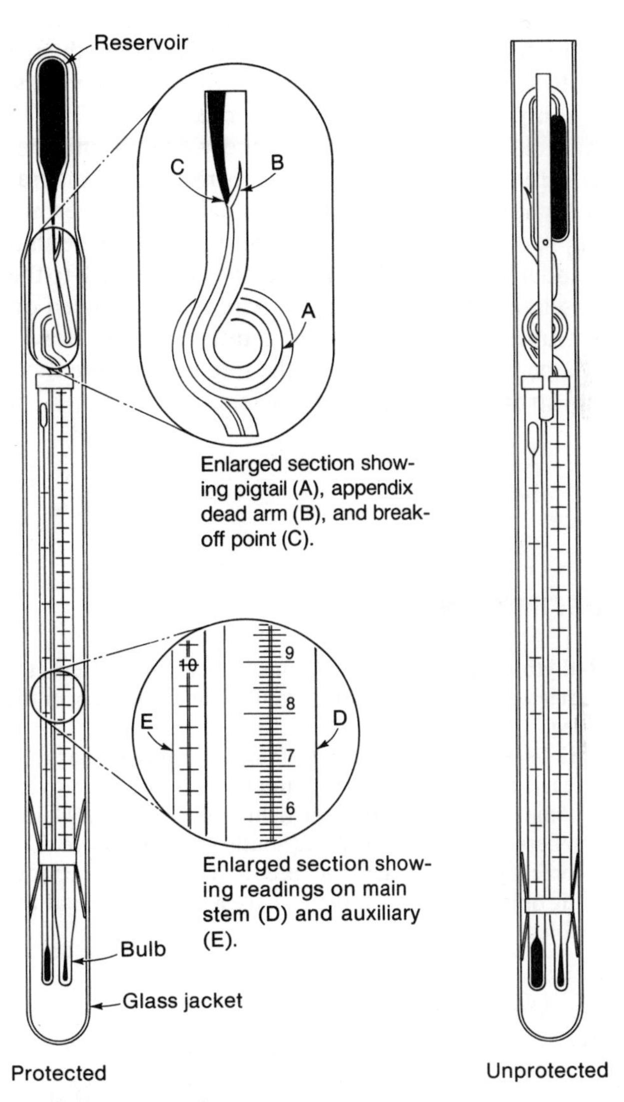

FIGURE 8-7 Protected and unprotected deep-sea reversing thermometers in the reversed position. In the upright position the main thermometer has a large reservoir of mercury at the lower end connected by a thin capillary to a small bulb at the upper end. The capillary is branched at the appendix (B) and goes through a 360-degree loop at the pigtail (A). In the upright position the mercury fills the reservoir, the capillary, and maybe even part of the bulb, depending on the temperature. When the thermometer reverses, the mercury column breaks at (C), descends into the bulb, now at the bottom, and fills the bulb and part of the stem, thus indicating the temperature on the main stem (D) and the auxiliary stem (E) at the time of reversal. No more mercury can enter the stem until the thermometer is returned to the upright position. [Hydrographic Office Publication No. 607, U.S. Naval Oceanographic Office, 1955.]

depth whereas the other is not. Since pressure compresses the reservoir, more mercury flows into the unprotected thermometer upon reversal than into the protected one. A rule of thumb is that the temperature reading from the unprotected thermometer is about 1 Celsius degree higher than that from the protected one for each 100 meters of water column, and thus the depth from which the temperature was obtained can be determined. For example, an unprotected thermometer reading 10 Celsius degrees higher than a protected one would indicate a depth of 1000 meters.

Temperature measurements made with DSRTs may take an hour or more, and individual thermometers are fragile and costly. For this reason oceanographers now prefer to use an expendable bathythermograph, or XBT. Dropped into the water while the ship is under way, the XBT relays temperatures back to the vessel via a threadlike copper wire that is spooled on instrument. Continuous temperature profiles drawn by a pen recorder on board the ship provide the thermal structure down to 500 meters.

DEFINITIONS

Bathythermograph (BT). A mechanical instrument for measuring depth and temperature simultaneously in the sea.

Coriolis effect. An apparent "force" on moving particles resulting from the earth's rotation. It causes moving bodies to be deflected to the right in the Northern Hemisphere and to the left in the Southern Hemisphere. The force is proportional to the speed and latitude of the moving object. It is zero at the equator and maximum at the poles.

CTD recorder. An electronic device for simultaneously measurement electrical conductance (*C*), temperature (*T*), and depth (*D*). From conductance it is possible to determine salinity; hence the device measures salinity, temperature, and depth of water.

Density. Defined as the mass per unit volume of a substance. In the metric system, the units for liquids are grams per cubic centimeter. For purposes of comparison only, seawater density may be taken as 1.0250 grams per cubic centimeter whereas fresh water is 1.000.

Downwelling. A downward movement (sinking) of surface water caused by onshore Ekman transport, converging currents, or when a water mass becomes more dense than the surrounding water.

Ekman spiral. A theoretical representation of the effect of wind blowing steadily over an ocean: The surface layer of water drifts at an angle of 45 degrees to the right in the Northern Hemisphere. Water at successive depths drifts in directions more and more to the right, until at some depth it moves in the direction opposite to the wind. Water velocity decreases with depth throughout the spiral. The net water transport is 90 degrees to the right in the Northern Hemisphere, and just the opposite in the Southern Hemisphere.

Isotherm. A line connecting points of equal temperature, either at the sea surface or with depth.

Surface, or mixed, layer. The temperature zone of water above the thermocline where winds and currents mix the surface waters and convey heat downward.

Thermocline. Literally, a temperature gradient or rapid decrease of temperature with depth in a body of water; also the layer in which such a gradient occurs. The permanent thermocline in the oceans occurs between the levels of about 200 meters and 1000 meters, and separates an almost uniformly warm upper layer from very cold dense bottom waters.

Upwelling. The process by which water rises from a lower to a higher depth, usually as a result of offshore water flow. It is most prominent where persistent wind blows parallel to a coastline, so that the resultant Ekman transport moves surface water away from the coast.

EXERCISE 8

REPORT

SEAWATER TEMPERATURES

NAME ______________________

DATE ______________________

INSTRUCTOR ______________________

1. Figure 8-8 represents a group of surface temperatures taken as part of a program known as Organization of Persistent Upwelling Structures (OPUS). The investigation centered on a region of intensified upwelling between Point Conception and Point Arguello, California. It involved a team of scientists from eight different institutions who shared a common interdisciplinary interest in ocean and atmospheric processes, three ships, and numerous data-collecting buoys and instrument packages, and it was managed by the University of Southern California.

Surface temperatures were recorded at a depth of 2 meters by lowering a CTD (see Definitions) from the Scripps Institution of Oceanography ship R. V. *New Horizon.* Data were plotted by computer and contoured manually.

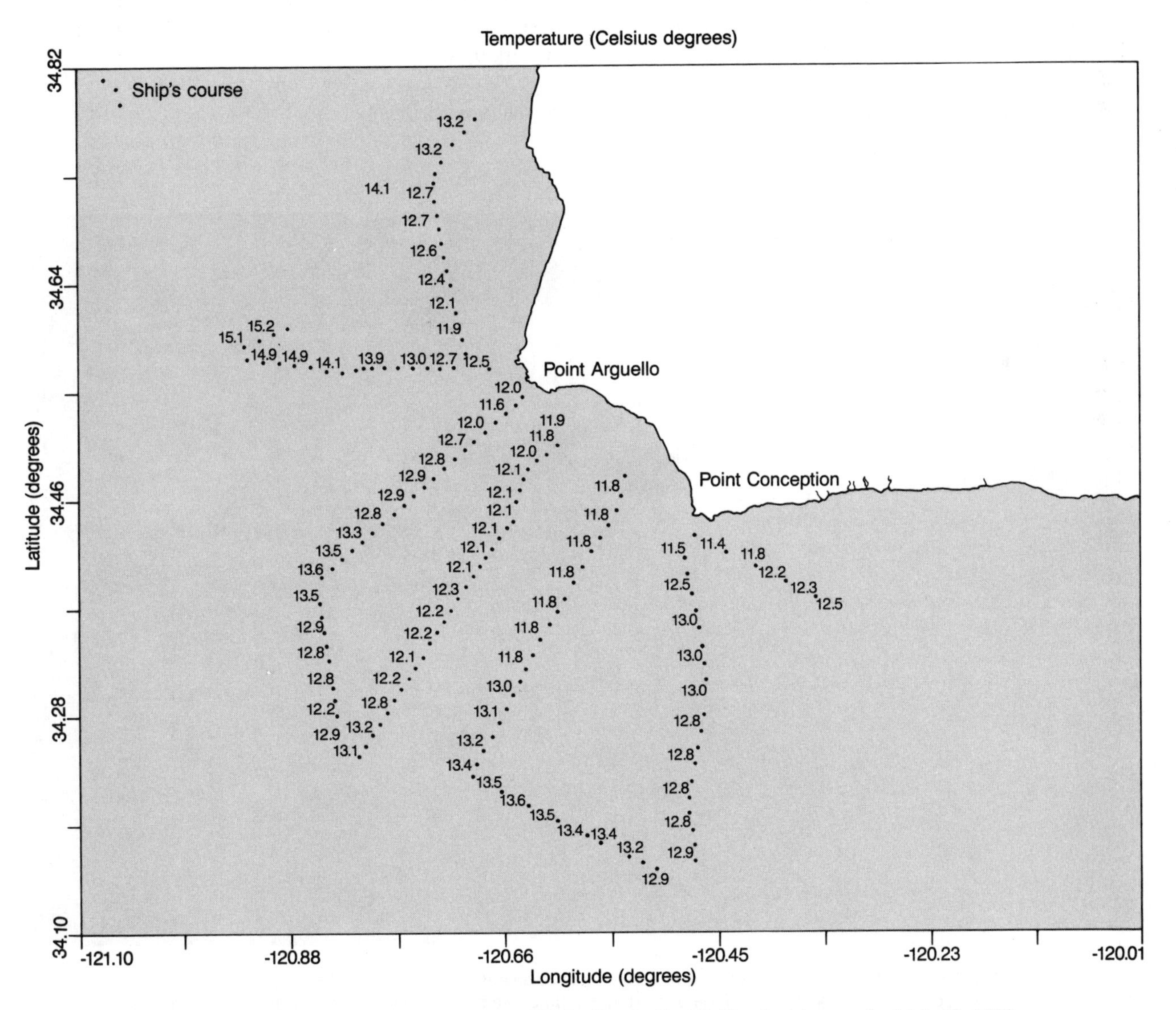

FIGURE 8-8 Surface temperature pattern off Point Conception, California, taken on April 14–15, 1983. [Courtesy Dr. Burton Jones, Allan Hancock Foundation, University of Southern California.]

(a) Contour the surface temperatures at a 0.5 Celsius-degree contour interval. It is suggested that you contour whole degrees to start, then fill in half-degree intervals by interpolation.

(b) Explain the temperature pattern found off Point Conception.

(c) Compare your isotherm map with the relative temperatures shown in the satellite image of Figure 8-9.

2. Figure 8-10 is a temperature cross section off Point Conception obtained from a vertical cast using a CTD recorder (see Definitions). Contour the profile down to the 8°C isotherm using a 0.5-Celsius-degree contour interval. It is suggested that you begin contouring at the bottom (8°C isotherm) and work upward.

(a) What process is indicated by the shape of the contours at shallower depths?

FIGURE 8-9 A satellite image of relative sea-surface temperature on April 16, 1982. Point Arguello and Point Conception are at the top of the photograph. Lighter areas represent coldest waters and darker areas are relatively warmer water. Land is black. [Courtesy Dr. Richard E. Pieper, Institute for Marine and Coastal Studies, University of Southern California.]

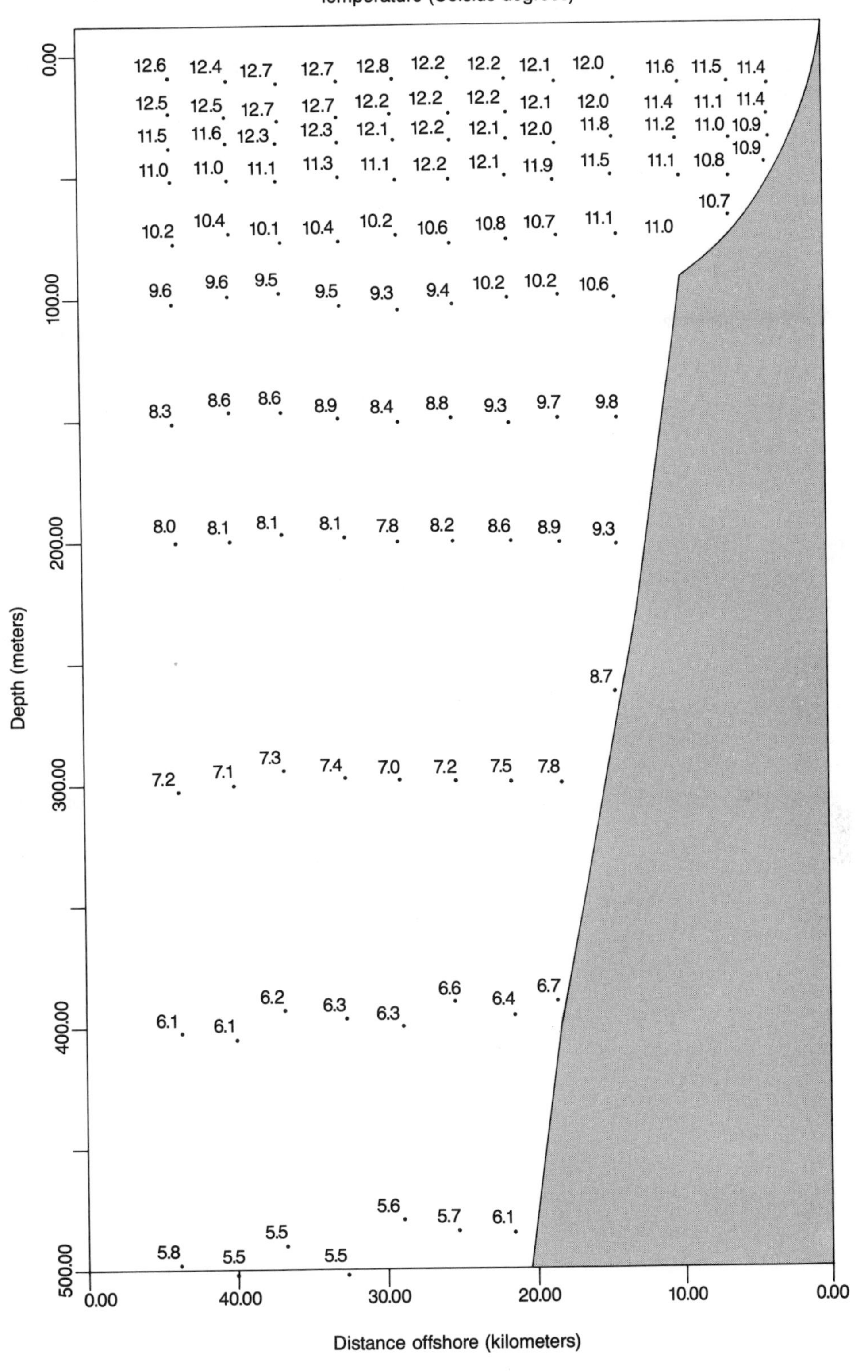

FIGURE 8-10 Typical vertical temperature profile off Point Conception during same time frame as Figure 8-9. [Courtesy Dr. Burton Jones, Allan Hancock Foundation, University of Southern California.]

(b) Judging from the trend of the contours, what seems to be the motion of the deeper water?

(c) Compare the temperature map and cross section to upwelling processes shown in Figure 8-4. What type of upwelling best describes what is taking place at Point Conception?

3. The map in Figure 8-2 shows the distribution of surface temperatures in August in the major ocean basins. Refer to it for your answers to the following questions.

(a) Where is water of the greatest density formed? Explain your answer.

(b) How would a depth–temperature (BT) curve from the Arctic regions differ from one at mid-latitudes or one from equatorial regions? Explain your answer.

(c) Why is water so much cooler along the coast of North America than at equivalent latitudes on the opposite side of the Pacific Ocean?

EXERCISE 9

THE SALINITY OF SEAWATER

The **density,** or mass per unit volume, of seawater depends upon two properties: temperature and salt content, or **salinity.** As water cools, its density increases. As its salt content increases, its density also increases. Because water of high density tends to sink, and that of low density tends to rise above or settle below water that is at the average density of the oceans, the change in density is one process whereby water motion is generated. Therefore we are interested in the distribution of both salinity and temperature of water, since these are the two factors that determine the circulation that is caused by density changes. In Exercise 8 we discussed seawater temperatures; in this one we will be concerned with salinity.

THE DISTRIBUTION OF SALINITY

The oceans get their salt from the weathering and dissolution of minerals on land and from volcanic emanations. The mobile constituents of minerals are carried in solution by streams to the sea where they accumulate and are recycled by various processes. Salinity is a "conservative" property; that is, one that remains constant for the ocean as a whole for long periods of time, even though the local salinity varies within limits over the surface of the oceans. The average salinity for the oceans as a whole is 34.73 parts salt per 1000 parts water (34.73‰), but concentrations between 33.0‰ and 37.0‰ have been measured in the open ocean. High salinity, or dilution, is found only in coastal waters or in partially enclosed seas. Such extremes are due largely to excessive runoff from the land, or to high evaporation rates and little mixing with other waters, as in the Red Sea and the Mediterranean Sea.

General variations in salinity are zoned from the equator to the poles. Values are low at the equator, highest in subtropical regions and at mid-latitudes, and lowest in the polar regions. The major processes responsible for this distribution are evaporation, precipitation, and mixing. Where evaporation exceeds precipitation, salinity values are high, and in areas of high rainfall, as at the equator, they are lower (Figure 9-1). The distribution of surface salinity in the major oceans for the months of August is shown in Figure 9-2. Points of equal salinity are connected by **isohalines.**

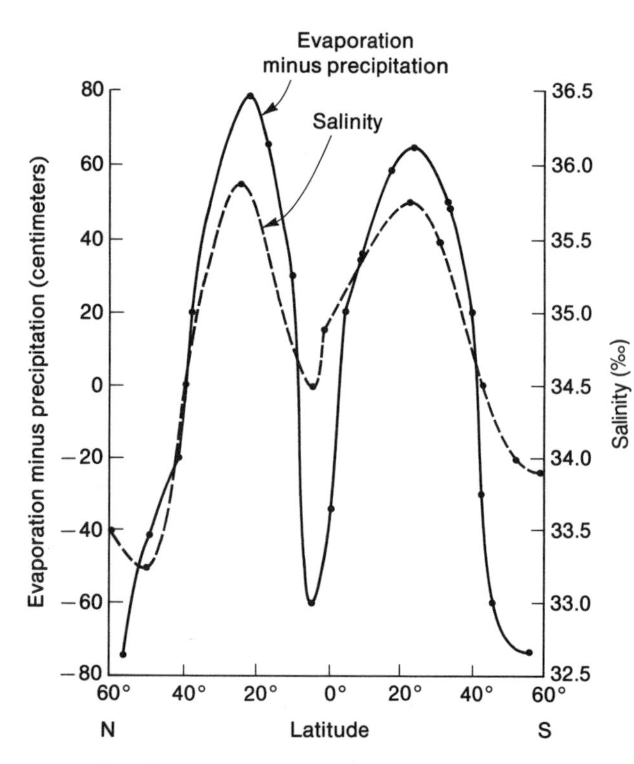

FIGURE 9-1 Distribution of surface salinity plotted against the evaporation minus precipitation. [After S. Defant, *Physical Oceanography,* Vol. 1. Pergamon Press. Copyright © 1961.]

DETERMINATION OF SALINITY

The salinity of seawater is not a difficult property to determine. One reason is that regardless of the absolute concentration of salts in solution, the major dissolved constituents exist in virtually a constant ratio to one another. This fact was first recognized by Johann Forchhammer and later confirmed in 1884 by Wilhelm Dittmar, who carefully analyzed 77 samples collected on the *Challenger* expedition (1872–1876). Modern analytical techniques have enabled refinement of Dittmar's ratios; however, the importance of his work is not the accuracy of its numerical values, but rather its demonstration of the constancy of the ratios of about a dozen dissolved constituents (Table 9-1). In theory, if you determine the concentration of a major dissolved ion in a sample, you should be able to calculate the concentration of the other major constituents. In practice this is not quite so simple because of the analytical problems in distinguishing between several of the elements. Because chloride is the most common dissolved ion and one of the easiest to determine precisely, its concentration is determined, usually by a procedure known as the **Knudsen titration,** and from that measurement the salinity is calculated:

$$\text{Salinity } (‰) = 1.80655 \times \text{chlorinity } (‰)$$

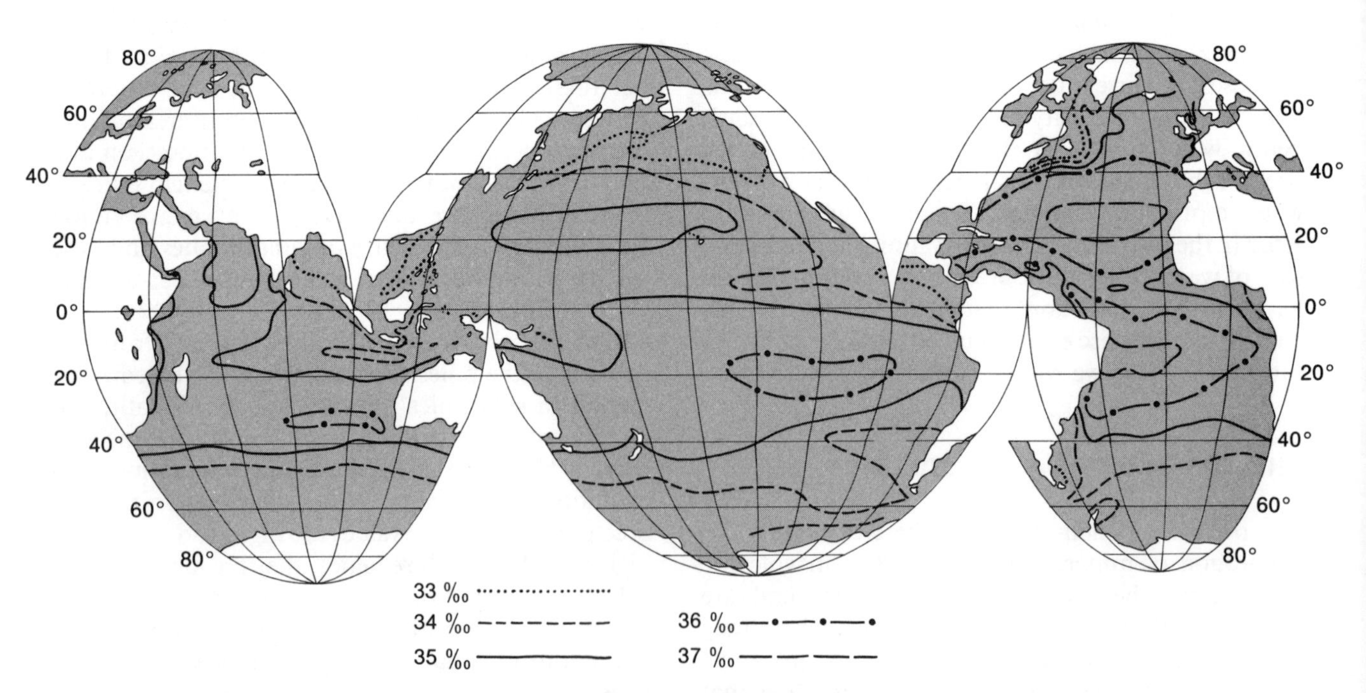

FIGURE 9-2 Salinity distribution in the surface waters of the oceans in August. [Goode Base Map Series. Copyright by the University of Chicago Department of Geography.]

TABLE 9-1
Major dissolved constituents in seawater with a chlorinity of 19‰ and a salinity of 34.32‰.

Dissolved substance	Concentration (grams per kilogram)	Ratio of dissolved salt to chlorinity (‰)	Percentage by weight
Chloride	18.980	0.99895	55.04
Sodium	10.556	0.55557	30.61
Sulfate	2.649	0.13942	7.68
Magnesium	1.272	0.06695	3.69
Calcium	0.400	0.02105	1.16
Potassium	0.380	0.02000	1.10
Bicarbonate	0.140	0.00737	0.41
Bromide	0.065	0.00342	0.19
Boric acid	0.026	0.00137	0.07
Strontium	0.013	0.00008	0.04
Fluoride	0.001	0.00005	0.00
Totals	34.482		99.99

Note in Table 9-1 that the salinity (34.32‰) calculated from the chlorinity of 19.00‰ is less than the salinity determined from the ratios of the elements to chloride ion (34.482). The reason is that bromine and iodine behave as if they were atoms of chlorine in the chemical analysis. However, the definition of salinity specifies that bromine and iodine should be converted to chlorine equivalent (weight) and carbonate converted to the oxide. When these mathematical manipulations have been completed, the chloride concentration increase, carbonate and bromide contents are reduced, and the salinity calculated from a chlorinity of 19.00‰ and that determined from the ratios in Table 9-1 agree very well.

Another analytical method of determining salinity of a salt solution is to measure its ability to conduct an electrical current. **Conductivity** increases with increasing salt content and this property of seawater may be measured with an electrical salinometer. At present, salinity determinations by high-precision conductivity measurements are more standard than chemical methods.

RESIDENCE TIME

An important concept in oceanography and marine geochemistry is that of **residence time.** It is defined as being equal to the quantity of a substance, whatever that might be, divided by the influx (or deposition) per unit time. It is the mean length of time an atom, molecule, or particle of a given substance spends in its reservoir, be it the atmosphere, oceans, or a lake, and it is a measure of the reactivity of that substance. Highly reactive substances have short residence times, and vice versa. There are several primary assumptions in the concept: (1) that the substance is thoroughly mixed in the reservoir, and (2) that the rates of supply and removal are constant over at least several residence times. For example, water vapor in the lower atmosphere has a residence time of about 10 days. This means that if no vapor were added to the atmosphere, it would be totally depleted of water in 10 days. Conversely, if the lower atmosphere contained no water vapor, about 10 days of evaporation from oceans and lakes would be required for the water vapor content to reach equilibrium; that is, the amount added by evaporation is equal to the amount lost by precipitation. Water vapor has a much shorter residence time than most substances in the oceans. In this exercise we will be concerned with the residence times for total salt, water, and sodium in the oceans, given the rate of addition from rivers (in tons per year). Residence time, R, is equal to the total quantity of the substance, C, divided by the rate of addition, A, of that substance: $R = C/A$.

DEFINITIONS

Conductivity. The ability of a fluid to conduct electrical currents. The conductivity increases with increased salt content. Salinity may be determined by measuring the conductance of seawater with an electronic device called a *salinometer.*

Density. The mass per unit volume of a substance. The density of seawater is expressed in grams per cubic centimeter.

Isohalines. Lines connecting points of equal salinity in the oceans.

Knudsen titration. The classical method for determining the chlorinity. The halides (chlorine, bromine, and iodine) are precipitated from a standard volume of seawater by a silver nitrate solution. The analysis is calibrated against a standard, or normal, seawater sometimes also called *Copenhagen water.*

Residence time. The mean length of time a quantity of a given substance—for example, salt, sodium, or water—spends in the ocean before removal.

Salinity. The total amount of solid material in grams contained in kilograms of seawater when all the bromine and iodine have been replaced by an equivalent amount of chlorine, all the carbonate has been converted to oxide, and all organic matter has been oxidized. It is usually expressed in parts per thousand.

REPORT

THE SALINITY OF SEAWATER

NAME ______________________

DATE ______________________

INSTRUCTOR ______________________

1. Give the concentration of seawater with a salinity of 3.45 percent in the following units:

(a) ______________ parts per thousand.

(b) ______________ parts per million.

(c) ______________ grams per kilogram.

(d) ______________ kilograms per ton.

2. What is the salinity of seawater with a chlorinity of 19.10‰?

3. Study the isohalines on the map in Figure 9-2 and answer the following questions.

(a) How does salinity vary from the equator to polar regions in the Pacific Ocean? Give exact values and general latitudinal zones.

(b) Which of the two oceans shown is saltier and by what amount?

How might you explain this condition? (Hint: Relate to the major wind belts.)

(c) Explain the tongue of low-salinity water extending from Baffin Bay west of Greenland to the eastern coast of Canada. (Remember that it is summertime in the Northern Hemisphere.)

4. Recall that residence time, R, is the mean length of time a given element remains in the ocean. It is equal to the total amount of substance, C, divided by the rate of influx or addition, A (that is, $R = C/A$). The following table gives the total amounts of water, salt, and sodium in the oceans, and the totals contributed each year by the river.

	Water	Dissolved salt	Sodium
Oceans	1.4×10^{18} tons (C)	5×10^{16} tons (C)	14×10^{15} tons (C)
Rivers	26.4×10^{12} tons per year (A)	2.7×10^{9} tons per year (A)	15×10^{7} tons per year (A)

Using these figures, give the residence time for each of the following.

(a) Water

_______________ years.

(b) Dissolved salt

_______________ years.

(c) Sodium

_______________ years.

5. (a) A series of stations off the California coast have been plotted on the chart in Figure 9-3. Contour the values of the salinity at intervals of 0.5‰ (that is, at 32, 32.5, 33, and so on). Shade in the areas in which the salinity is greater than 33.80‰. A regional trend of slightly greater salinity to the south is present though not well defined. This variation is probably due to the wetter climate farther north and to the evaporation rates, which increase as we go southward along the coast.

(b) Briefly explain the high-salinity station (34.00‰) shown near the bay of San Francisco in Figure 9-3.

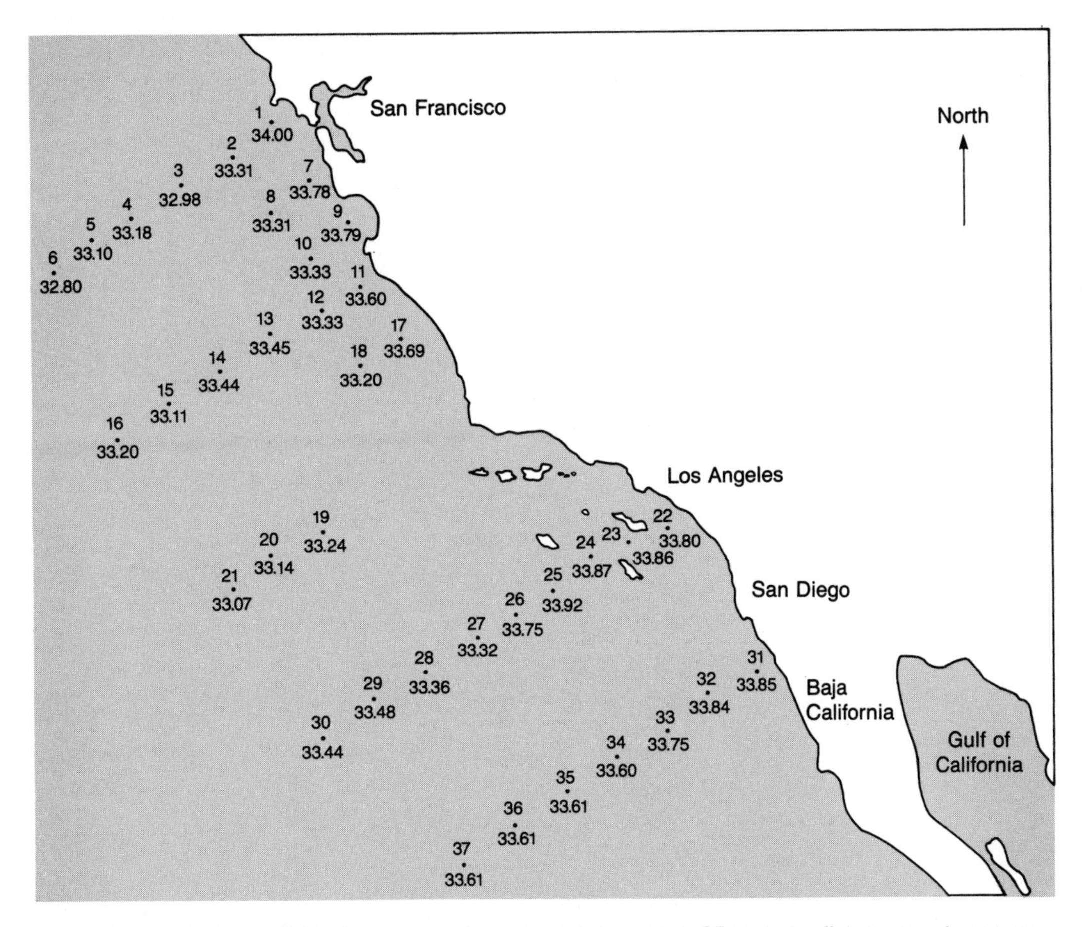

FIGURE 9-3 Surface salinity observations (in parts per thousand) at 37 stations off the coast of southern California and Baja California, summer 1964.

EXERCISE 10

WATER MASSES AND TEMPERATURE–SALINITY DIAGRAMS

A **water mass** is a large volume of water that can be identified as having a common origin or source area. Water masses are formed by an interaction of water with the atmosphere or by the mixing of two or more bodies of water. Once formed they sink to a depth determined by their *density* relative to the waters above and below them in the vertical column. The important determinants of the density are *temperature* and *salinity.*

IDENTIFICATION OF WATER MASSES

Because water masses mix with the surrounding waters only very slowly, they tend to retain their original temperature and salinity. Thus the distinctive temperatures and salinities (and sometimes the oxygen content) of these masses make it possible to identify them. The identification is important because it gives us information on their place of origin, deep circulation, and the rates at which waters of different densities mix.

Deep circulation — the motion of water at depth — is called *thermohaline* (temperature–salinity, hence density) *circulation* and is almost completely separate from that of the surface currents. Whereas surface circulation is largely in an east–west direction and moves warm water toward the poles, deep and bottom currents transport water in a north–south direction, returning cooler water along the meridians toward the equator. The cold water eventually returns to the surface at some point to be reheated and returned to the poles by surface currents, or to mix with other waters and return to the depths. The velocity of thermohaline currents is very slow, about 1 centimeter per second, whereas surface currents are 10 or 20 times faster. Using out concept of residence time — the average time that a given substance (deep water in this case) remains in the ocean before being recycled (see Exercise 9) — about 500–1000 years would be required to replace all the deep water in the Atlantic Ocean.

The identification of large water masses in the oceans is made possible by careful collection of oceanographic data. The most useful data for this purpose

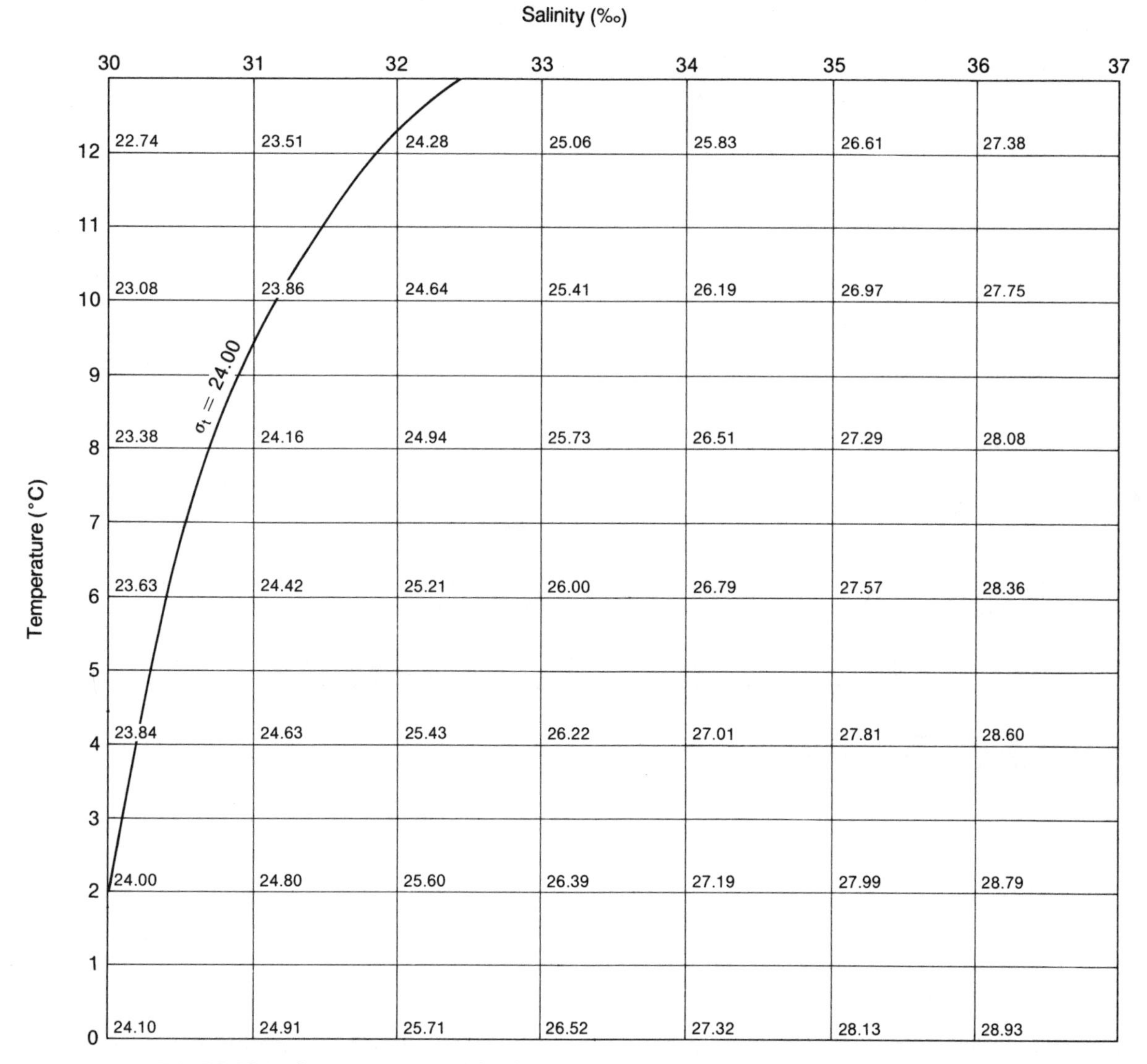

FIGURE 10-1 A temperature–salinity diagram show σ_t values. Here the contour is $\sigma_t = 24.0$.

are temperature, salinity, and oxygen content. The extremes—that is, the maxima and minima for each of these parameters in a vertical column of seawater—are important for identification. Because the number of possible combinations of temperature and salinity is limited, it follows that only a reasonably small number of water masses are formed in the oceans. However, the density alone of a water mass is not sufficient for its identification because various combinations of temperature and salinity can produce the same density.

THE DETERMINANTS OF DENSITY, AND THE DENSITY FACTOR

Temperature, salinity, and pressure are the determinants of density, which is measured in grams per cubic centimeter. Inasmuch as the density of seawater is always greater than 1.0 gram per cubic centimeter (the density of fresh water), and never as great as 1.1 grams per cubic centimeter it is more convenient to use a **density factor,** sigma-t, symbolized by the greek sigma and subscript "t," σ_t. The density factor most commonly used takes into account temperature and salinity but ignores pressure and is written as follows:

$$\sigma_t = (\text{density} - 1) \times 1000$$

Thus a seawater sample with a density of 1.02594 would have a $\sigma_t = 25.94$. Note that the mathematical manipulation involved in changing density to σ_t is simply dropping the 1 and moving the decimal point three places to the right.

TEMPERATURE–SALINITY DIAGRAMS

If we plot the density factors for a series of combinations of temperature and salinity on a **temperature salinity** (T–S) **diagram,** we see that contours of equal density, **isopycnals,** are curved lines (Figure 10-1). For example, water with a salinity of 32.00 parts per

TABLE 10-1
Density factor, σ_t values for various temperatures and salinities

Temperature (°C)	Salinity (‰)						
	30	31	32	33	34	35	36
0	24.10	24.91	25.71	26.52	27.32	28.13	28.93
2	24.00	24.80	25.60	26.39	27.19	27.99	28.79
4	23.84	24.63	25.43	26.22	27.01	27.81	28.60
6	23.63	24.42	25.21	26.00	26.79	27.57	28.36
8	23.38	24.16	24.94	25.73	26.51	27.29	28.08
10	23.08	23.86	24.64	25.41	26.19	26.97	27.75
12	22.74	23.51	24.28	25.06	25.83	26.61	27.38

thousand (‰) at a temperature of 10°C has a σ_t of 24.64 density = 1.02464). This value has been plotted in the $T-S$ diagram in Figure 10-1 where the 10°C and 32‰ lines intersect. Note the isopycnal. Every point on this line has a σ_t of 24.00; for example, this line crosses the 12°C temperature line at a salinity of about 31.85‰. Thus a water type with a temperature of 12°C and a salinity of 31.85‰ has a σ_t value of 24.00 (density = 1.0240 grams per cubic centimeter). Table 10-1 gives σ_t values for the range of temperature and salinity conditions commonly found on the open ocean. Note that the highest density, 1.02893 grams per cubic centimeter, is yielded by the water type with the highest salinity, 36‰, and lowest temperature, 0°C.

From Figure 10-2, we can see that the mixing of two **water types** of the same density, but of different temperatures and salinities, will produce a water mass that is denser than the two that originally mixed. This mixing process is known as **caballing.** In the figure, water types A and B, which have the same density,

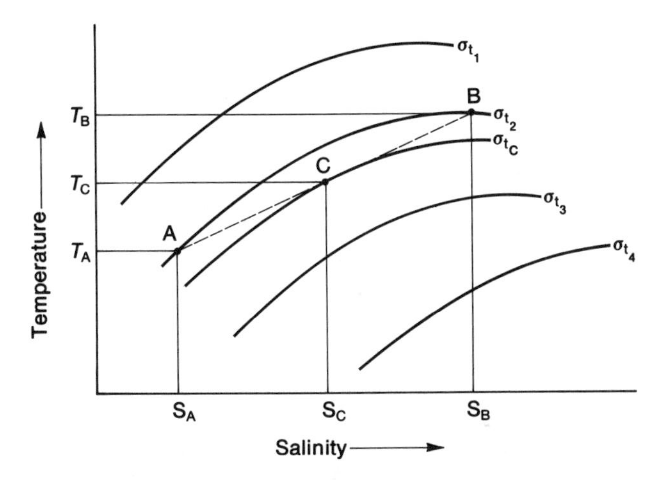

FIGURE 10-2 A temperature-salinity diagram showing the simple mixing of two water types, A and B, to form water mass C. Note that the density of C is greater than that of either of its end members A or B. The density increases from σ_{t_1} to σ_{t_4}.

mix to form water mass C. When mixed in equal quantities $T_C = (T_A + T_B)/2$, and $S_C = (S_A + S_B)/2$, but σ_{tC} is greater than $(\sigma_{tA} + \sigma_{tB})/2$, where T denotes the temperature, S the salinity, and σ_t the density factor. In general, the temperature and salinities that result from mixtures of water types can be computed by simple averaging, whereas density cannot. Caballing produces intermediate water masses (500–1500 meters), deep-water masses (1500–4000 meters), and bottom-water masses in the oceans. Figure 10-3 shows these water masses in the South Atlantic off

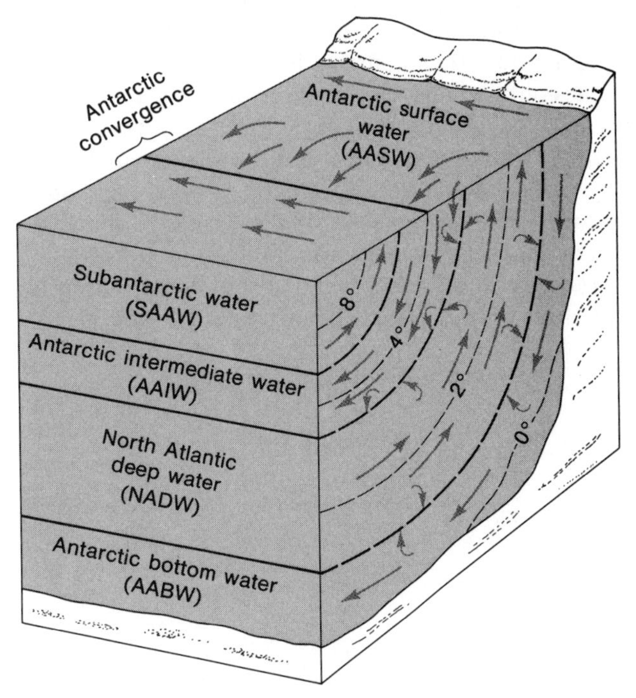

FIGURE 10-3 Complex thermohaline currents in the South Atlantic Ocean off Antarctica. Water cooled by sea and land ice of Antarctica sinks and flows along the bottom (AABW). It is replaced by an upward flow of warmer salty water (NADW), which mixes with surface water (AASW). Generalized isotherms show temperatures characteristic of the water masses. [After V. G. Kort, "The Antarctic Ocean." Copyright © 1962 by Scientific American, Inc. All rights reserved.]

TABLE 10-2
Subsurface-water masses in the North Atlantic Ocean

Water mass	Source	Identifying characteristic
Antarctic bottom water (AABW)	Weddell Sea	Bottom temperature minimum
North Atlantic deep water (NADW)	Area near Greenland	Intermediate salinity maximum, may show intermediate temperature maximum, low oxygen content in South Atlantic
Surface-water masses	Regional	Variable, generally warm
Mediterranean intermediate water (MIW)	Mediterranean Sea off Turkey	High salinity and temperature tongue at intermediate depths

Antarctica. Surface-water masses and types are generally formed by direct interaction and exchange between sea and air. The different water masses found in the North Atlantic Ocean are listed in Table 10-2.

When you plot the $T-S$ values for a real water station (as in Question 4 in the report form) you will note that they are in a stable position relative to each other because their densities increase as we go deeper in the column. The relative degee of stability is shown on a $T-S$ diagram by the angle the $T-S$ curves make with the contours of density. If the curves cut across the lines of equal density at large angles, the water column changes density rapidly. This means that minor changes in density will not cause water to sink or rise a great distance, and that the water column is therefore very stable. If the curves are at very small angles to the density contours, then the water changes density only slightly with increasing depth, so that minor changes in temperature and salinity (density) will cause water to move to different depths rapidly.

Also, a $T-S$ diagram in a well-known area of the oceans can be used to correct or find data points that are in error. The incorrect data appear as isolated points on the diagrams outside of the areas in which the water conditions are usually found.

DEFINITIONS

Caballing. A mixing of two water masses of identical *in situ* densities but different temperatures and salinities. The resulting water mass then becomes denser than either of its components.

Density factor(σ_t). A convenient numerical value for manipulating density data: $\sigma_t = (\text{density} - 1) \times 1000$. If density is 1.02386 then σ_t is 23.86.

Isopycnal, or isopycnic line. A line of equal or constant density.

Temperature–salinity ($T-S$) diagrams. Diagrams on which the characteristics of large masses of water are identified by plotting salinity and temperature data. Salinity is plotted with increasing values toward the right, and temperature is plotted with decreasing values downward. Depth at which each sample has been taken is usually indicated along the curve.

Water mass. A large volume of seawater that can be recognized as having a common origin. Water masses may be formed by interaction between air and sea or by mixing of two or more water types. A water mass is characterized on a $T-S$ diagram by a group of values that may be plotted as a curve or a straight line.

Water type. A homogeneous mass of water having well-defined temperature and salinity characteristics. It appears as a single point on a $T-S$ diagram.

EXERCISE 10 REPORT

WATER MASSES AND TEMPERATURE–SALINITY DIAGRAMS

NAME ____________________

DATE ____________________

INSTRUCTOR ____________________

1. Below is a duplicate of Figure 10-1 in which the σ_t value for 24 has been plotted. Plot the values from $\sigma_t = 25$ to $\sigma_t = 28$ on the T–S diagram.

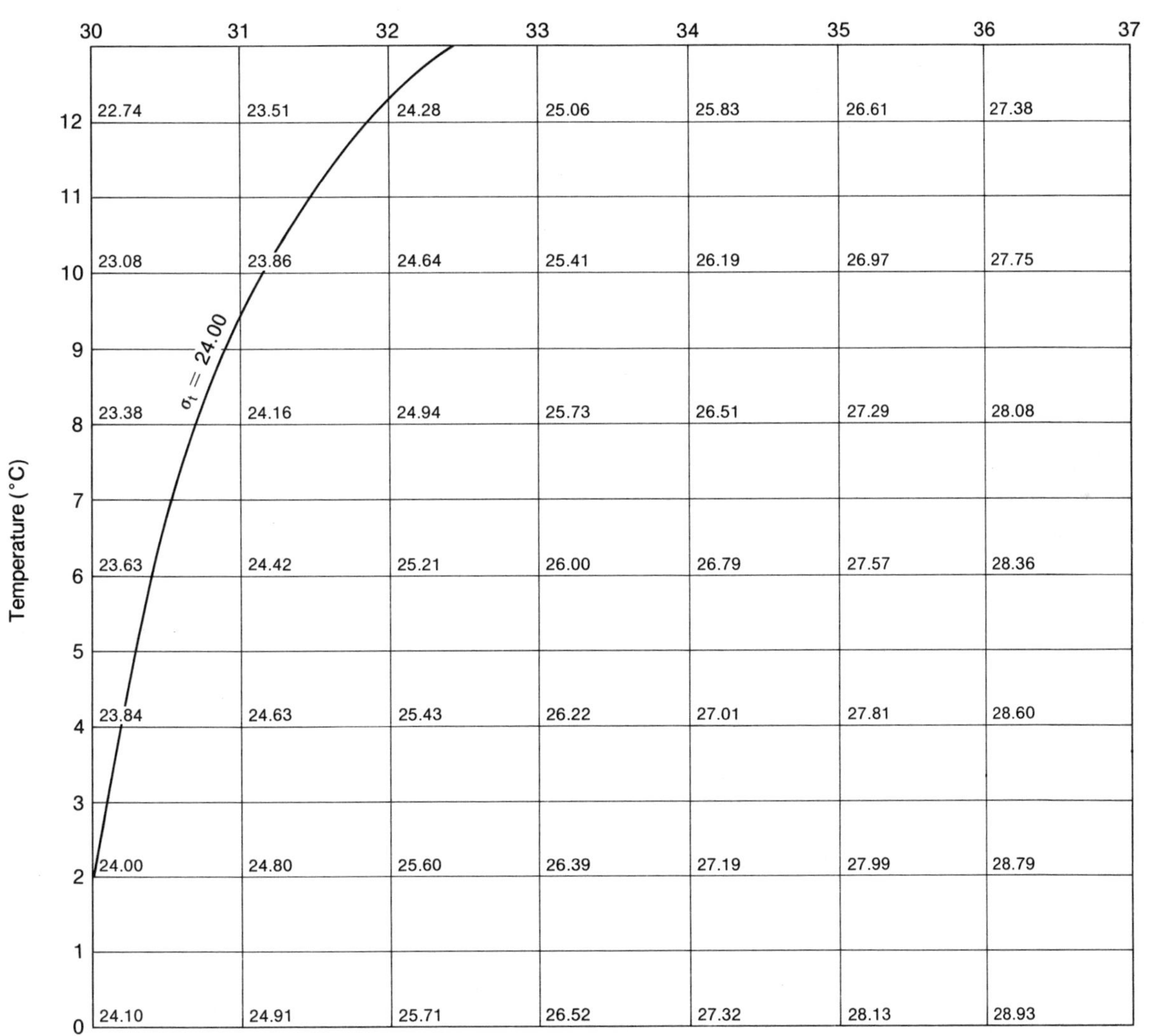

2. On the 24.00 σ_t contour, plot a water type A where the contour crosses the 32‰ salinity line, and a water type B where it crosses the 6°C temperature line. Connecting these points with a straight line yields a mixing line for these two water types. Plot a point in the middle of this mixing line. Note that it falls below and to the right of the 24.00 σ_t contour and is therefore denser. This midpoint is the temperature and salinity of a 50–50 mixture of water types A and B. The line itself represents all possible mixtures of water types A and B and thus the water mass that would form from such a theoretical mixture.

(a) What is the range of T and S that defines your water mass?

(b) What is the density of the 50–50 mixture of water types A and B? (Hint: Interpolate between the 24.00 and 25.00 isopycnals.)

3. The following table gives the temperature and salinity data taken at one station. Plot the points on the T–S diagram that you contoured in Question 1, and connect them with straight lines.

Depth (meters)	Temperature (°C)	Salinity (‰)
0	12	33.7
10	12	33.7
20	12	33.8
50	10	33.9
100	8	33.7
200	7	34.1
300	6	34.2
400	5	34.2

(a) What is the relative stability of the water column at this station; how deep is the mixed layer; and in what portion of the curve is the water most unstable (portion below the mixed layer)?

(b) In what portion of the curve is the water most stable?

4. The following are oceanographic data from a typical station in the North Atlantic at about La 20°N.

Depth (meters)	Temperature (°C)	Salinity (‰)
100	16.0	36.1
200	13.0	35.8
400	11.0	35.5
500	9.0	35.3
600	8.0	35.0
850	13.0	37.3
950	12.5	37.1
1200	11.0	36.7
1500	4.8	34.8
2000	4.0	35.0
2200	3.5	34.9
2500	2.0	34.8
3000	0.0	34.7
4000	−1.9	34.6
5000	−2.0	34.6

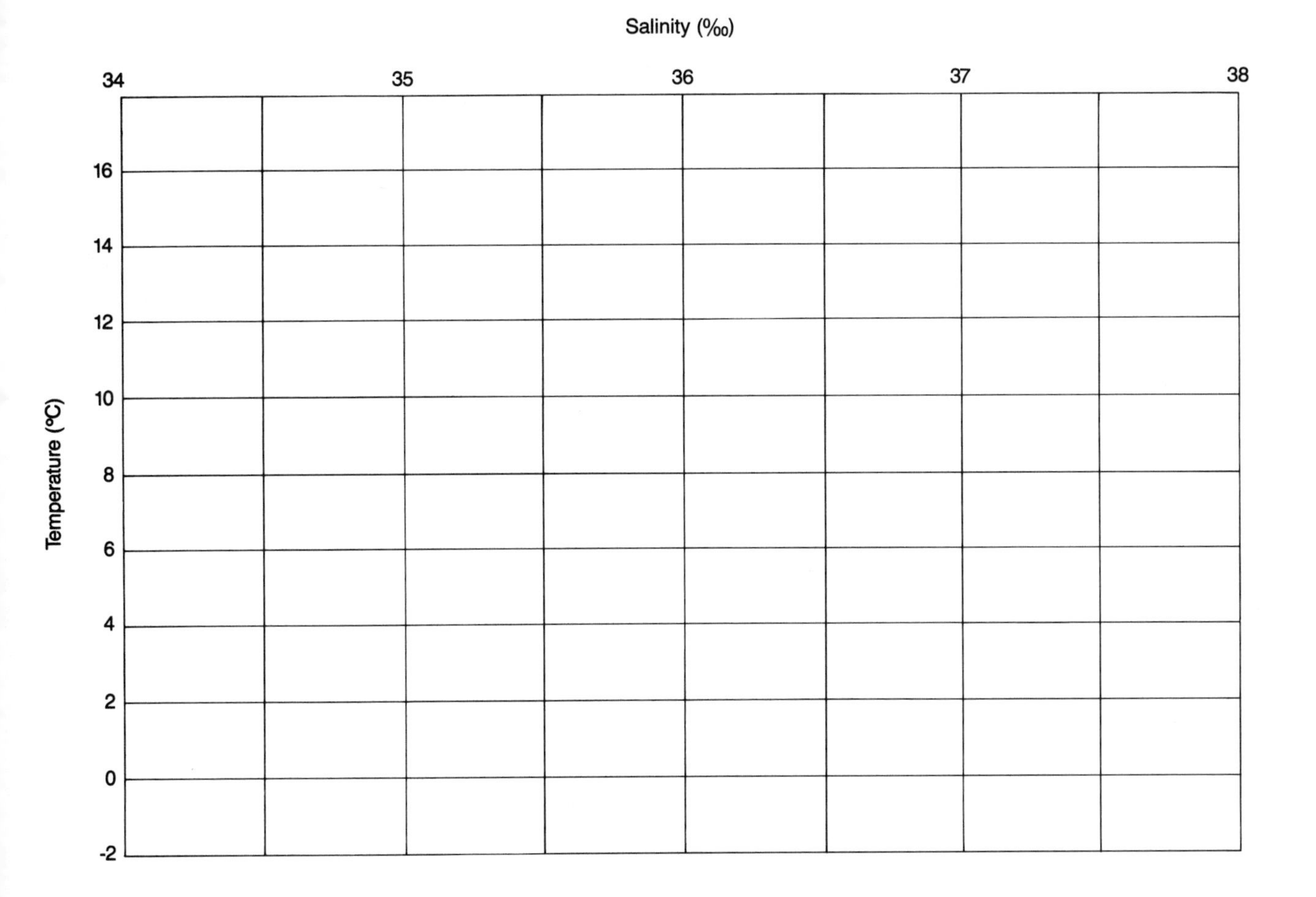

Plot the data on the blank T–S diagram above and draw a line connecting all points in order of depth. Using the diagnostic parameters provided in Table 10-2, name the major water masses represented in the diagram.

5. (a) It is estimated that the flow southward of North Atlantic deep water is 6×10^6 cubic meters per second. If the volume of the Atlantic Ocean is 3.24×10^{17} cubic meters, what would be the length of time required to circulate the entire Atlantic through the deep water?

(b) If the oceans are at least 3×10^9 years old, and we assume that the mixing rate has been constant, how many times has the ocean been "stirred" through the deep water?

(c) How long would you have to run a household blender spinning at 1000 revolutions per minute to equal the stirring of the oceans?

(d) Is the ocean a well-stirred solution?

EXERCISE 11

SURFACE CURRENTS

Surface oceanic circulation is the result of several processes, including wind stress acting on the water surface and differences in density due to solar heating. If we assume that observed **current** systems are simply the result of wind stress then they should very closely follow the major wind belts on the earth (see Figure 11-1). In fact they do, but you will note that there is a slight clockwise divergence (to the right) from the wind direction in the Northern Hemisphere and a counterclockwise one in the Southern Hemisphere. This is due to the earth's rotation, and its influence on ocean currents was not appreciated until 1835 when Gaspar de Coriolis, while studying equations of motion in a rotating frame of reference (the earth), discovered what is now called the **Coriolis effect** or, not quite accurately, the Coriolis force. The word "force" is misleading because the effect is due not to force or acceleration, but to the earth's eastward rotation, and any moving object not attached to the earth's surface shows the apparent deflection described above. The amount of deflection depends on the velocity of the object and its latitude. It is zero at the equator and maximum at the poles, and fast-moving objects are deflected more than slow-moving ones. It should be noted that the Coriolis effect has no influence on the energy of motion and modifies only direction, that is, to the right in the Northern Hemisphere and to the left in the Southern Hemisphere.

MAJOR CURRENTS OF THE WORLD'S OCEANS

Figure 11-2 is a map of the major wind-driven currents in the world's oceans. Note that in the Northern Hemisphere the currents form large clockwise-rotating gyres or rings. Note also that the gyres are not exactly centered in the ocean basin but are slightly offset to the west. Thus the currents along the western sides of the northern oceans are narrow, fast, and deep, whereas those on the eastern sides are wide, shallow, and sluggish. As an example, the Kuroshio Current in the western Pacific Basin has about six times the flow of the California Current on the eastern side but is only one-fourth as wide. Velocities within the Kuroshio Current can attain 10 kilometers per hour, whereas the California Current generally moves with a velocity of less than 2 kilometers per hour. This phenomenon is known as *westward intensification,* and is due to the earth's rotation and the necessity of balancing or conserving angular momentum. The same phenomenon may be observed in the Canary–Gulf Stream current system in the Atlantic Ocean.

The currents in the southern oceans are essentially mirror images of those in the northern parts. They rotate in a counterclockwise direction, as predicted by Coriolis, and show westward intensification. Note that the plane dividing the northern and southern

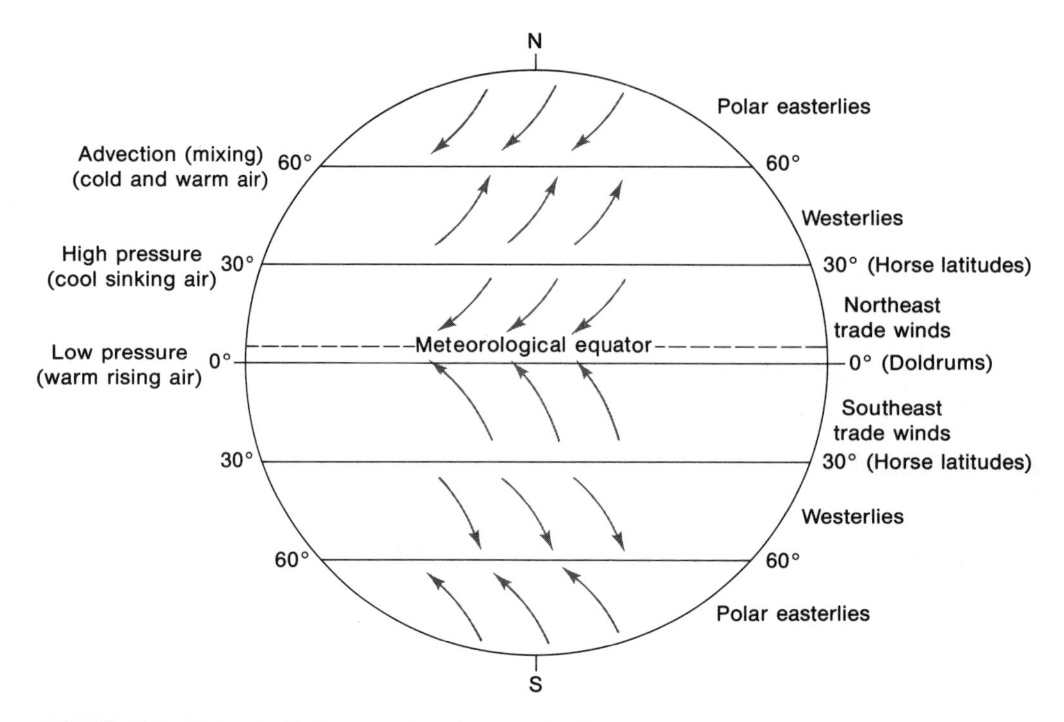

FIGURE 11-1 Major wind belts on earth and zones of high and low pressure. Note that the meteorological equator is 5–10° north of the geometric equator.

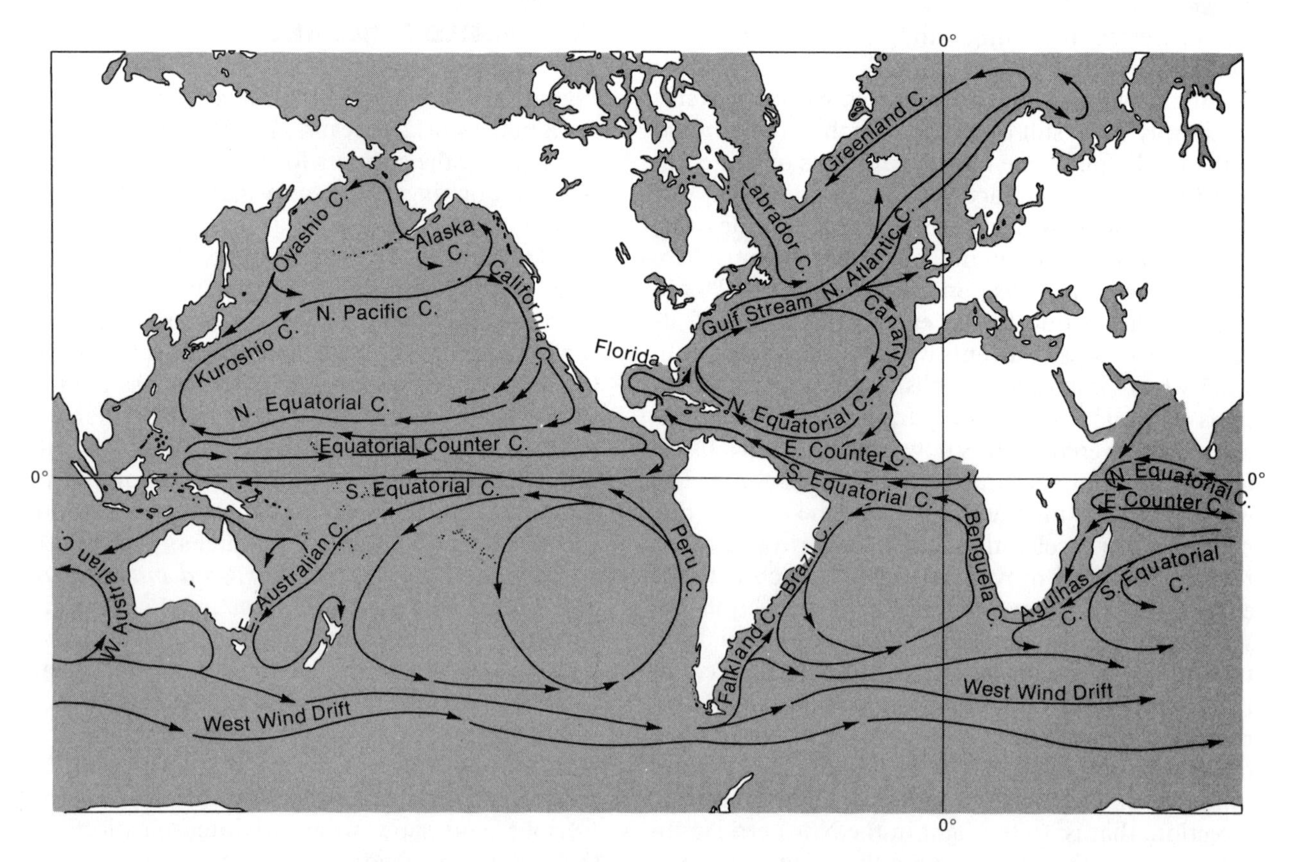

FIGURE 11-2 Major surface currents of the oceans of the world.

circulation lies a few degrees north of the equator. This plane is known as the *meteorological equator* and also marks the convergence between the northeast and southeast trade winds. Known as the *doldrums* to mariners (Figure 11-1), the meteorological equator is a zone of erratic winds and rising warm air (low pressure), and it also contains the Equatorial Countercurrent. This current results from warm light water piling up on the west side of the Pacific Basin to produce an eastward slope of the water surface of about 4 centimeters per 1000 kilometers. The countercurrent flows eastward between the west-flowing North and South Equatorial currents.

Periodically, around Christmas time, the coast of Peru is washed by a current of warm water that the local fishermen call *El Niño,* after the infant Christ. Oceanographers now use the term to describe anomalous warming events that involve the entire equatorial Pacific Ocean. These sometimes catastrophic events are episodic, and no less than seven severe El Niños have occurred over the past 100 years. The largest recent one took place in 1982–1983 and had an effect on the entire Pacific rim and Indian Ocean. The normal coastal current along South America is the Peru, or Humboldt, Current, which sets northward with cold waters from the Antarctic regions. However, the southeast tradewinds (Figure 11-1), combined with the Coriolis effect, deflect this current offshore to be replaced by cooler, nutrient-rich upwelled water (see Exercise 8). This colder upwelled inshore water is so distinctive it is called the *Peru Coastal Current* to distinguish it from the Peru Current offshore. It constitutes one of the richest fisheries in the world.

In El Niño years the usual low atmospheric pressure of the western equatorial Pacific becomes high pressure. The pressure differential that drives the trade winds (high pressure to low pressure) lessens and the trades die down or even reverse direction. Without winds to produce persistent upwelling along the coast, the huge mass of piled-up warm water in the western Pacific surges back toward South America in the Equatorial Countercurrent. When it strikes the coast of South America it moves southward over the cooler coastal waters, wreaking ecological havoc. Many fish species disappear, driven away by the warm, dilute, nutrient-poor water blanketing the region; others simply die. Millions of seabirds that ordinarily feed on these fish and sustain the valuable guano fertilizer industry also vanish. As a result, coastal communities experience economic hardship and sometimes even famine.

The increase in atmospheric pressure in the western Pacific that triggers El Niño is periodic and is called the *southern oscillation,* a designation reflecting the cyclical nature of the pressure difference involved. So closely related are the two phenomena (which affect global weather and eastern Pacific oceanography) that atmospheric scientists now refer to a combined El Niño–southern oscillation (ENSO). The 1982–1983 ENSO was responsible for flooding and droughts in 12 countries, thousands of deaths, and billions of dollars in property damage. In southern California the rainfall nearly tripled and the coast was battered by destructive winds and high waves.

VOLUMES OF WATER FLOW

Major currents are so large that tremendous volumes of water flow are transported by them. A flow unit, called the sverdrup (Sv), is used to indicate a flow of 1,000,000 cubic meters per second and was named after Harald U. Sverdrup, a famous oceanographer. The volumes of transport of water of the major currents of the Pacific Ocean are shown in Figure 11-3.

Data for the Atlantic Ocean are more difficult to obtain but it is estimated that the flow of the Gulf Stream is at least 55 Sv, and for the Canary Current a range of 2–16 Sv is estimated. The westward intensification is clearly demonstrated in these figures.

SPEED OF FLOW AND DYNAMIC TOPOGRAPHY

The greatest sustained speeds are found in the Kuroshio Current, which is known to move as fast as 3 meters per second. Speeds as great as 50 centimeters per second are reported in the westward flow near the equator and the West Wind Drift.

An ideal way to measure velocity and direction of currents would be to establish current-meter stations in all parts and at all depths of the oceans. The exorbitant expense and difficulty involved in mooring and maintaining such meters has limited the use of such direct methods to relatively small areas. The common procedure used to determine large-scale current motion is to measure the distribution of density (therefore pressure) in the oceans or of a particular current and convert the readings into profiles or sections showing horizontal density changes or anomalies. The patterns of these anomalies reflect the deflection of the water surface from horizontal and enable the investigator to construct a map showing **dynamic topography,** or irregularities in the sea surface. The topography so illustrated in dynamic meters reflects horizontal pressure gradients, and it is then possible to obtain useful approximations of real currents (to

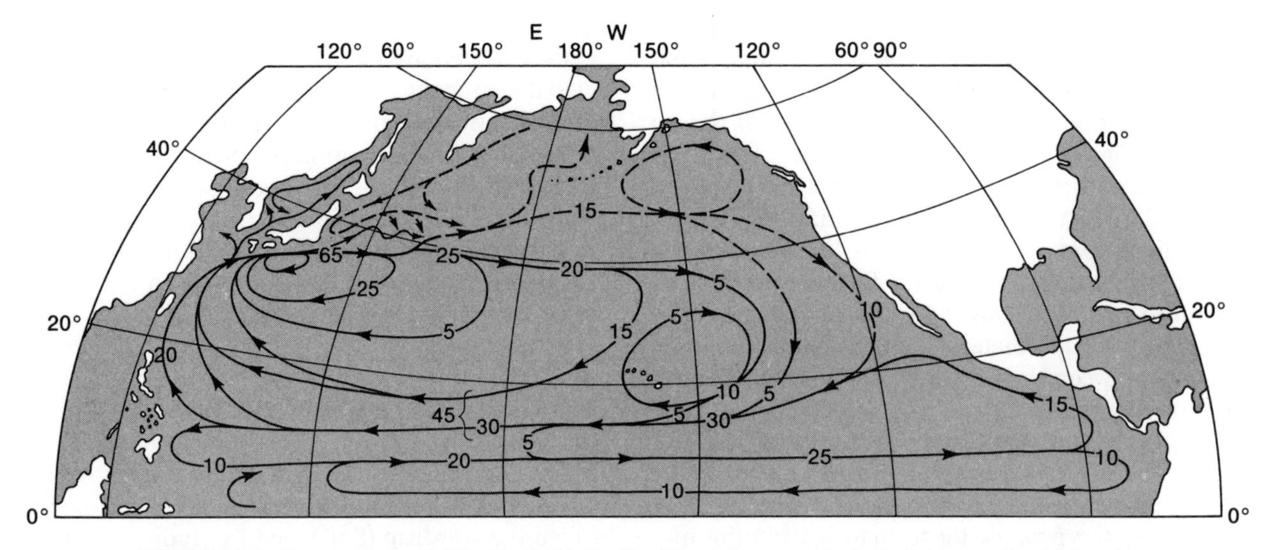

FIGURE 11-3 Transport chart of the North Pacific. The lines with arrows indicate the approximate direction of transport above 1500 meters, and the numbers show transported volumes in sverdrups. Broken lines show cold-water currents; solid lines warm ones. [After Harald U. Sverdrup, Martin W. Johnson, and Richard H. Fleming, *The Oceans: Their Physics, Chemistry, and General Biology.* Copyright © 1970. Redrawn by permission of Prentice-Hall, Inc., Englewood Cliffs, New Jersey.]

within about 15 percent). To do this a series of temperature and salinity measurements are made to a depth of about 1 kilometer at a number of locations (called *stations*). From these data the horizontal pressure gradients can be computed and a chart of dynamic topography prepared. If we contour values of equal dynamic topography and assume that only the Coriolis effect applies, then the flow will theoretically parallel the contours. When we put arrowheads on the contours pointing in the proper direction, they indicate the direction of flow and are called *streamlines.* Figure 11-4 is such a chart, showing dynamic topography and streamlines for the Kuroshio Current in the western Pacific Ocean. The spacing of the streamlines gives us an idea of the speed of flow. Where they are closely spaced the horizontal pressure gradient is large and the flow will be fastest.

THE ORIGIN OF DYNAMIC TOPOGRAPHY

How are the irregularities in the sea surface formed and maintained? To answer this question we must examine the interaction between wind, gravity, and the Coriolis effect. In 1905 V. W. Ekman provided a theoretical explanation that laid the cornerstone for all subsequent studies of wind-driven currents. Observers had noted that icebergs in the northern oceans moved 20–40° from the wind direction. Ekman showed that, given a steady wind and homogeneous sea, surface water will flow 45° to the right of the wind direction in the Northern Hemisphere and 45° to the left of it in the Southern Hemisphere. However, this surface layer, a few meters or tens of meters thick, sets in motion an underlying layer that is deflected farther to the right (in the Northern Hemisphere) but with lesser velocity because of friction. Subsequent layers downward are set in motion, each being deflected farther to the right until at depth there is a weak current flowing in the opposite direction. This rotation of the wind-driven current with depth is known as the **Ekman spiral** (Figure 11-5). The base of the wind-driven water column varies with wind velocity and persistence but is usually taken to be at a depth of 100 meters. This theory itself was revolutionary, but more important is the fact that the average direction of flow throughout the entire water column is 90° to the right of the wind direction in the Northern Hemisphere, and 90° to the left in the Southern Hemisphere. This net motion of water at right angles to the wind is known as *Ekman transport.*

Now we are able to see how dynamic topography develops. The northeast trades and the westerly winds in the Northern Hemisphere push water into a bulge near the center of the current gyres, owing to Ekman transport (Figure 11-6). This bulge of warm surface water is unstable and the water attempts to return to a level stable condition with a uniform distribution of warm water near the surface and cooler water below. As a water parcel starts to flow downslope by gravity (to a region of lower pressure) the Coriolis effect deflects its motion to the right after it reaches an appreciable velocity. As it continues downslope it is turned

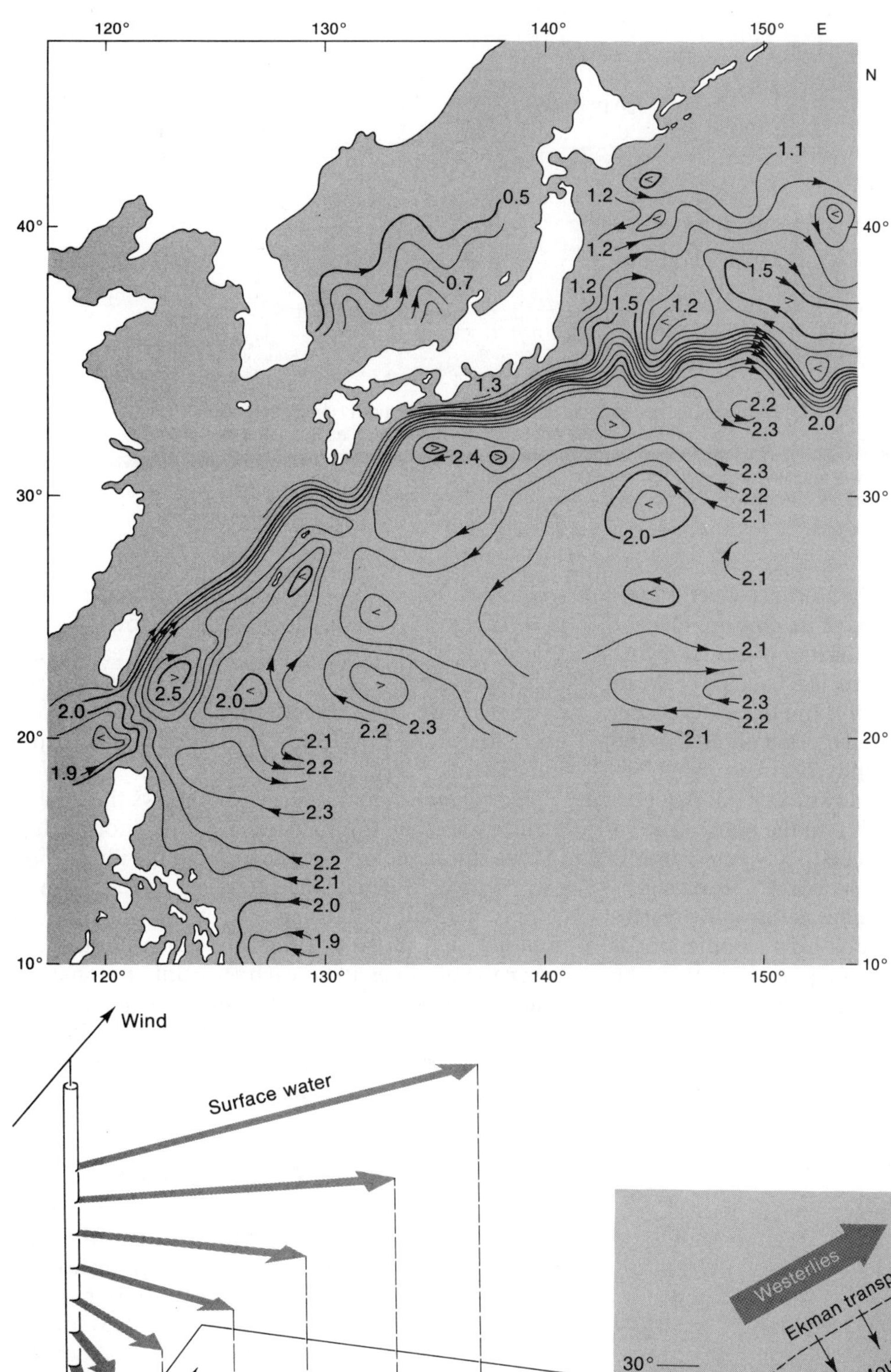

FIGURE 11-4 Chart of dynamic topography for the Kuroshio Current. Numerals along the contours give the elevation of the sea surface in meters, and the streamlines show direction of flow. Note that the current flows in a clockwise direction from high-pressure areas toward low-pressure areas. The closer together the streamlines are, the faster the current. [After R. A. Barkley, "The Kuroshio Current," *Science Journal*, Vol. 6, 1970, pp. 54–60.]

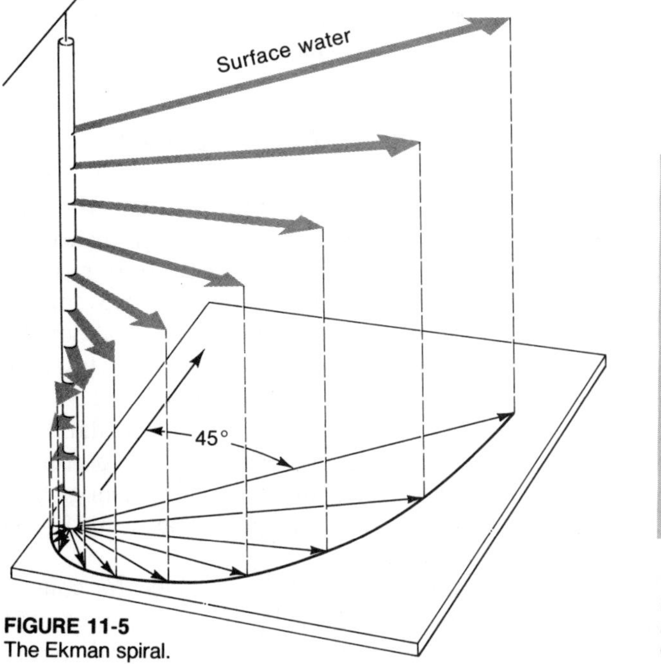

FIGURE 11-5
The Ekman spiral.

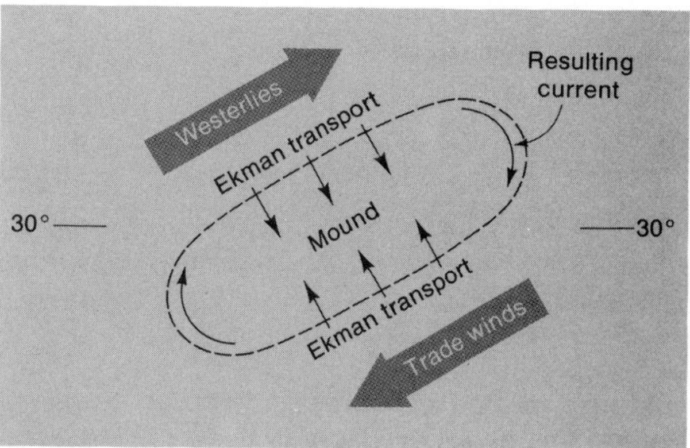

FIGURE 11-6 The creation of an oceanic high-pressure area at 30° N by prevailing winds. [After J. Williams, J. J. Higginson, and J. D. Rohrbough, *Sea and Air: The Naval Environment.* Copyright © 1968, U.S. Naval Institute Press.]

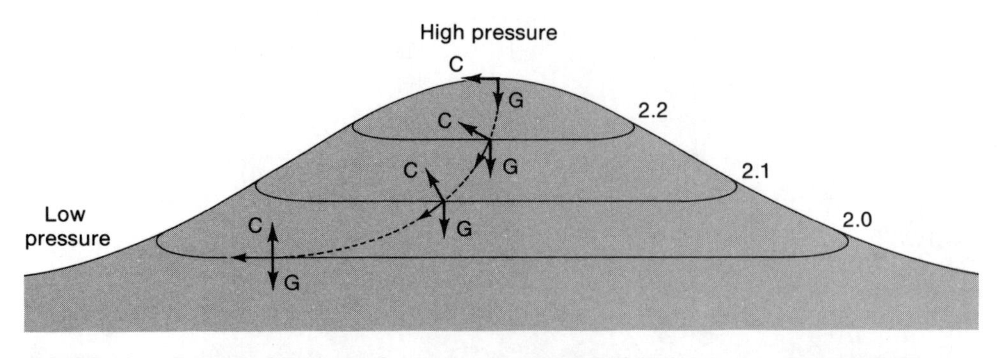

FIGURE 11-7 Diagram of the forces for geostrophic currents of the Northern Hemisphere. Note the perfect balance that is achieved between the Coriolis effect, C, and gravity, G, as the water parcel moves around the mound parallel to the contours. Numerals represent the height in dynamic meters. [After W. Anikouchine and R. Sternberg, *The World Ocean*. Copyright © 1973. Redrawn by permission of Prentice-Hall, Inc., Englewood Cliffs, New Jersey.]

farther to the right because the Coriolis effect is continually operating to the right of its motion. Once it has turned 90° it cannot turn farther without flowing uphill. At this point there is a perfect balance between the Coriolis "force" and gravity, and the water parcel continues to move around the mound parallel to the contours of dynamic topography, or streamlines (Figure 11-7). If it turns slightly downhill it gains speed, and the Coriolis effect deflects it to the right again. In this way a balance is reached between Coriolis "forces" and pressure forces (gravity). Currents generated in this manner are called **geostrophic** (earth-turned) **currents.** A schematic of this phenomenon for the North Atlantic is shown in Figure 11-8. It can be appreciated that once the mound of water is elevated by the wind it is not relieved by geostrophic flow because the water simply moves around the hill. However, in nature this is not exactly true, because some friction occurs in fast-moving currents and a water parcel eventually spirals down back to a low. It is said that if the wind were to stop, then the ocean would become perfectly flat in about 3 years and surface currents as we know them would cease to exist; however, density differences would continue to cause movement at depth.

The current map in Figure 11-9 shows the flow of surface waters off California and Baja California during the summer of 1939. The scales in the illustration can be used to determine the speed of the current: Simply measure the distance between two adjacent contours and move this distance along the gap between the two lines above each scale until they match. The speed can then be read from the scale. You are actually using a graphic method for calculating the horizontal pressure (gravity) changes and the flow rate produced by these changes.

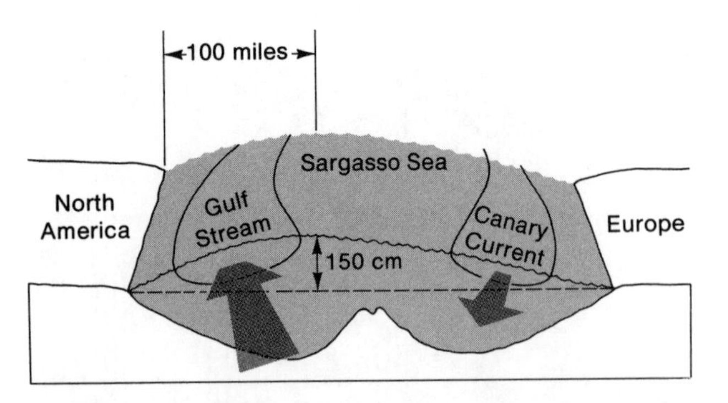

FIGURE 11-8 The mound in the North Atlantic Ocean between the Gulf Stream and the Canary Current. The Sargasso Sea is a region of warm, clear surface water, deep blue in color and having large quantities of sargassum, or seaweed. [After J. Williams, J. J. Higginson, and J. D. Rohrbough, *Sea and Air: The Naval Environment*. Copyright © 1968, U.S. Naval Institute Press.]

SATELLITES FOR OCEANOGRAPHY

Satellites have the great advantage of covering the surface of the globe in a matter of days instead of months or years, as shipboard investigations require. Satellite-obtained imagery depicts the sea surface only; it conveys little direct information about the water at depth. But because satellites move so fast they record data in a quasi-synoptic manner and give us a better overall picture of the ocean at a particular time. Furthermore, responses at the sea surface to perturbations in temperature, salinity, or dynamic topography are significantly briefer than those that occur in the interior of the ocean.

The first satellite to be used exclusively for collecting ocean data was *Seasat.* It successfully measured surface winds, significant wave heights, the direction

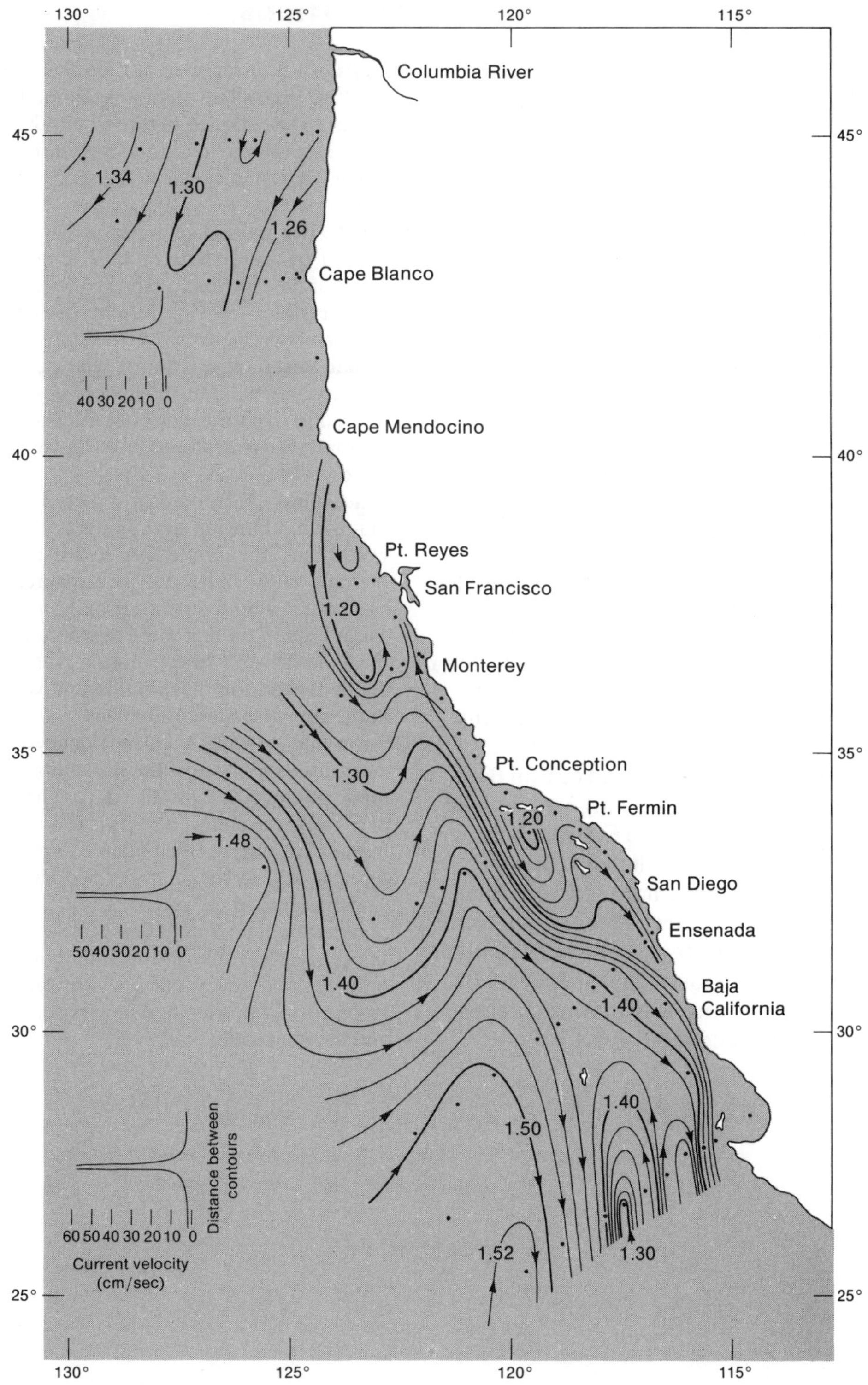

FIGURE 11-9 The dynamic topography of the sea surface relative to an arbitrary level of surface at depth. Note that the direction of flow alternates to and from the coast. The contours represent the height in dynamic meters. Contour intervals are 0.02 dynamic meter. [After Harald U. Sverdrup, Martin W. Johnson, and Richard H. Fleming, *The Oceans: Their Physics, Chemistry, and General Biology.* Copyright © 1970. Redrawn by permission of Prentice-Hall, Inc., Englewood Cliffs, New Jersey.]

and frequency of surface waves, and sea-level changes. Subsequent satellite systems have been used to map the distribution of sea ice and to measure temperature, chlorophyll content, and other biological properties of seawater. In addition to collecting scientific data, satellites have practical value as an aid to shipping, since they can provide advance information about winds and waves (which can be measured by satellite-borne radar techniques) along given sea routes.

An example is the imagery of El Niño reproduced on the cover of this book. Taken by advance very high-resolution radiometry (AVHRR) aboard the *NOAA-T* satellite in 1982–1983, these color-coded pictures show the warmest areas in red and orange, intermediate-temperature areas in yellow and green, and the coldest regions in blue. Clouds appear white. The left-hand photo (taken in January 1982) depicts normal conditions, with cool California Current water flowing southward past Point Conception, which corresponds to the area labeled 1 in the photo at the right. The cool coastal waters (green) are typical of southern California and Mexico, where winds from the north cause upwelling along the coast. The enriched coastal waters give rise to plankton blooms and high biological productivity.

Los Angeles is at 3, San Diego at 4, and central Baja California at 2 in the right-hand photo, which was taken in January 1983. The impact of El Niño is obvious: South of Point Conception the surface waters average 2 Celsius degrees above normal. Abnormal onshore winds have caused warm water to pile up along the coast, raising the average sea level 20 centimeters above normal. In addition, the 1982–1983 El Niño brought with it exotic species of marine life such as small pelagic red crabs and large squid.

DEFINITIONS

Coriolis effect. An apparent "force" on moving particles resulting from the earth's rotation. It causes moving bodies to be deflected to the right in the Northern Hemisphere and to the left in the Southern Hemisphere. The "force" is proportional to the speed and latitude of the moving object.

Current. The motion of water as it flows down a slope, pushed by wind stress or tidal forces. The velocity or speed of flow is usually expressed in centimeters per second, or for fast-moving currents in meters per second or kilometers per hour.

Dynamic topography. The irregularities in the sea surface produced by wind or differences in density, usually expressed in dynamic meters or fractions thereof. This topography is dynamic and thus changes with time and the seasons.

Ekman spiral. A theoretical representation of the effect that a wind blowing steadily over a large body of water would cause the surface layer to drift at an angle of 45° to the right in the Northern Hemisphere. Water at successive depths would drift more and more to the right in a spiral fashion until at some depth, known as the *base* of the wind-driven current, motion is essentially zero. This depth depends on the duration and velocity of the wind but is approximately 100 meters.

Geostrophic current. A current defined by assuming an exact balance between the horizontal pressure gradient (density) and the Coriolis effect. The usual manner of deriving geostrophic currents is to prepare a chart of dynamic topography based upon observations of temperature and salinity for a number of oceanographic stations. The direction of the current is indicated by the contours of dynamic topography, and its speed by the spacing of the contours. Although the underlying assumptions are only approximately correct, the direction and speed computed by this method are very close to the direction and speed actually observed.

EXERCISE 11

REPORT

SURFACE CURRENTS

NAME ______________________

DATE ______________________

INSTRUCTOR ______________________

1. By inspection of Figure 11-2, describe the following currents as either warm or cold, and fast or slow.

Current	Relative temperature	Relative speed
Peru	________	________
Kuroshio	________	________
California	________	________
Gulf Stream	________	________
Agulhas	________	________
Canary	________	________
West Wind Drift	________	________

2. Describe briefly the function of ocean currents in the distribution of heat on the earth.

3. Capetown at the tip of South Africa has a cool mild climate, whereas Durban a few hundred miles to the east is very hot and humid. Why is this so?

4. Why is the Equatorial Countercurrent in the Atlantic Ocean so poorly defined in comparison with the same current in the Pacific Ocean?

5. What is the only current that completely circumscribes the earth?

6. Using the scales on the California Current map in Figure 11-9, determine the maximum and minimum velocities off Point Conception. Use the scale closest to the speed to be determined.

Maximum ______________ centimeters per second. Minimum ______________ centimeters per second.

Convert these values to the following units (1 meter per second = 3.6 kilometers per hour; 1 kilometer per hour = 0.54 knot).

Maximum ______________ meters per second; ______________ kilometers per hour; ______________ knots.

Minimum ______________ meters per second; ______________ kilometers per hour; ______________ knots.

Compare these results with the Kuroshio Current, where speeds of 2 meters per second are common.

______________ kilometers per hour; ______________ knots.

7. From observation of the Kuroshio and California currents (Figures 11-4 and 11-9), what generalization can you make about major ocean currents with respect to boundaries, shape, and uniformity of velocity over long periods of time?

8. A subsurface temperature section from San Francisco, California, to Honolulu, Hawaii, for the period of April 2–5, 1972, is shown in Figure 11-10. The section is taken through the California Current and the eastern North Pacific central water to a depth of 500 meters. Surface temperature, salinity, and temperature–depth profiles were taken at each of 21 stations (Table 11-1 lists the surface temperature and salinity readings). The California Current is recognizable by colder surface waters, low salinities, and irregular isotherms; eastern North Pacific central water is much warmer, with salinities in excess of 34.8‰. A transition, with salinities in the range of 34.0–34.8‰ exists between the two water masses.

(a) Plot the surface salinity data at each station on the graph above the section, using the salinity scale at the right-hand side. Connect the points for each station with a line.

(b) Plot the surface temperature at each station in the same manner, using the temperature scale at the left-hand side. Connect these data points with a dashed line.

(c) How far from San Francisco does the region of eastern North Pacific central water appear in the section?

Is there a well-defined surface mixed layer in this water mass, and to what depth does it extend?

(d) How wide is the California Current in this section?

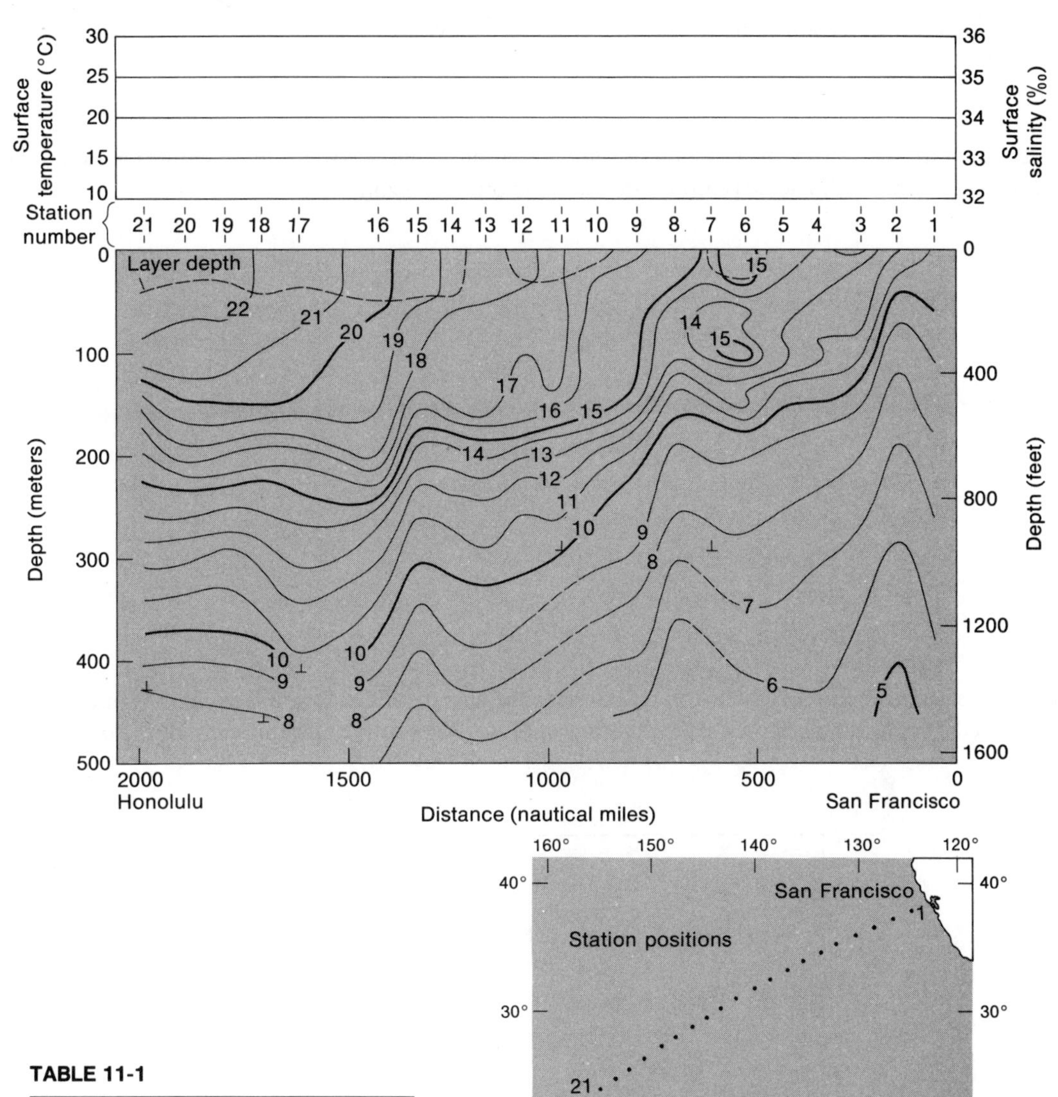

FIGURE 11-10 Cross section of temperature to a depth of 500 meters from San Francisco to the Hawaiian Islands. Temperatures in Celsius degrees are indicated along the isotherms. The symbol ⊥ shows the depth of the bathythermograph drop. (From *Fishing Information.* U.S. Department of Commerce, National Oceanographic and Atmospheric Administration, April 1972.)

TABLE 11-1

Station number	Surface salinity (‰)	Surface temperature (°C)
1	33.20	11.2
2	32.90	12.5
3	33.10	14.8
4	33.20	14.5
5	—	15.0
6	33.50	15.8
7	33.45	15.2
8	33.40	15.7
9	33.41	16.0
10	34.65	17.0
11	34.95	17.5
12	35.10	18.5
13	35.20	18.9
14	35.20	19.0
15	35.20	19.8
16	35.25	20.5
17	35.10	22.0
18	35.15	22.4
19	35.10	22.8
20	35.10	23.2
21	35.15	23.7

(e) How wide is the transition zone between the two water masses?

(f) Describe briefly the shape of the isotherms and variation in surface-water temperatures between the California Current, eastern North Pacific central water, and the transition zone.

9. Compare the satellite photos of El Niño (on the cover) to the map of the California Current in Figure 11-9. Based on the data presented in both of these figures, sketch the flow of El Niño along the California coast on the outline map in Figure 11-11.

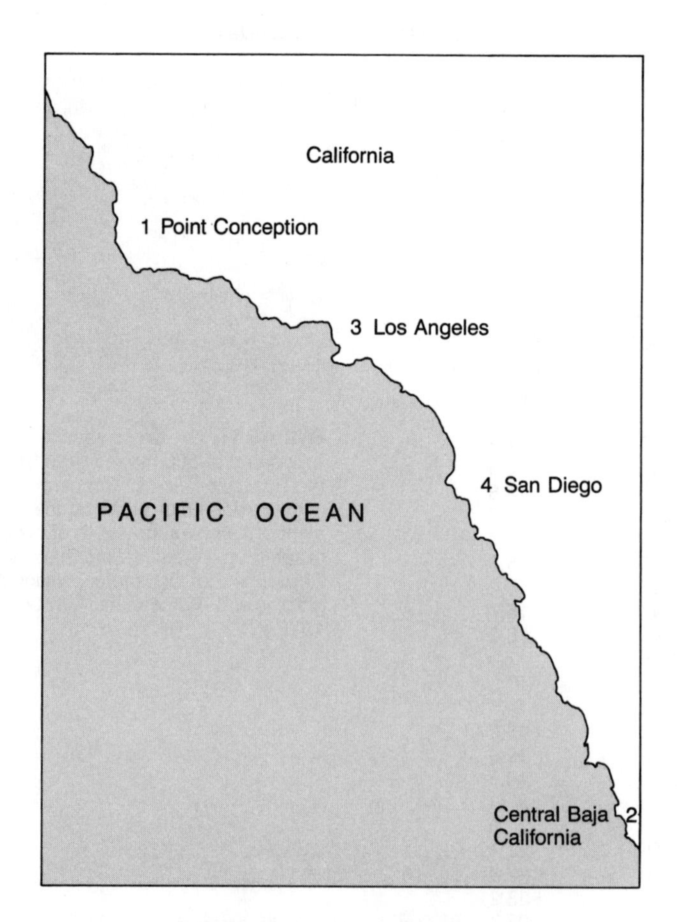

FIGURE 11-11 The California coast.

10. In a few sentences describe the flow of El Niño that you determined in answering Question 9 as compared to normal current flow.

11. On the outline map in Figure 11-11 draw arrows showing wind directions that produce upwelling and normal conditions along this stretch of coastline. Using dashed lines and arrowheads, show the wind directions that caused destruction and severe beach erosion along the the coast during the ENSO of 1982–1983.

EXERCISE 12

TIDES

To the casual observer the most obvious change in the level of the sea is that of the tides. They are caused mainly by the moon and the sun exerting forces on different parts of the rotating earth. The tidal wave, or bulge, is the result of *gravitational attraction* and *centrifugal force,* which act in combination to produce a regular variation in water level in the course of a day. Consider only the earth – moon system. Although the moon appears to revolve about the earth, the two bodies are actually rotating about a common center of mass. They are held together by gravity and kept apart by an equal and opposite centrifugal force. Thus, on the side of the earth closest to the moon, the tide-producing force is gravitational, whereas on the opposite side centrifugal force dominates.

We can illustrate this point by showing the water motion resulting from only the horizontal tide-generating forces (Figure 12-1a). When the moon is over the equator water is drawn by gravitational attraction toward the side nearest the moon, and toward the opposite side by centrifugal force, so that high tides with a low tide belt in between result. Because the forces are of equal magnitude symmetrically about the equator, two high and two low tides of equal magnitude should be experienced, at least in theory, at any given latitude on the earth. As the moon shifts north or south of the equator (that is, when it is at north or south declination), the forces are as shown in Figure 12-1b. A point at the equator is still subject to highs and lows of equal magnitude, but points at higher latitudes will experience strong diurnal inequalities; in other words, they will experience high tides of unequal heights or perhaps only one high tide.

TIDE LEVELS AND DATUM PLANES

Because the ocean basins vary in size and shape, and because land masses interfere with the tidal bulge, the tides do not assume a simple regular pattern. Although a purely mathematical solution of the tidal phenomenon is still beyond the limits of marine science, it is possible to predict the tide level at least 1 year in advance by careful analysis of tide records from stations at which observations have been made for long periods of time. Most of the averages are based on at least 19 or 20 years of records, and quite accurate prediction is routine.

The tide level is usually measured in reference to a local base level, or **datum** plane, which is an average of many years' observations. A typical datum in the United States is mean lower low water (MLLW), which is the average of the lowest tide each day. Another common datum is mean low water (MLW). This is the average of all low-tide levels at the station,

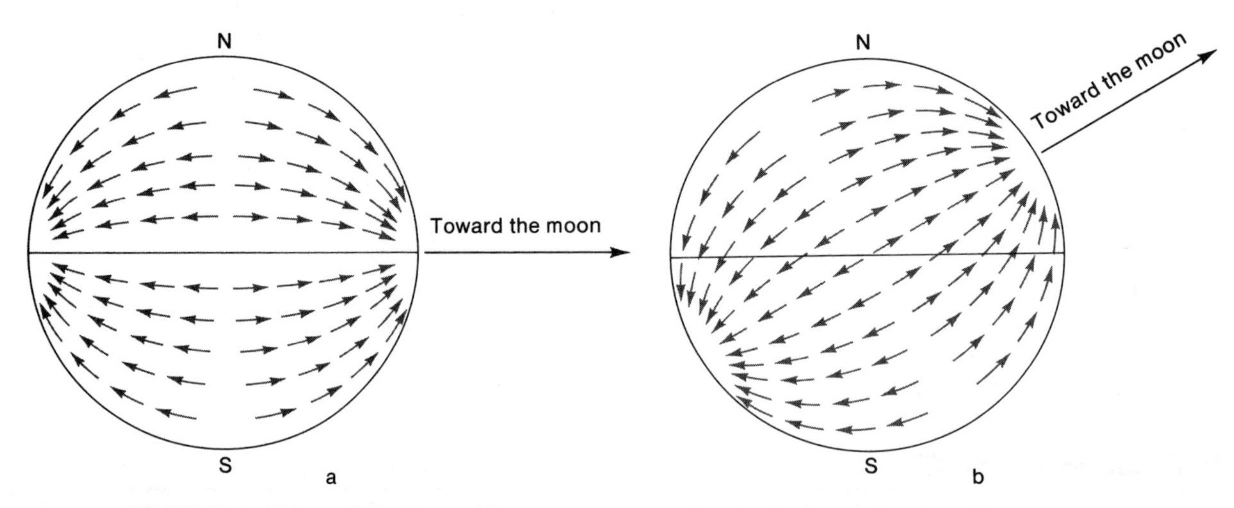

FIGURE 12-1 Tide-producing forces. The arrows represent the magnitude and direction of the horizontal tide-generating forces on the earth's surface. The force pulling toward the moon is gravitational attraction, and the force pulling away from the moon is the centrifugal force produced by the rotation of earth and moon about their common center of mass. (a) When the moon is in the plane of the earth's equator, the forces are equal in magnitude at the two points on the same parallel of latitude on opposite sides of the earth. (b) When the moon is at north or south declination, the forces are unequal at such points and tend to cause an inequality in the two high waters and the two low waters of a tidal day. [After N. Bowditch, *American Practical Navigator*. Hydrographic Office Publication No. 9, U.S. Naval Oceanographic Office, 1966.]

but is not as safe a point of reference for navigational purposes as mean lower low water, since at least half of the lows during a month will be lower than the datum. Other datum planes are mean sea level (MSL), mean high water (MHW), and mean higher high water (MHHW). Inasmuch as mariners depend on the charted depths, and since these are established in reference to the tide datum, it is obvious that the best datum will be the lowest normal level that the tide will reach. The relationship between levels of the sea and datum planes for the Orange County coast of California are shown in Table 12-1 and Figure 12-2.

TABLE 12-1
Sea levels and datum planes for the coast of Orange County, California

Sea level	Datum mean sea level (MSL)	Datum mean lower low water (MLLW)
Highest tide	4.8	7.5
Mean higher high water	2.6	5.3
Mean high water	1.9	4.6
Mean sea level	0.0	2.7
Mean low water	−1.8	0.9
Mean lower low water	−2.7	0.0
Lowest tide	−5.2	−2.5

TYPES OF TIDES

Three major types of tides can be recognized on the basis of frequency of occurrence and symmetry of the tidal curve. **Diurnal tides** occur once daily, meaning that there is one high and one low tide of about equal height in the course of a **tidal day.** A tidal day is 24 hours and 50 minutes long because the moon, which exerts the greatest tidal influence, advances 50 minutes each day in its orbit around the earth. The **semidiurnal tides** occur twice daily and are also of about equal height. The **mixed tides,** also known as *irregular semidiurnal tides,* occur twice daily but exhibit two highs and two lows of significantly unequal height. The type of tide that occurs on a given coast and its variation in amplitude, or height, depend on a number of factors. Among them are the shape of the basin in which the tide occurs, natural oscillations of the water (seiches) within the basin, declination of the sun and moon, and relative position of the sun and moon. Tides on the East Coast of the United States are representative of the semidiurnal type, whereas those on the West Coast are mixed tides. However, either coast may exhibit both types at certain times of the year. Dirunal tides typically occur in partially enclosed basins such as the northern Gulf of Mexico, the Java Sea, and the Gulf of Tonkin off the Vietnam–China coast (Figure 12-3).

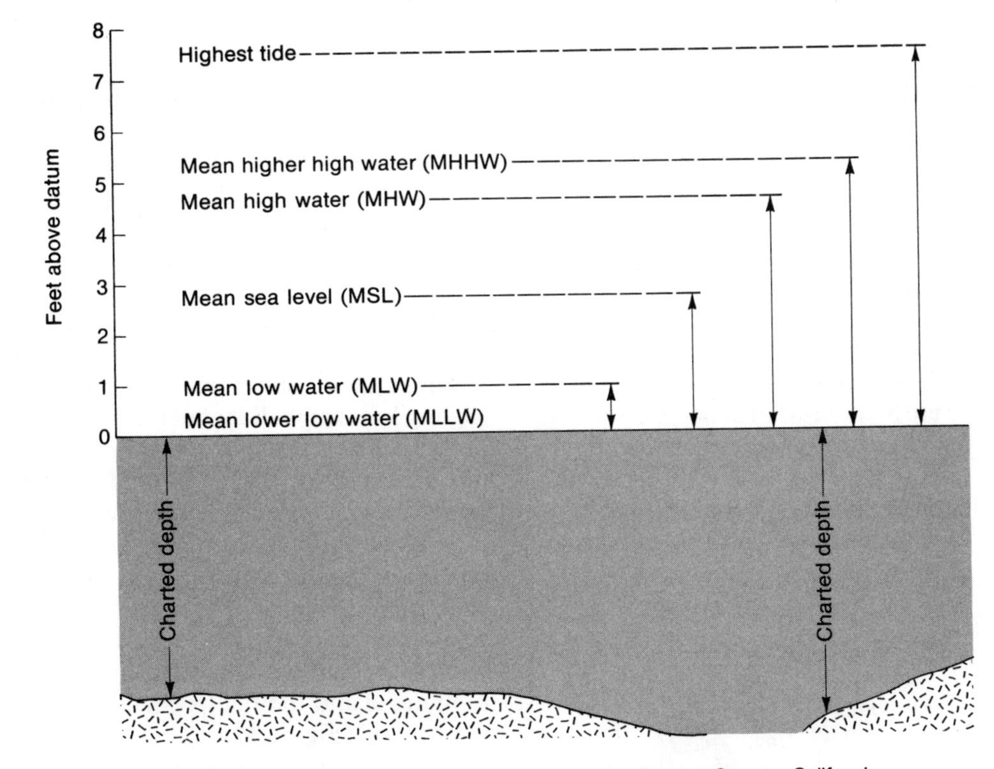

FIGURE 12-2 Depths at various stages of the tide, Orange County, California.

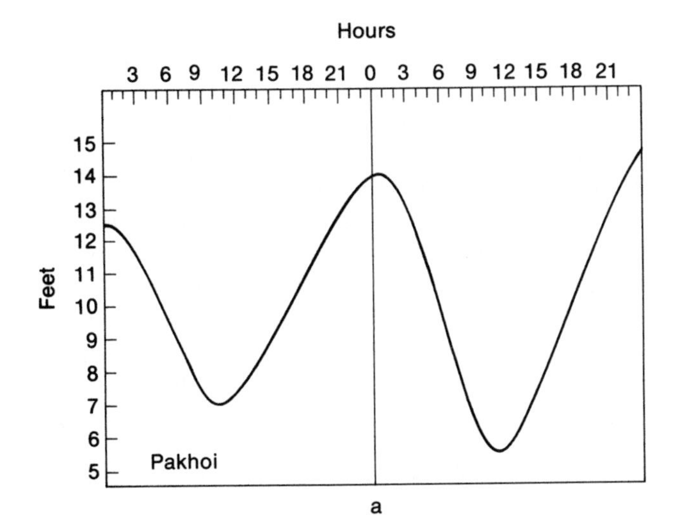

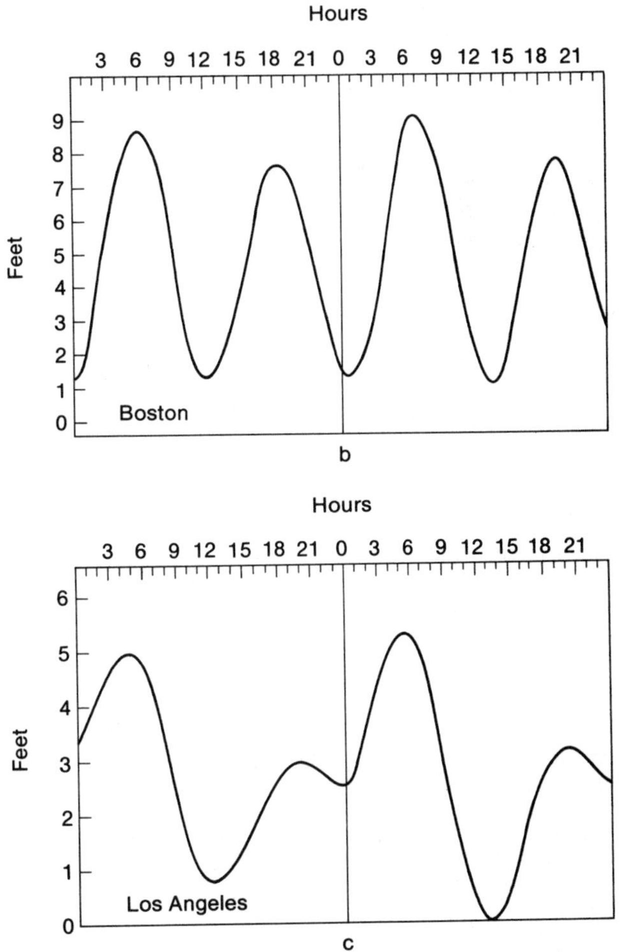

FIGURE 12-3 Types of tides from the Atlantic and Pacific ocean basins: (a) diurnal type; (b) semidiurnal type; (c) mixed type.

MONTHLY TIDAL CYCLES

Tides are also identified by their **tidal range:** Those having, in the course of a lunar month, the largest difference in level between high and low are called **spring tides;** those having the smallest range of the month are the **neap tides.** Spring tides occur twice monthly at or near the time of new moon and of full moon. At these periods, tides are at their highest and lowest levels in relation to their mean level. Perhaps the best datum to use for navigational purposes would be the average of all low spring tides or mean lower springs. Neap tides—the tides of lowest range (the lowest high tides and the highest low tides)–are also influenced by the lunar cycle and occur twice a month, at or near the first-quarter and third-quarter phases of the moon (the half moon).

Other fluctuations in the tidal range occur in response to the elliptical nature of the moon's orbit. When the moon is at a point closest to the earth (the *perigee*) once every 27.5 days in its orbit about the earth, the range of the tide is increased; when it is at the point farthest away from the earth (the *apogee*), smaller tidal ranges occur, other factors being equal.

The sun, too, causes tides, the solar influence being slightly less than half that of the moon. The solar tidal cycle occurs in the course of a 24-hour period and is not synchronous with the lunar tidal period of 24 hours and 50 minutes. However, when the sun, moon, and earth are in alignment, as they are at new and full moons (Figure 12-4a,b), the lunar and solar components reinforce one another and spring tides result. When neap tides occur, at first- and third-quarter moons, the sun–earth–moon system forms a right angle and the tide-producing forces are greatly diminished (Figure 12-4c).

The typical tidal curves given for various localities in Figure 12-5 show the three major types of tides and the effects of proximity and alignment of the moon and sun. Note the tidal curve for New York for September 22–26. The tidal range is high because the sun, moon, and earth are lined up; the sun and moon are at the equator, producing a high degree of symmetry, and the moon is at perigee, causing the higher spring tides to occur at this time rather than at the new moon phase when the moon was at apogee. The other curves may be explained in the same way, except the one for Port Adelaide, where the solar and lunar tides are about equal, so that they nullify one another at neap tides (at the other localities the lunar tide-producing force is about twice that of the sun).

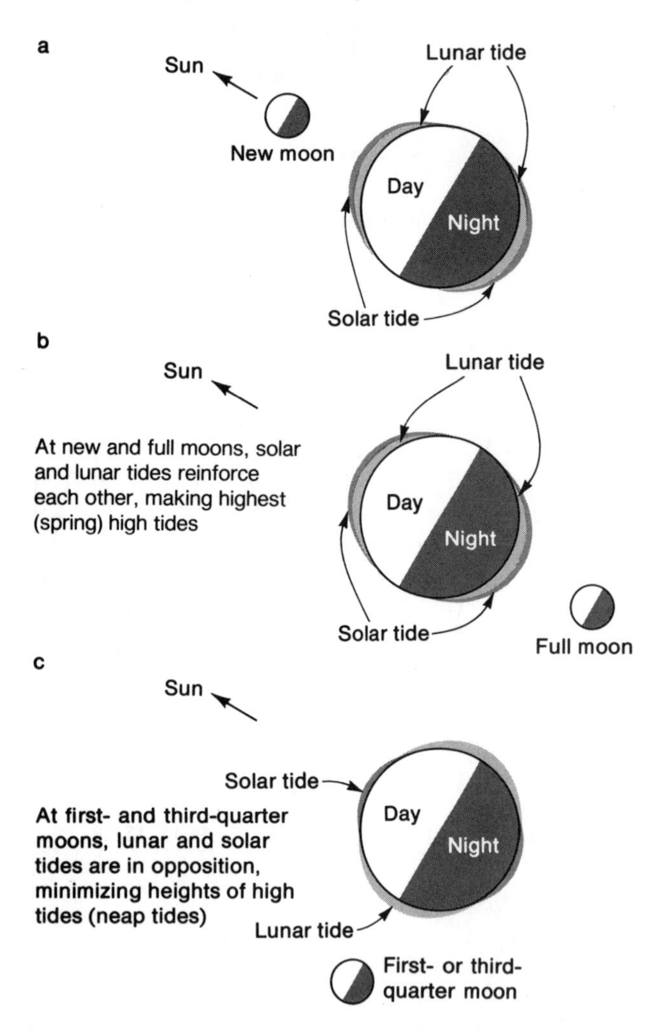

FIGURE 12-4 The relative positions of earth, moon, and sun determine the heights of high tide during the lunar month. The highest high and lowest low tides (spring tides) come at new and full moons; the lowest high and highest low tides (neap tides) come at first- and third-quarter moons. [After F. Press and R. Siever, *Earth,* 4th ed. W. H. Freeman and Company. Copyright © 1986.]

UNUSUAL TIDES

There are places in the world where tidal ranges exceed 10 meters and may reach as much as 16 meters. These occur in bays or harbors open to the ocean that are very long with respect to their depth. Natural oscillations, known as *seiches,* in these basins cause water to slosh back and forth much like what you might observe in a coffee cup. Physicists refer to these waves as *standing waves,* or *forced oscillations,* because the water stands first high and then low through one cycle. If the fundamental period (one up–down cycle) of the tidal basin or harbor is equal to the tidal period of 12 hours and 25 minutes, then resonance results and extreme tidal ranges may occur.

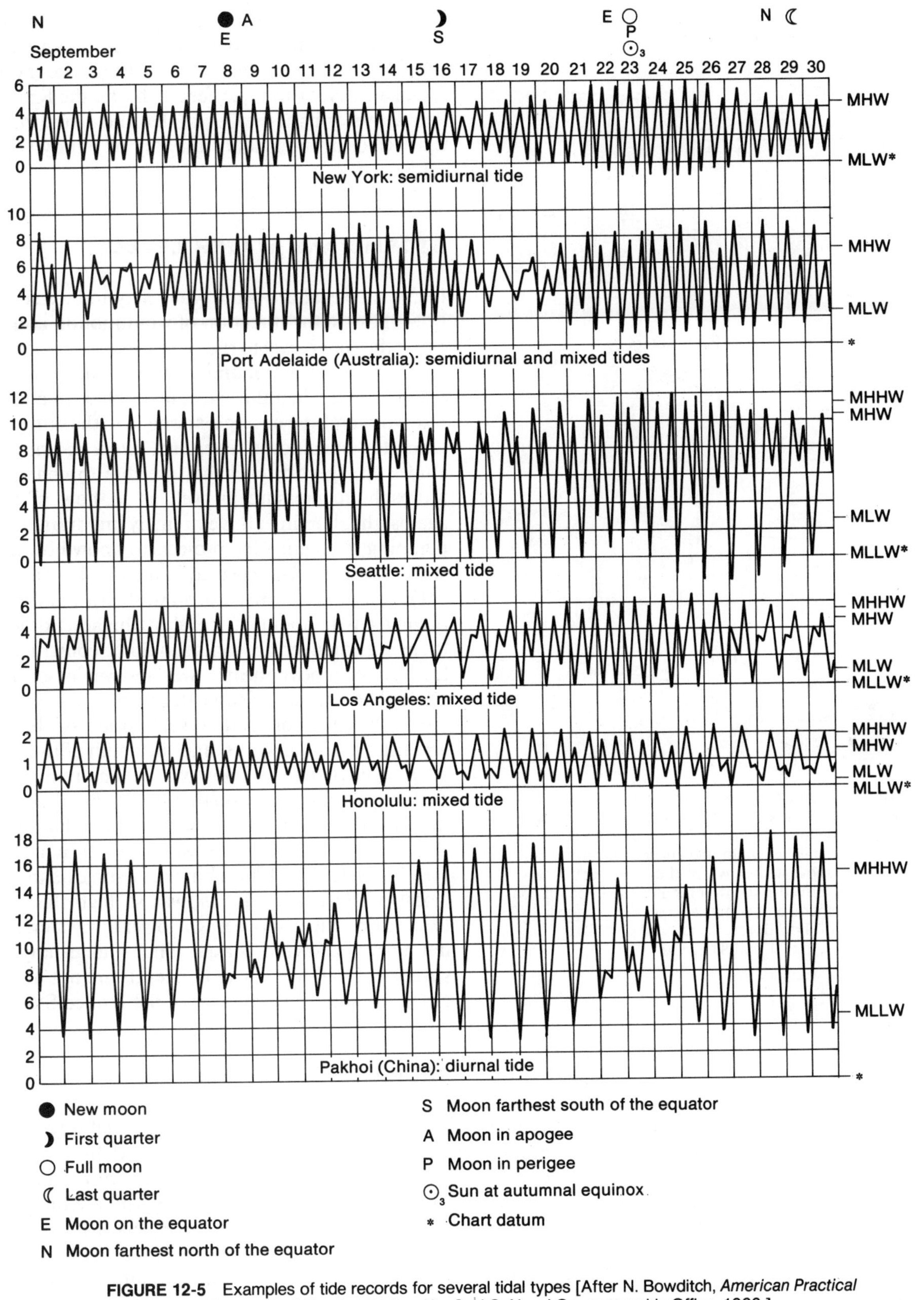

FIGURE 12-5 Examples of tide records for several tidal types [After N. Bowditch, *American Practical Navigator.* Hydrographic Office Publication No. 9, U.S. Naval Oceanographic Office, 1966.]

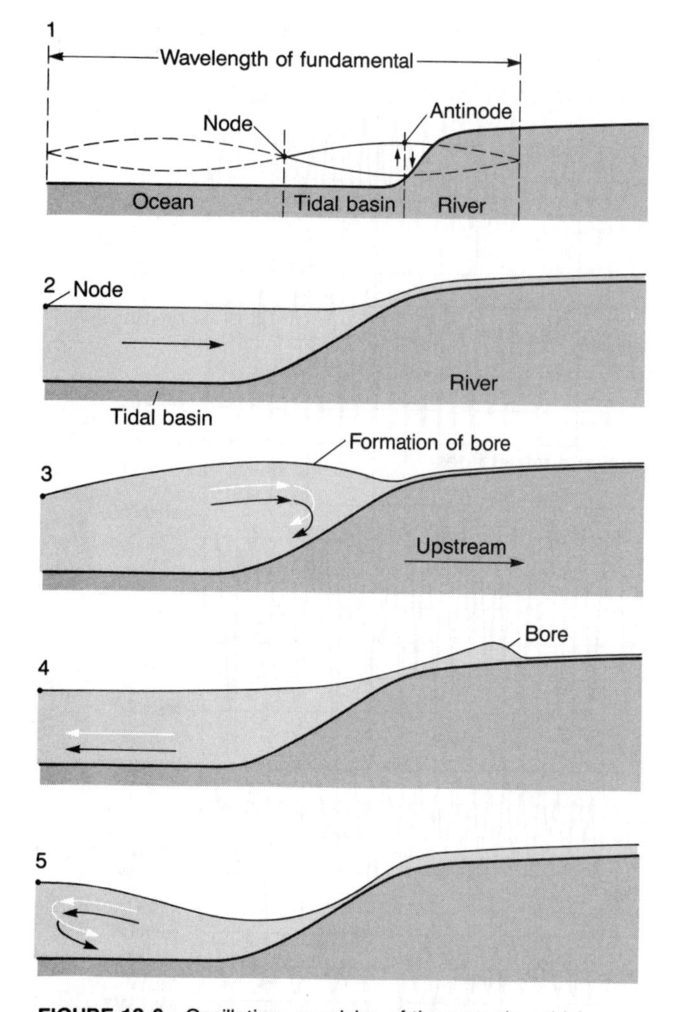

FIGURE 12-6 Oscillation, or seiche, of the water in a tidal basin. the time it takes the standing wave to make one complete oscillation is the fundamental period. The node is a point of little or no vertical movement of water; the antinode is a point of maximum vertical movement of water. If the tide rises fast enough, a bore is formed in the river mouth. The effect of constructive reinforcement, or resonance, is shown in panels 2–5. [After D. K. Lynch, "Tidal Bores." Copyright © 1982 by Scientific American, Inc. All rights reserved.]

Figure 12-6 shows the way in which the natural frequency of oscillation of a tidal basin and the tidal frequency may be in phase to produce resonance. This constructive reinforcement creates large tidal ranges—as great as 16 meters in the Bay of Fundy, 10 meters in the upper reaches of the Gulf of California, and 10 meters at Anchorage, Alaska. In some narrow funnel-shaped estuaries a *tidal bore* develops. This phenomenon, an abrupt solitary wave that moves upstream with the incoming tide, can be quite dangerous, since bores range in height from a few inches to as much as 25 feet. The most famous is on the river Severn in England; about 4 feet high, it can pass an observer on the run (Figure 12-7). Bores occur on the Amazon River (up to 25 feet high), on the Knik and Turnagain arms of the Cook Inlet, and on the Petitcodiac River at the head of the Bay of Fundy (Figure 12-8).

STORMS AND WATER LEVEL

In most coastal areas the wind may induce surface-water flow in the direction of wind motion and thus cause the water level to rise or fall above or below that level owing to astronomical tides. The term **wind setup** is used when this effect takes place in a lake or reservoir, and **storm surge** is applied to the same effect along the open coast. The term *hurricane surge* includes changes in level that are due to changes in atmospheric pressure as well as to flow of water against the shoreline. It is extremely important for the planning of engineering projects to know how much storm surge can be expected in a coastal area.

For example, storm surge in southern California is predicted to be about 2.5 feet above highest tide levels; therefore engineering works should be constructed at least 10 feet above mean lower low water (see Table 12-1). The amount of surge depends on the wind velocity, the length of open sea surface across which the wind can generate waves, and the depth of the water; surge is greater for shallow water. The influence of shallow water is the reason that storm-surge values are higher on the Gulf Coast than on the Atlantic (and that surge on the Atlantic Coast is higher than on the Pacific). Indeed, in 1900 a hurricane surge on the Gulf Coast of Galveston, Texas, was so high that water levels rose 15 feet above mean lower low water, inundating much of the coastal land; and in any year surges of 5–10 feet above tide levels are not unusual along the southeastern coast of the United States.

TSUNAMIS AND TIDES

Although sometimes referred to as tidal waves, tsunamis are actually sea waves caused by submarine landslides, earthquakes, or volcanic eruptions. Like the tides, tsunamis in the open ocean are only a few feet high and travel at speeds in excess of 400 miles per hour. In certain circumstances they can be very destructive. For example, the tsunami set off by the eruption of Krakatoa in 1883 rose to 30 meters in the shallow waters of the Sunda Strait between Java and Sumatra, killing 36,000 people in that coastal region.

FIGURE 12-7 Tidal bore on the Severn River is large enough for surfers to ride upstream for miles. [From D. K. Lynch, "Tidal Bores." Copyright © 1982 Scientific American, Inc. All rights reserved. Photo by C. G. Kershaw, Severn-Trent Water Authority.]

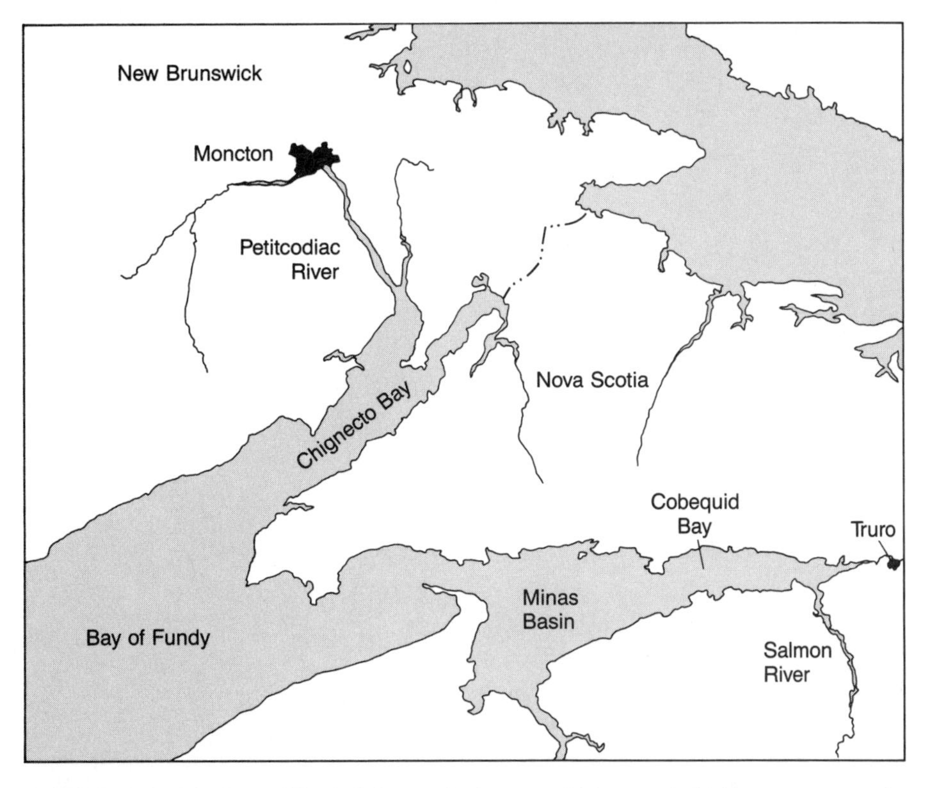

FIGURE 12-8 The Bay of Fundy is known for its great tidal range. A tidal bore moves up the Petitcodiac River as the high tide is funneled into Chignecto Bay and then into the river. [After D. K. Lynch, "Tidal Bores." Copyright © 1982 by Scientific American, Inc. All rights reserved.]

DEFINITIONS

Datum. The reference level to which tide levels are compared. The datum planes commonly used are mean low water or mean lower low water, which are the average levels of low tides taken over a 19-year period. These are also the datum planes (0′) used in constructing bathymetric charts.

Diurnal tides. Tides occurring once daily, one high and one low tide per tidal day.

Mixed tides. Complex tide curve, usually with two highs and lows of unequal height per tidal day.

Neap tides. The tides of lowest range, occurring twice monthly when the moon is at quadrature (so that the sun and moon are 90° apart).

Semidiurnal tides. Tides occurring twice daily. There are two high and two low tides per tidal day.

Spring tides. The tides of highest range, occurring twice monthly when the lunar and solar tides are in phase.

Storm surge. A rise above normal water level on an open coast due to strong winds blowing onshore. Storm surge resulting from a hurricane or other intense storm also includes the rise in level due to atmospheric pressure reduction as well as that due to the winds. A storm surge is most severe when it occurs in conjunction with a high tide.

Tidal day (or lunar day). The time between two successive transits or passings of the moon over a local meridian. It is derived from the rotation of the earth relative to the movement of the moon about the earth. As the earth rotates once on its axis (24 hours) the moon has advanced in its orbit about the earth about 50 minutes; therefore the tidal day is 24 hours and 50 minutes long.

Tidal range. The difference in height between successive high- and low-tide levels.

Wind setup. The vertical rise in the water level on the leeward, or downwind, side of a body of water due to strong winds. Wind setup is similar to storm surge but the term is usually applied to reservoirs and smaller bodies of water.

EXERCISE 12

REPORT

TIDES

NAME ____________________

DATE ____________________

INSTRUCTOR ____________________

1. Table 12-2 at the bottom of the page is a sample tide table for a period of 4 days. In this question, you will plot the data given in Table 12-2 and construct a relatively simple tide curve in Figure 12-9 on page 120. Straight lines are used to connect points rather than a smooth sine curve, as would ordinarily be done, but the record is still adequate for seeing the tidal variation. You may plot smooth curves also if you wish.

   (a) On the graph in Figure 12-9, plot the tide heights at the proper times, and connect them with a straight line to produce a tide curve.

   (b) During what days do the tide curves exhibit the following characteristics:

   Semidiurnal ______________.

   Diurnal ______________.

   Mixed ______________.

   (c) What is the smallest range and on what day does it occur?

   Range ______________ feet. Day ______________.

   (d) What is the elevation of mean high water for the 4 days?

   ______________ feet.

   Of mean low water?

   ______________ feet.

   What is the mean range?

   ______________ feet.

TABLE 12-2

Day	Time 24-hour day	Height (feet)
1	0115	1.0
	0730	4.1
	1600	1.0
	2015	4.0
2	0200	1.0
	0830	6.5
	1415	2.0
	2100	5.0
3	0415	0.0
	1600	7.0
4	0430	1.0
	1000	3.0
	1600	1.0
	2200	2.9

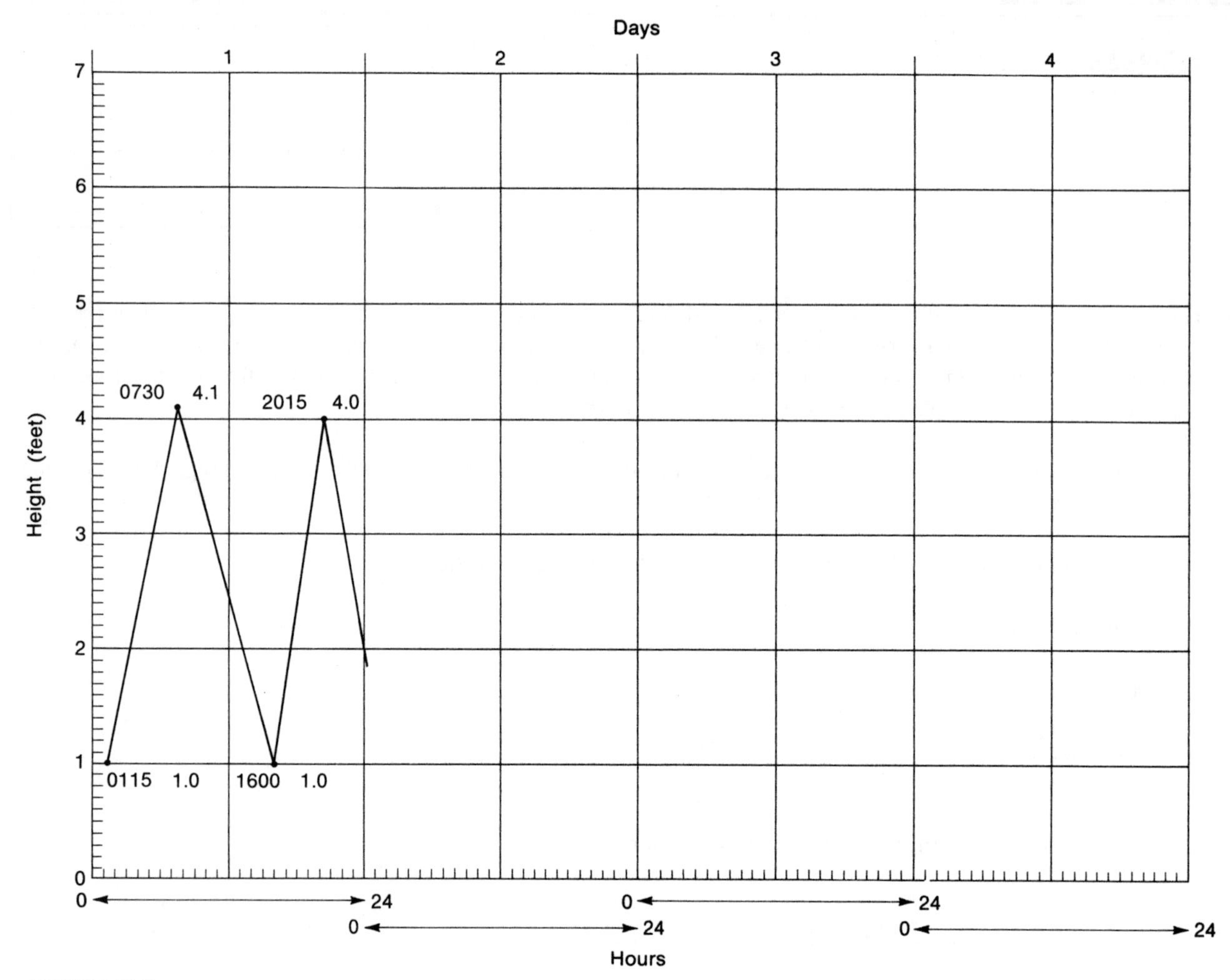

FIGURE 12-9

(e) On what days and at what times could a person sail a boat that draws 5 feet through a passage underlain by a reef that is exposed at mean lower low water? Assume that the datum for the graph is MLLW.

2. Figure 12-10 shows the contours of water levels that were computed on Lake Erie during a storm on November 8, 1957.

(a) In the grid at the bottom of the figure, draw a profile of the computed water levels from Toledo (gauge out of water) to Buffalo.

(b) The nodal point is the point of no vertical change and thus represents the mean level of the lake. What is the value of the wind setup at Buffalo?

What is the maximum difference in water level between Buffalo and Toledo?

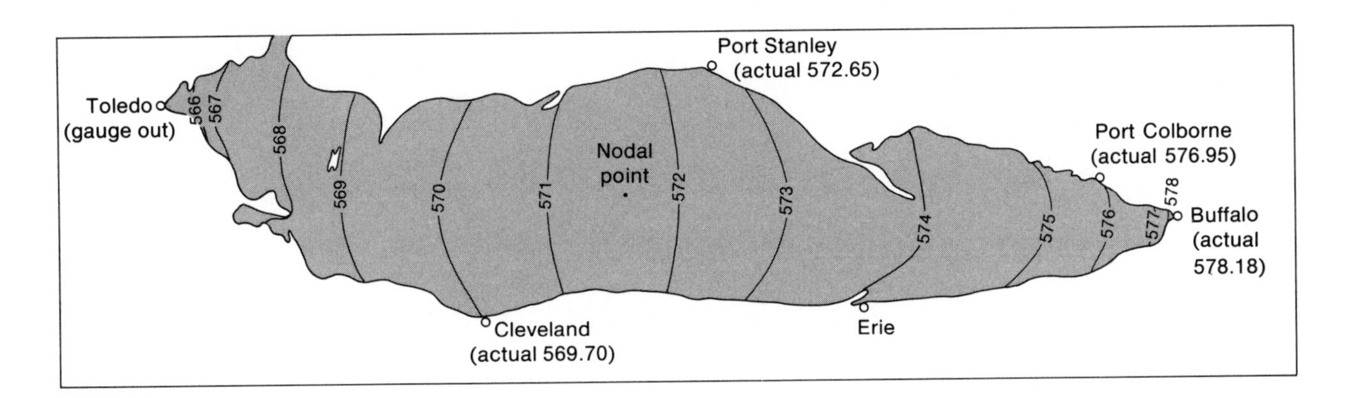

FIGURE 12-10 Effects of wind on surface-water level of Lake Erie. The contours show the water-level computer at 11:00 P.M. on November 8, 1957. [After I. A. Hunt, Jr., *Winds, Wind Set-up and Seiches on Lake Erie.* U.S. Lake Survey, U.S. Army Corps of Engineers, 1959.]

3. If the charted depth of the water at a given point is 10 feet (MLLW), what is the actual depth of water at the following tide levels? (Use Table 12-1.)

Highest tide ________________ feet.

Mean sea level ________________ feet.

Lowest tide ________________ feet.

OPTIONAL QUESTION

4. In the waters off southern California a small smeltlike fish, *Leuresthes tenuis* (the grunion), exhibits an interesting reproductive strategy finely timed to the tidal cycle. During spring tides from April to August, grunions come ashore shortly after the highest tides (which occur at night), and the female deposits eggs a few inches deep in the damp beach sand. The eggs are then fertilized by male grunions and are ready to hatch in 9–10 days, but only when the tidewater reaches them and they are agitated by surf action. When will grunion eggs deposited at full moon on July 3 have their first opportunity to hatch?

EXERCISE 13

WAVES AT SEA

Waves at sea are created by winds blowing across the water surface and transferring energy to the water by the impact of the air. Small ripples develop first and frictional drag on their windward side causes them to grow larger or collapse and contribute part of their expended energy to larger waves. Consequently the large waves capture increasing amounts of energy and continue to develop as long as the wind maintains sufficient strength and a constant direction. Generally, high winds of long duration produce large waves with long **wavelengths** and **wave periods.** Thus as more and more energy is transferred to the water surface, waves become higher and longer, and travel with increasing **wave velocities.** A mariner's rule of thumb is that in an area of foul weather, **wave height** in feet will be approximately half the wind velocity in knots.* That is, winds of about 50 knots would develop waves about 25 feet high. Whether or not this approximation is reliable, the fact remains that 50-foot waves are not uncommon in the open ocean and waves more than 100 feet high have been reported.

* Recall that the knot is defined as 1 nautical mile per hour; 1 nautical mile is equivalent to 1.15 statute miles.

WATER MOTION

It has long been observed that floating objects on the sea surface simply bob up and down or move with a slight rotary motion as waves pass underneath them. This is because water particles respond to the passing wave and move in circular orbits that decrease in diameter with depth (Figure 13-1). At a depth equal to about one-half the wavelength, the orbital diameters of the water particles are only 1/25 of those at the surface and, for all practical purposes, we may consider this level as the maximum depth of wave motion. In water deeper than half the wavelength, the orbiting particles do not contact the ocean bottom, whereas at depths shallower than half the wavelength the orbits are flattened by frictional resistance, they lose energy, and the wave is said to "feel the bottom." Geologists recognize this depth, which is called the **wave base,** as the maximum at which waves can move particles and erode fine sediment on the sea floor.

THE VELOCITY OF DEEP-WATER WAVES

The velocity of deep-water waves (where water depth is greater than half the wavelength) is a function of their wavelengths, that is, the longer a wave, the faster

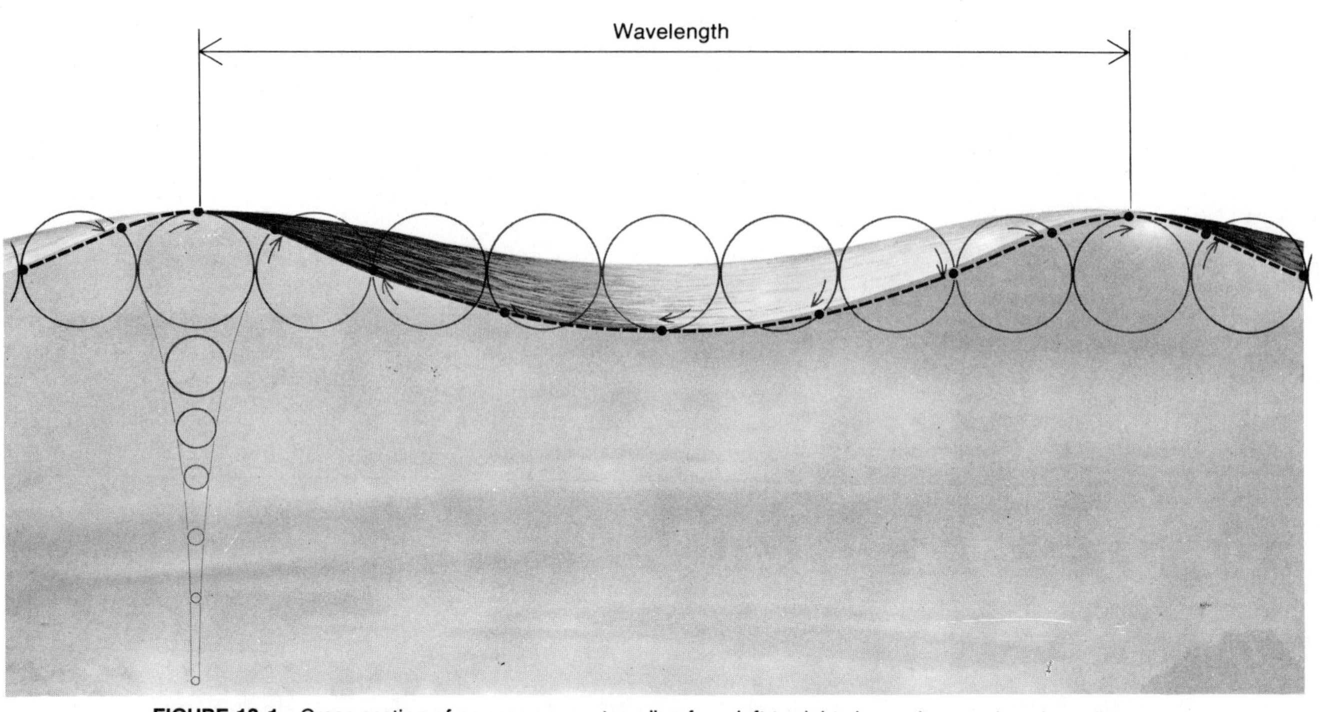

FIGURE 13-1 Cross section of an ocean wave traveling from left to right shows the wavelength as distance between successive crests. The time it takes for two crests to pass a point is the wave period. The circles are orbits of water particles in the wave. At the surface their diameter equals the wave height. At a depth of half the wavelength (left), the orbital diameter is only 4 percent of that at the surface. [After W. Bascom, "Ocean Waves." Copyright © 1959 by Scientific American, Inc. All rights reserved.]

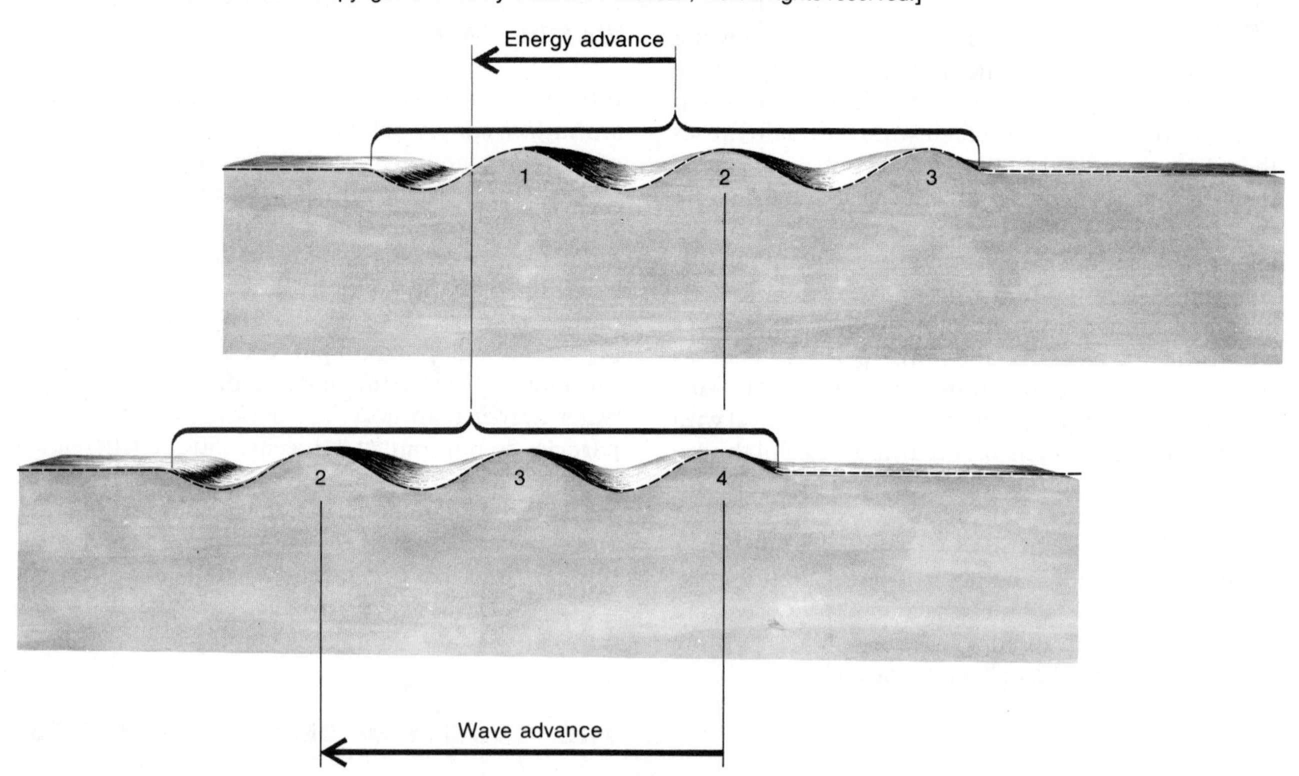

FIGURE 13-2 A moving train of waves advances at only half the speed of individual waves. At the top is a wave train in its first position. At the bottom, the train and its energy have moved only half as far as wave 2. Meanwhile wave 1 had died, but wave 4 has formed at the rear of the train to replace it. Waves arriving at shore are thus remote descendants of waves originally generated. [After W. Bascom, "Ocean Waves." Copyright © 1959 by Scientific American, Inc. All rights reserved.]

it travels. However, the energy contained in a group of waves is transmitted at half the velocity of the individual wave. The reason is that waves at the front of a wave group decay and lose energy as they raise the water surface, and they are replaced by waves from behind (see Figure 13-2). It is somewhat analogous to a marching band in which the individual members walk at 5 miles per hour, but because the players in front leave the row to execute some other maneuver, the front row advances at only 2.5 miles per hour.

TABLE 13-1
Minimum fetch and duration required for selected wind speeds to set up fully developed seas

Wind speed (knots)	Fetch (nautical miles)	Duration (hours)*
10	10	2
20	75	10
30	280	23
40	710	42
50	1420	69

* Duration times rounded off to the nearest hour.

FULLY DEVELOPED SEA

The development of waves in deep water is complex but may be attributed to three primary factors. These are the wind speed, the duration of wind, and the **fetch** of the wind (the distance of open surface across which the wind blows). In a discussion of wave development, the term **sea** refers to the occurrence on the sea surface within the fetch area of irregular waves of many periods coming from many directions. A **fully developed sea** is formed when the speed of a given wind lasts long enough, and the wind has enough open water to work upon, to produce the maximum wave height that can be maintained by the wind. The necessary combination of sufficient duration and sufficient fetch rarely occurs for winds of high speed, but it is possible for most light winds. Table 13-1 shows the minimum fetch and duration required for various wind speeds to set up fully developed seas.

If conditions of duration and fetch permit a fully developed sea to form, we can predict the characteristics of the resulting waves as shown in Table 13-2. Since the data in this table are based on field observations, they are reasonably good estimates. The highest waves can be estimated statistically but cannot be predicted exactly. Therefore, for a 30-knot wind developing a full sea, we can expect that the highest wave in ten will be about 28 feet (8 meters) high, but we cannot predict when it will occur.

The diagram in Figure 13-3 shows the characteristics of waves from different source areas that strike the California coast. Such diagrams are useful for planning purposes because they aid in predicting the kind of waves that would strike a given part of the shoreline with a specific orientation, such as a south- or west-facing coast. Note that the waves with the longest period, and thus the ones containing the most energy, come from distant storms with large unobstructed fetch areas. The waves that have traveled out of their source area are called **swell.** They are more regular, and have greater wavelengths and flatter crests, than the seas that generated them. Swell is usually denoted by source direction, such as *south* or *north swell.*

Table 13-3 shows the criteria for fully developed seas at differing wind speeds and the parameters of the waves that are produced. For example, for a wind speed of 30 knots, a fetch of 280 nautical miles and a storm duration of 23 hours are required for a fully developed sea. The resulting waves would have an average height of about 14 feet (4.1 meters) and a period of 8.6 seconds. The table shows that for storms that cover a large area, with only moderate subhurricane wind velocities, waves more than 30 feet high can develop. Thus fetch and storm duration are almost as important as wind velocity in producing large waves.

TABLE 13-2
Characterteristics of waves resulting from selected wind speeds in a fully developed sea

Wind speed	Average height		Average length		Average period	Highest 10 percent of waves	
(knots)	(feet)	(meters)	(feet)	(meters)	(seconds)	(feet)	(meters)
10	.9	.27	28.0	8.5	2.9	1.8	.55
20	5.0	1.5	111.0	33.8	5.7	10.2	3.1
30	13.6	4.1	251.0	76.5	8.6	27.6	8.4
40	27.9	8.5	446.0	135.9	11.4	56.6	17.2
50	48.7	14.8	696.0	212.2	14.3	98.9	30.2

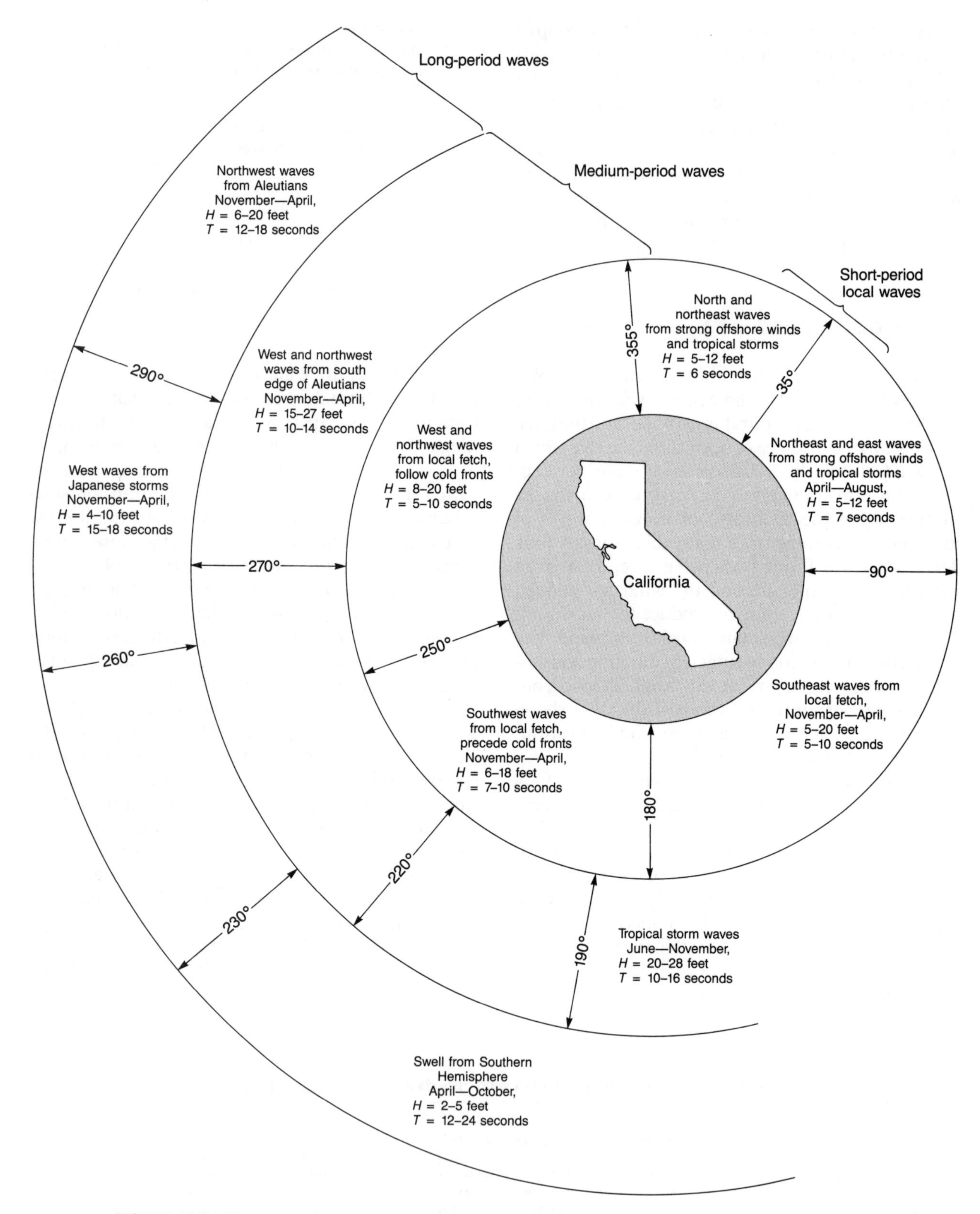

FIGURE 13-3 Diagram for the California area showing the characteristics of waves from different directions: H refers to the wave height, T to the wave period. [After D. L. Inman, U.S. Army Corp of Engineers Beach Erosion Board Technical Memorandum.]

TABLE 13-3
Conditions necessary for a fully developed sea at given wind speeds and the parameters of the resulting waves

Wind speed (knots)	Fetch (nautical miles)	Duration (hours)	Average height (feet)	Average height (meters)	Average length (feet)	Average length (meters)	Average period (seconds)
10	10	2	.9	.27	28	8.5	3.0
12	18	4	1.4	.43	40	12.2	3.4
14	28	5	2.0	.61	55	16.8	4.0
16	40	7	2.8	.85	71	21.6	4.6
18	55	8	3.8	1.2	90	27.4	5.0
20	75	10	4.9	1.5	111	33.8	5.7
22	100	12	6.3	1.9	135	41.2	6.3
24	130	14	7.8	2.4	160	48.8	7.0
26	180	17	9.5	2.9	188	57.3	7.4
28	230	20	11.4	3.5	218	66.4	8.0
30	280	23	13.6	4.1	251	76.5	8.6
32	340	27	16.0	4.9	285	86.9	9.0
34	420	30	18.6	5.7	322	98.2	9.7
36	500	34	21.4	6.5	361	110.1	10.3
38	600	38	24.5	7.5	402	122.6	10.9
40	710	42	27.9	8.5	446	136.0	11.4
42	830	47	31.5	9.6	491	149.7	12.0
44	960	52	35.4	10.8	540	164.6	12.6

DEFINITIONS

Fetch. The length of unobstructed open sea surface across which the wind can generate waves.

Fully developed sea. The waves that form when wind blows for a sufficient period of time across the open ocean. The waves of a fully developed sea have the maximum height possible for a given wind speed, fetch, and duration of wind.

Sea. Local irregular waves of many periods and from many directions. A sea forms within storm areas or when local winds are blowing over the sea surface.

Swell. Waves that have traveled a long distance from the generating area and have been sorted out by travel into long waves of the same approximate period.

Wave base. The plane or depth to which waves may erode the bottom in shallow water.

Wave height. The difference in elevation between the crest and trough of a wave.

Wave period. The length of time, in seconds, required for a wave to pass a fixed point.

Wave velocity. The wavelength divided by the wave period (in feet per second or meters per second).

Wavelength. The distance, in feet or meters, between equivalent points (crests or troughs) on waves.

EXERCISE 13

REPORT

WAVES AT SEA

NAME ______________________

DATE ______________________

INSTRUCTOR ______________________

1. What is the maximum wind that can set up a fully developed sea in a channel or strait 20 miles wide? (Refer to Table 13-1 for your answer.)

2. If the maximum fetch off Florida is from the southeast, and if the usual storm duration is 48 hours, what is the maximum wind speed that can form a fully developed sea?

   What will be the average dimensions of the waves generated by this storm? (Refer to Table 13-3.)

   Wave height ______________. Wavelength ______________. Wave period______________.

3. Typical afternoon breezes off southern California are 10–20 knots from the northwest. How long must they blow and over how much sea to produce a fully developed sea? (Refer to Table 13-3.)

   Duration ______________. Fetch ______________.

4. Figure 13-3 in the discussion shows the characteristics of California waves coming from various directions in the course of a year.

   (a) From what sector do the largest waves strike California during the winter?

   (b) From what sector does the largest swell come in the summertime?

   What is the range of potential periods and heights?

   (c) Waves of what period (long, medium, short) produce the highest waves?

5. (a) What is the group velocity of storm waves with an individual wave velocity of 100 kilometers per hour?

   (b) How long would it take such waves to reach Hawaii from a storm center in the northeastern Pacific Ocean 1500 miles away?______________ hours.

   (c) Why are long waves the first to arrive at the coast from an open-ocean distant storm?

EXERCISE 14

SHALLOW-WATER WAVES AND COASTAL PROCESSES

In Exercise 13 we discussed the origin and nature of deep-water waves in the open ocean. When these waves move into shallow water and strike the shoreline a transformation takes place, producing breakers, or surf, which can create or destroy beaches. Whereas most of us regard a beach as a place of recreation where we can relax, watch the surf, and get a tan, geologists and engineers recognize beaches as the last rampart of protection of the land against the sea. If we do not manage this resource wisely, the contest between human ingenuity and the ocean's relentless pounding will be won by the sea. Beaches along both coasts of the United States have been deteriorating badly in the course of the past few decades, and in some places they have completely disappeared. Where this occurs valuable recreational land is lost and coastal-zone property endangered.

SHALLOW-WATER WAVES

Shallow-water waves are defined as those that are traveling in water whose depth is half the **wavelength** or less. Recall that the water motion in a deep-water wave is circular and that the diameter of the circles decreases downward until, at a depth equal to about half the wavelength, water motion ceases. However, when a wave moves into shallow water a drastic change takes place. The orbits of water particles at depth become flattened and those in contact with the sea floor simply move back and forth (Figure 14-1). The wave is said to "feel the bottom" and, as a result, the wave velocity and length decrease, and the wave

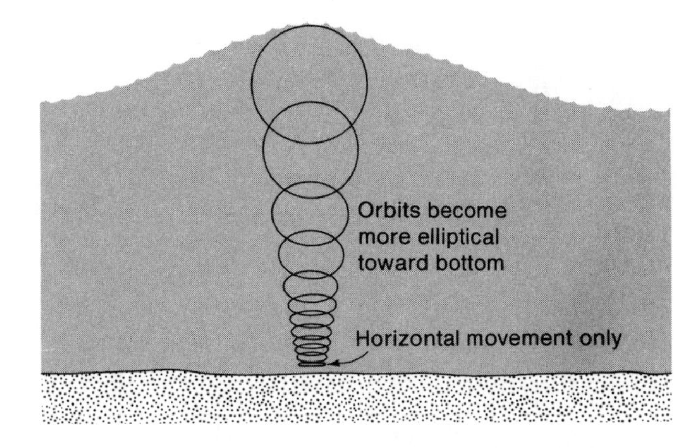

FIGURE 14-1 Orbits of water particles become elliptical as they approach a shallow bottom. At the bottom, particles move back and forth only. [After F. Press and R. Siever, *Earth,* 4th ed. W. H. Freeman and Company. Copyright © 1986.]

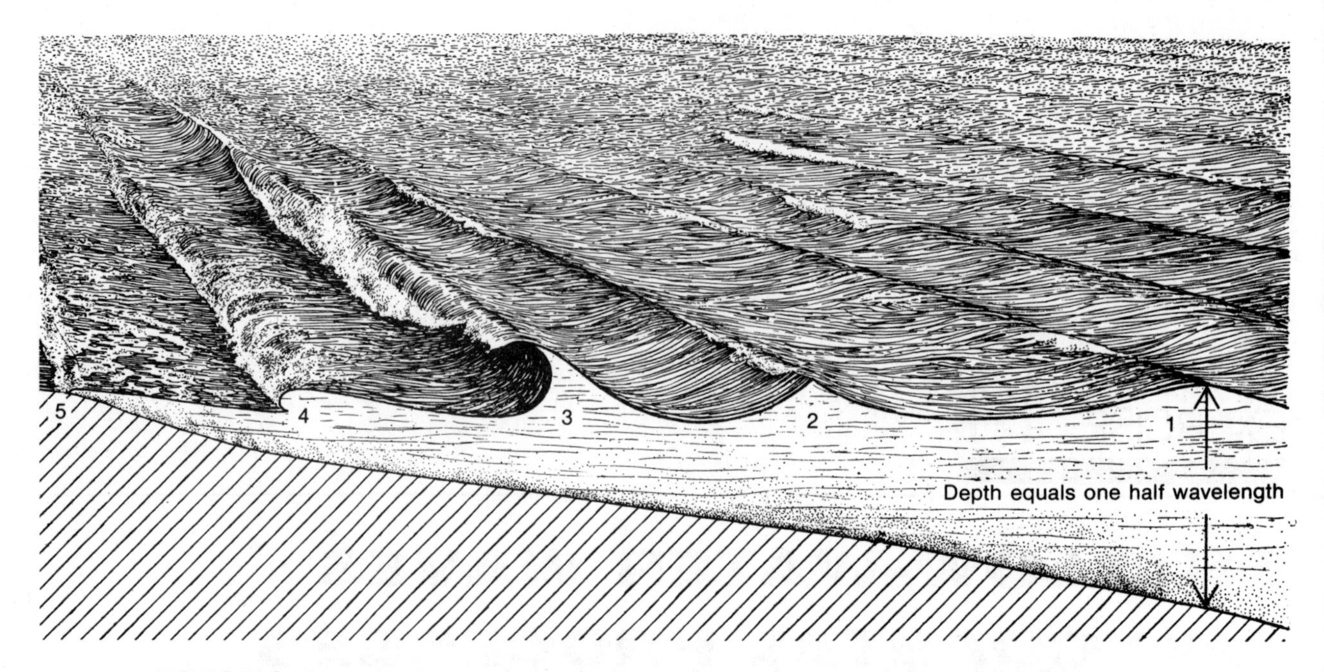

FIGURE 14-2 A wave breaks up at the beach when swell moves into water shallower than half the wavelength. (2) The shallow bottom raises wave height and decreases length. (3) At a water depth of 1.3 times the wave height, water supply is reduced and the particles of water in the crest have no room to complete their cycles; the wave forms and breaks. (4) A foam line forms and water particles, instead of just the wave form, move forward. (5) The low remaining wave runs up on the beach face as "swash," or uprush. [From W. Bascom, "Ocean Waves." Copyright © 1959 by Scientific American, Inc. All rights reserved.]

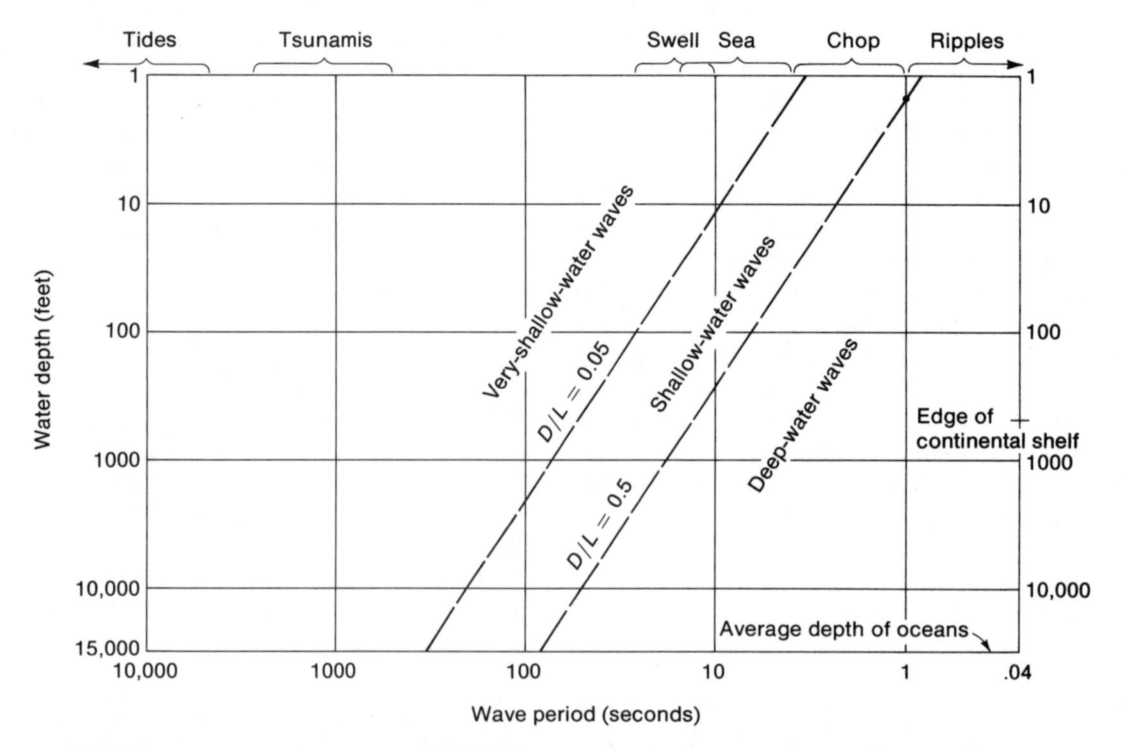

FIGURE 14-3 Wave characteristics. [After W. Bascom, *Waves and Beaches.* Doubleday and Company, New York, 1964.]

form steepens until it becomes unstable and spills against the shoreline as breakers, or surf. At this point the oscillations of the water particles cease and the water motion is all in one direction toward the beach. Figure 14-2 shows this transformation from **swell** to shallow-water wave to surf. Whether a wave is classified as a shallow-water one or not depends on both the wave itself and the basin within which it is traveling. Figure 14-3 shows the relationship between **wave period** and water depth. Because the length of a wave is proportional to the square of its period, the diagram also shows the relationship between wavelength and depth of water. We see that very short-period waves, such as ripples, are deep-water waves even in relatively shallow water. **Sea** and swell are also deep-water waves in water from about 100 to 1000 feet deep. However, in water shallower than 100 feet, sea and swell become shallow-water waves. In the deep ocean, with an average depth of about 15,000 feet, all waves with periods greater than 80 seconds are shallow-water waves. In this category we find the tides and **tsunamis,** or so-called tidal waves. Tsunamis are long-period sea waves produced by submarine earthquakes, volcanic eruptions, or landslides. They may travel unnoticed across the ocean for thousands of miles from their point of origin and build up to great heights over shallow water. We will treat the nature and travel times of these waves in more detail at the end of this discussion and in the report section.

Table 14-1 shows the relationship between selected wave periods, the calculated wavelength and wave velocity, and the water depth at which the wave feels the bottom or becomes a shallow-water wave. Note that wind-generated waves with periods greater than 14 seconds are capable of moving sediment at depths as great as the edge of the continental shelf. Most wind waves have periods between 5 and 25 seconds.

WAVE REFRACTION

Shallow-water waves are subject to **refraction** over humps or depressions of the sea floor, and to **reflection** from seawalls or breakwaters. Refraction occurs when a wave moves into shallow water at some angle other than parallel to the shoreline. The part of the wave crest in the shallowest water is slowed the most, whereas the part of the wave in deeper water moves forward at a higher velocity. The result is a bending of the wave crest, known as refraction, and a concentration or dissipation of energy at the shoreline. An example of a refraction pattern at sea is shown in Figure 14-4. We can determine the relative amount of concentration or dissipation of wave energy by drawing lines perpendicular to the wave crests, known as **orthogonals,** on a diffraction diagram. The wave's energy, or its ability to do work on a shoreline, is the same between orthogonals drawn at equally spaced intervals along the wave crest. By tracing the orthogonals shoreward on the crests of successive waves of selected periods or lengths, we can determine how wave energy is concentrated or dissipated at the **surf zone** (Figure 14-5). The **wave energy coefficient,** e, which is the relative amount of concentration or dissipation of energy, may be calculated by simply dividing the distance between two orthogonals at the shoreline by the distance between the same two orthogonals in deep water. Where e is greater than 1, wave energy is dissipated and we would expect little erosion; where it is less than 1, energy is concentrated on a shorter stretch of shoreline and accelerated erosion would be predicted (Figure 14-5). The height of a breaker is approximately equivalent to 0.78 of its depth. Thus we would predict that a breaker 7.8 feet high would occur in water 10 feet deep, provided that swell of sufficiently long period were approaching the

TABLE 14-1
The relationships between selected wave periods, calculated wavelengths and velocities, and water depths at which wave feels bottom

Wave period (seconds)	Wavelength* (feet)	Approximate velocity (miles per hour)	Water depth† (feet)
6	184	21	92
8	326	28	163
10	512	35	256
12	738	42	369
14	1000	49	500
16	1310	56	655

* Length is equivalent to 5.12 times the wave period squared.
† Depth is equivalent to half the wavelength.

FIGURE 14-4 Wave refraction associated with a small rocky island 1 mile north of Prouts Neck, Maine. [From John Shelton, *Geology Illustrated.* W. H. Freeman and Company. Copyright © 1966.]

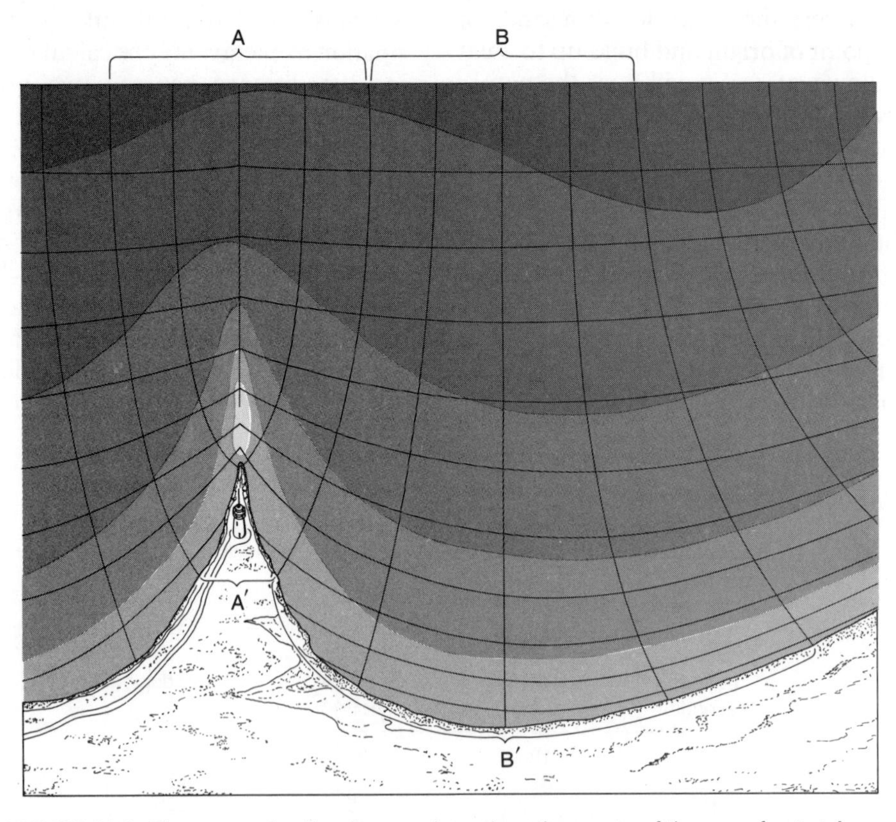

FIGURE 14-5 The wave refraction diagram shows how the energy of the wave front at A is concentrated by refraction at A′ around the small headland area. The same energy at B enters a bay but is spread at the beach over wide area B′. The horizontal lines are wave fronts; the vertical lines (orthogonals) divide the energy into equal units for purposes of investigation. Such studies are vital preliminaries to design of shoreline structures. [After W. Bascom, "Ocean Waves." Copyright © 1959 by Scientific American, Inc. All rights reserved.]

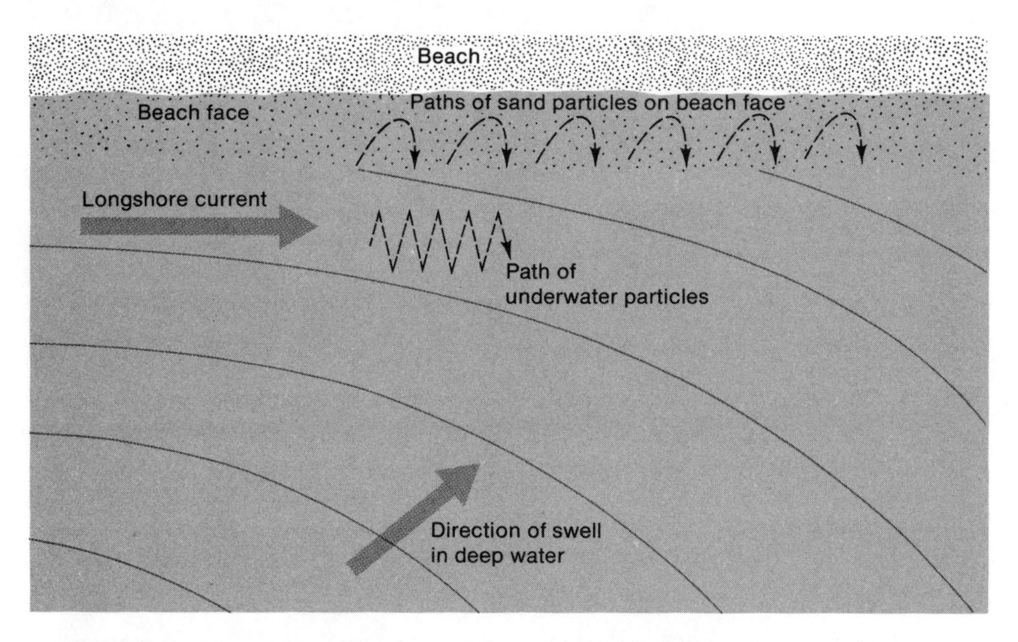

FIGURE 14-6 A longshore, or littoral, current is set up parallel to the beach when waves move toward the beach at an oblique angle. Sand is also transported parallel to the beach by the current. [After W. Bascom, "Beaches." Copyright © 1960 by Scientific American, Inc. All rights reserved.]

shoreline. This type of approximation is important in determining the height that piers should be built above mean high water or some other datum. Obviously fishing or commercial activity from a pier 15 feet high would be impeded if the pier were exposed to 18-foot waves (water depth about 23 feet).

FIGURE 14-7 Groins—which are dams constructed of wood, concrete, or stone—are built perpendicular to the beaches, as shown in this photograph. They are widely used to trap sand.

LONGSHORE CURRENTS AND LITTORAL DRIFT

As a rule waves approach the shoreline at an angle and are refracted; however, because refraction is usually incomplete, the waves strike the shore at a slight angle. Consequently some of the water is transported parallel to the beach and a weak **longshore current** (which flows parallel to the shore) is created (Figure 14-6). The current is like a river on land, and is capable of moving sand along the beach, a process known as *beach drifting.* The earth material moved along the beach by the longshore current is known as **littoral drift.** When an obstruction, such as a groin or jetty, is placed in the path of the current, a buildup, or accretion, of littoral drift results on the upstream side and erosion occurs on the downstream side (Figure 14-7). The extent of this accretion and erosion depends on the velocity and persistence of the current and the supply of sand. Also, like a river, the longshore stream may overflow. If we view the banks of the river as the beach and the surf zone, we realize that as water builds up in the longshore current, it must eventually "overflow" and return seaward. It does so in the form of **rip currents,** improperly called rip tides, which can pose a formidable hazard to swimmers when the waves are high. They can be identified by the presence of a gap in the wave forms of the breakers, white foamy water extending seaward beyond the turbulence of the surf zone, and streaks of sediment or floating objects moving seaward in the rip (Figure 14-8). A diagram of the nearshore system of circulation and currents is shown in Figure 14-9.

The longshore current, unlike a river, may reverse its direction as a consequence of differing wave approaches, but it never stops flowing. Thus, where the

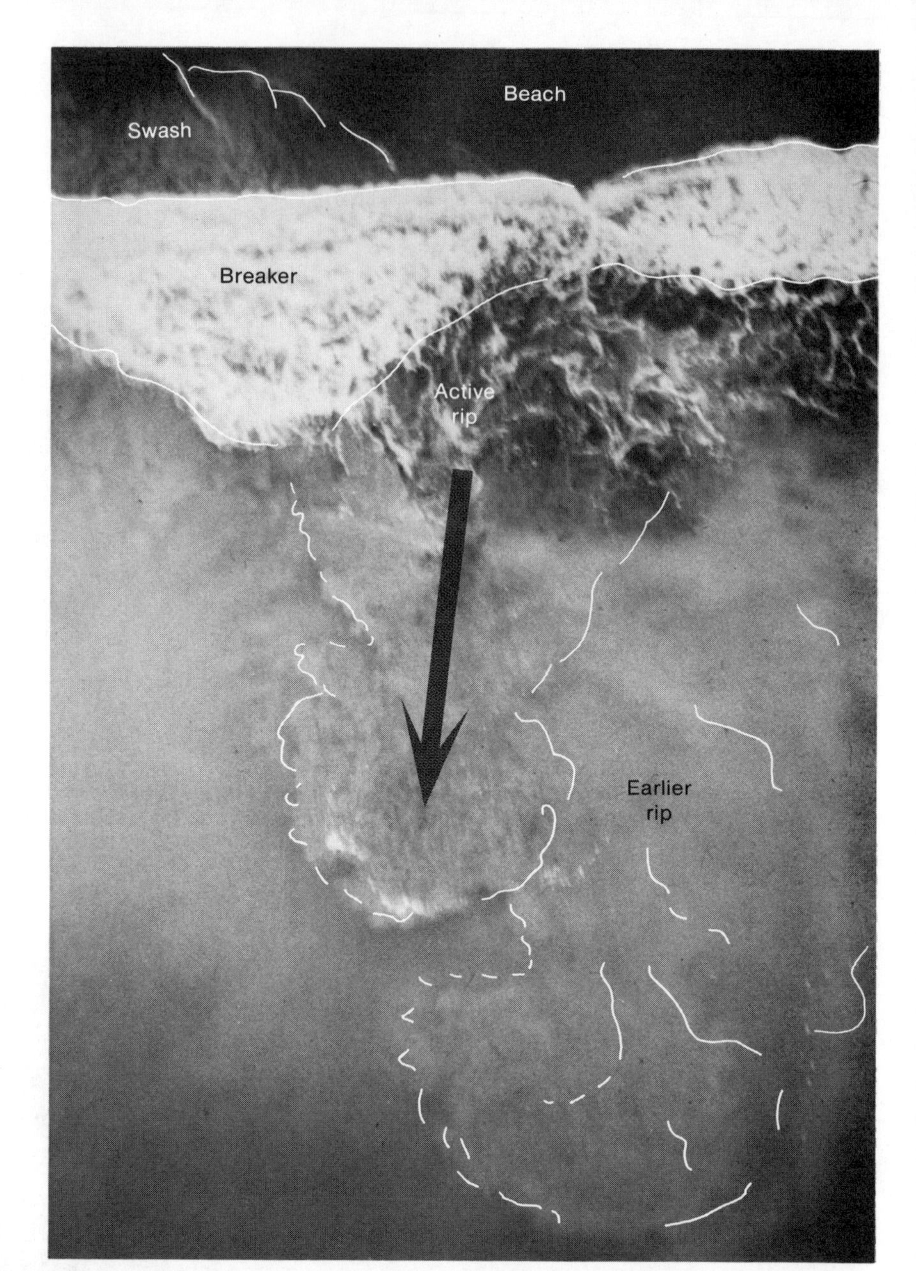

FIGURE 14-8 Rip currents in the breaker zone near Carpinteria, California, in February 1969. Two pulses of flow can be seen: An earlier jet is seaward and deeper than the active jet. Each pulse is probably the product of a series of breakers. The older pulse is the result of rip generated a few minutes earlier.

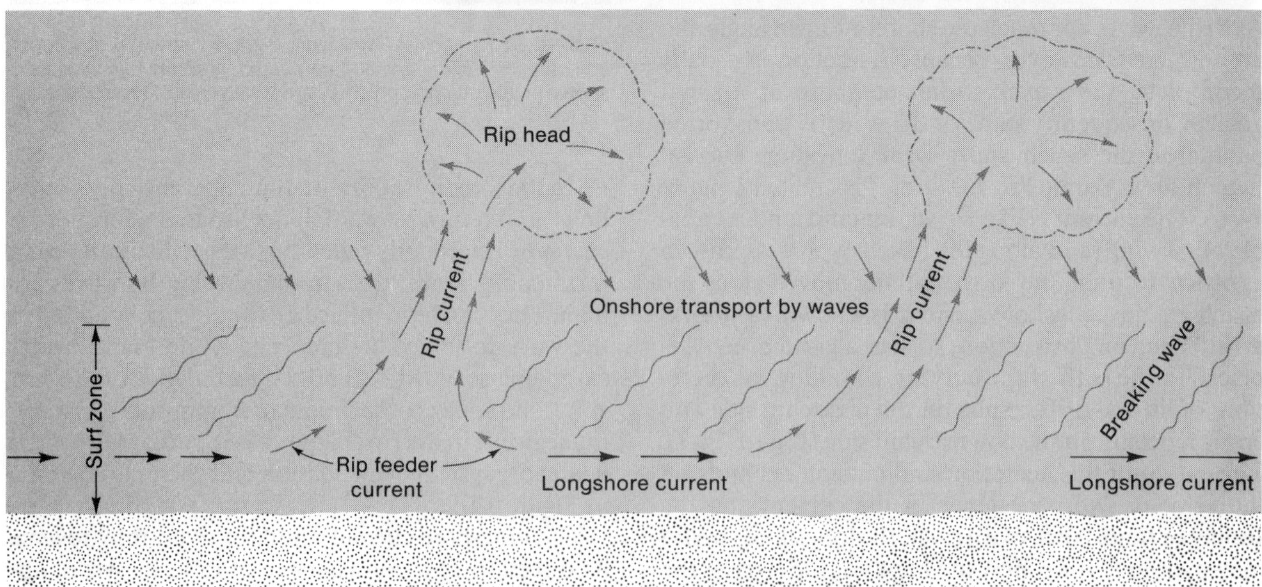

FIGURE 14-9 Diagram of the nearshore current system.

FIGURE 14-10 Breakwater at Santa Barbara, California, causes sand transported by longshore currents to be deposited as a sandbar in the lee of the breakwater. Sand must be periodically dredged from the harbor to beaches farther east (top of the photograph) to prevent closure of the harbor and beach erosion downcurrent. Dark line along the beach in the harbor and the streaky appearance of the water due to oil that was accidently spilled from a drilling operation in the Santa Barbara Channel.

FIGURE 14-11 The breakwater at Santa Monica, California, runs parallel to the shoreline. The designers thought the longshore current flowing inside the breakwater would carry sand through, but instead sand was deposited in the "shadow" created by the structure.

supply of sand to a given point along the beach is less than the amount removed by longshore currents, erosion and loss of beach ensue. Although beach erosion may result from natural causes—such as drought, which decreases sand supply from rivers—coastal engineering works are more often directly responsible.

Structures of considerable height that extend seaward for some distance are effective barriers to littoral drift. But short groins are usually constructed for shoreline stabilization and only temporarily disrupt the longshore transport of sand (see Figures 14-10 and 14-11). Table 14-2 indicates the magnitude of the problem in southern California. Keep in mind that for every cubic yard of sand trapped or stored by a groin or breakwater, the beaches that are downcurrent, are deprived of an equivalent amount, and the loss of beach width by erosion inevitably occurs. The same problem is also prevalent in the Great Lakes and along the East and Gulf Coasts of the United States. The figures shown for accretion are even more striking when we consider that a cubic yard is roughly equivalent to 1 square foot of beach. That is, if

TABLE 14-2
Rates of littoral drift and accretion at shoreline structures for selected southern California localities

Location	Accretion rate (10^3 cubic yards per year)
Santa Barbara breakwater (Figure 14-10)	280
Santa Monica breakwater (Figure 14-11)	259
Redondo Beach	30
Anaheim Bay jetties	175
Balboa Bay jetty	72

100,000 cubic yards of sand are lost, this volume is equivalent to beach retreat of about 1 foot along a length of 100 feet. It may be seen that if loss of sand continues for many years it can seriously diminish a recreational area. Finally, reparation of the damage is costly: To replace sand on the beach costs roughly between $2.00 and $4.00 per cubic yard, and more than 80 million cubic yards have been artificially placed on California beaches in the course of several decades.

The natural transport of longshore drift is shown dramatically in three photographs of Sand Beach, New Jersey, taken over a period of 23 years (Figure 14-12a, b, and c). Note the growth of the spit within that period of time and the direction of the longshore current.

TSUNAMIS

Tsunamis are impulsively generated waves that are very destructive along certain shorelines. They have wavelengths of about 80–100 miles and travel with velocities in excess of 400 miles per hour. They are refracted and exhibit the same shoaling effects that wind waves do. As they enter shallow water their velocity and wavelength decrease, whereas their heights increase. In deep water their heights are less than 10 feet and they are unnoticeable there. Figure 14-13 shows tide-gauge records at various ports for the destructive tsunami of April 1, 1946. It was generated by an earth movement on the sea floor off the Aleutian Islands and sped from there across the Pacific. We can see why tsunamis generated in that location are so damaging along the exposed Hawaiian coast, whereas they have much less impact on the West Coast of the United States.

DEFINITIONS

Littoral drift. The transport of sand, gravel, and other materials along the beach face by longshore (or littoral) currents.

Longshore (littoral) current. A current running parallel to the beach and generated by waves striking the shoreline at an angle.

Orthogonal. A line drawn perpendicular to wave crests so that refraction or bending can be visualized more clearly.

Reflection. The process by which the energy of the wave is returned seaward.

Refraction. The process by which the direction of a wave moving in shallow water at an angle to the bottom contours is changed. The part of the wave moving shoreward in shallower water travels more slowly than that portion in deeper water, causing the wave to turn or bend to become parallel to the contours.

Rip current. A current flowing seaward from the shore through gaps in the surf zone. The strength of the current is proportional to the height of the breakers striking the shoreline.

Sea. Local irregular waves of many periods and from many directions. A sea forms in storm areas or when local winds are blowing over the sea surface.

Surf zone. The nearshore zone along which the waves become breakers as they approach the shore.

Swell. Waves that have traveled a long distance from the generating area and have been sorted out by travel into long waves of the same approximate period.

Tsunami. A large, high-velocity wave generated by an earthquake, volcanic explosion, or landslide on the sea floor.

Wave energy coefficient *(e)*. Calculated as follows:

$$e = \frac{\text{width of orthogonals at shore}}{\text{width of orthogonals in deep water}}$$

Wave period. The length of time required for a wavelength to pass a fixed point.

Wavelength. The distance between equivalent points on waves.

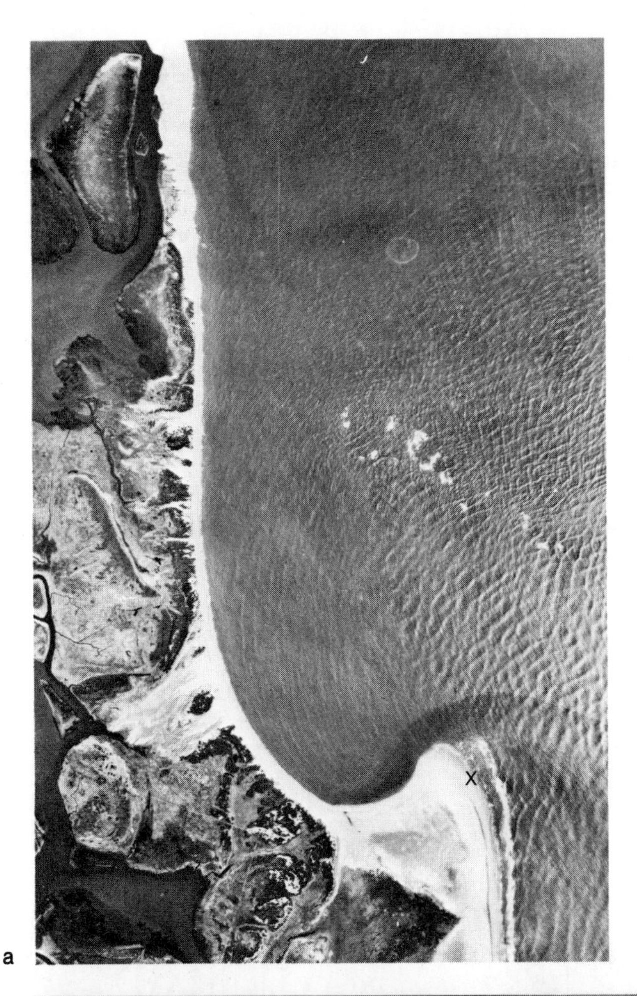

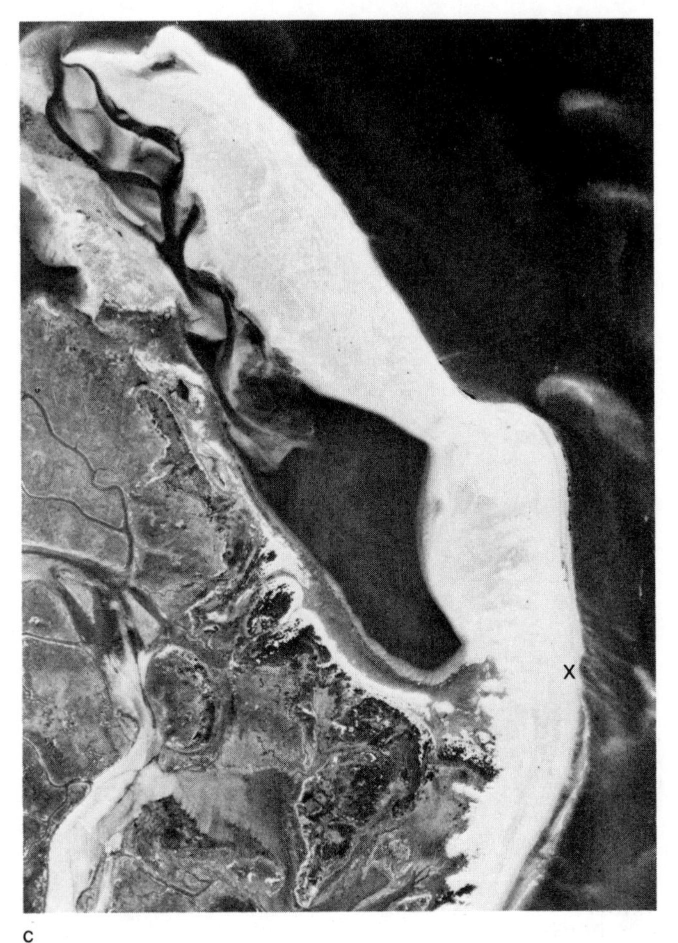

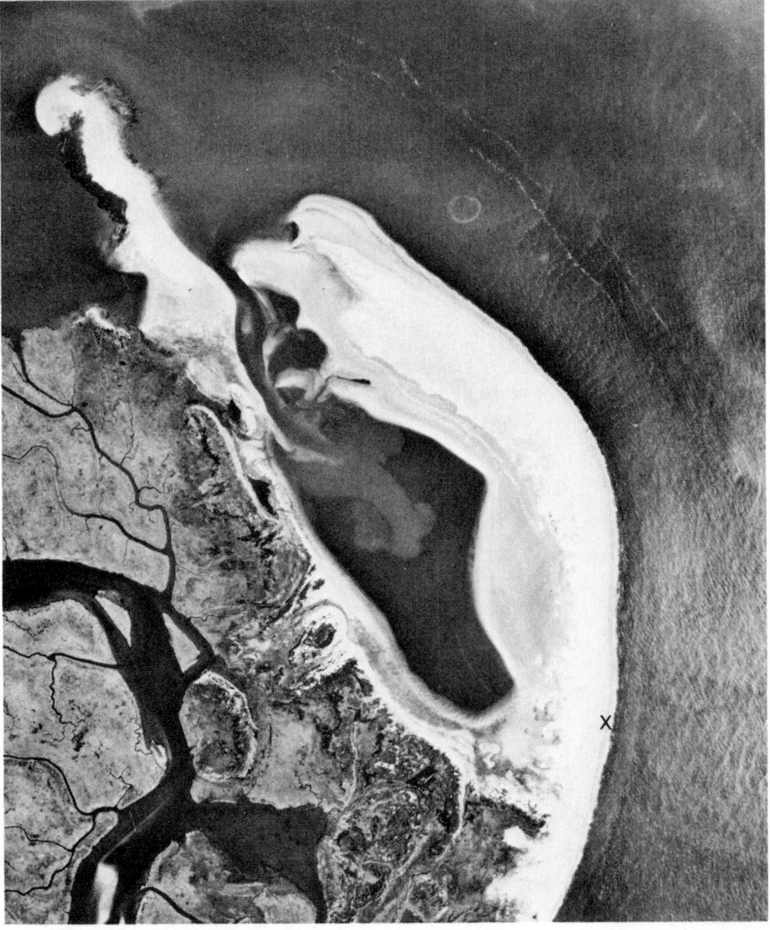

FIGURE 14-12 The natural transport of longshore drift over a 23-year period at Sandy Beach, New Jersey (designated by the cross in all photos). (a) 1940: The first stage in the development of a spit; the longshore current runs toward the north. (b) 1957: The spit extends with time in the direction of longshore transport of sediment. (c) 1963: The spit joins the land and the recurved spit grows around the tip. Also note that the earlier shoreline downdrift (east) of the main spit has extended and erosion has cut back the older islands until the main spit has screened the shore from wave action. [From C. S. Denny et al., *A Descriptive Catalog of Selected Aerial Photographs of Geologic Features in the United States.* U.S. Geological Survey Professional Paper 590, 1968.]

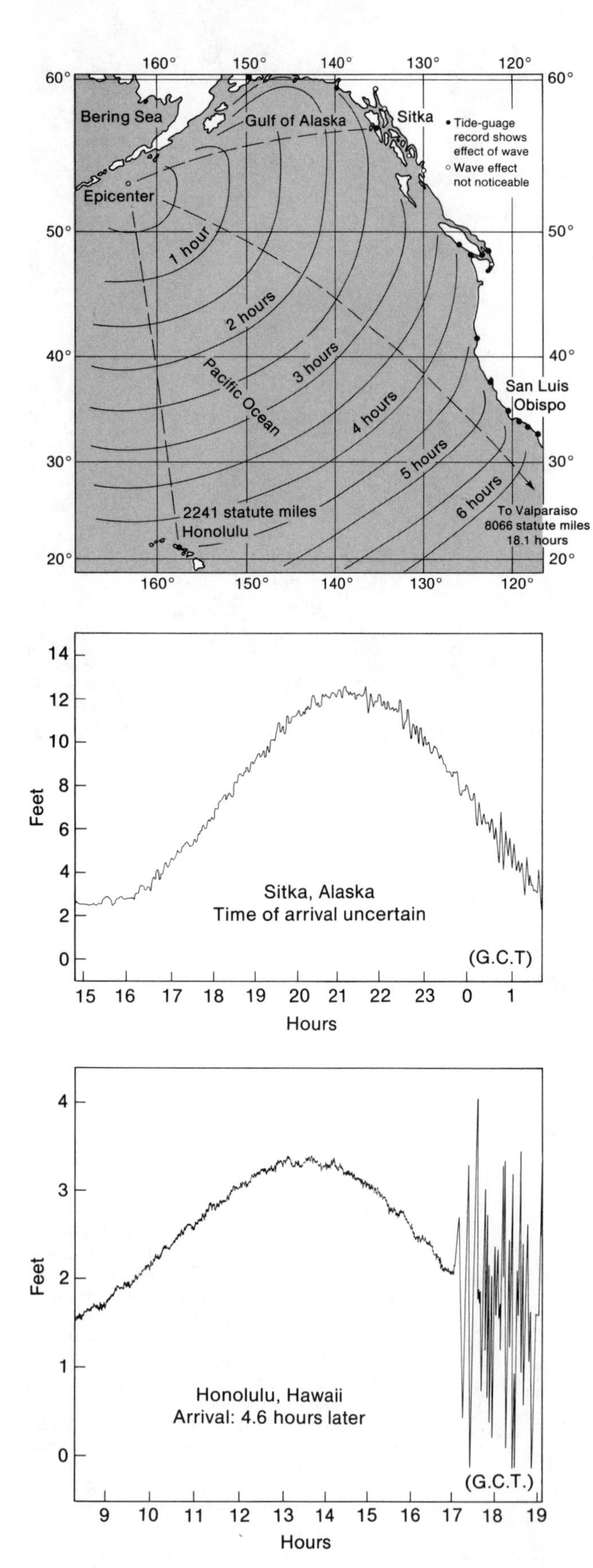

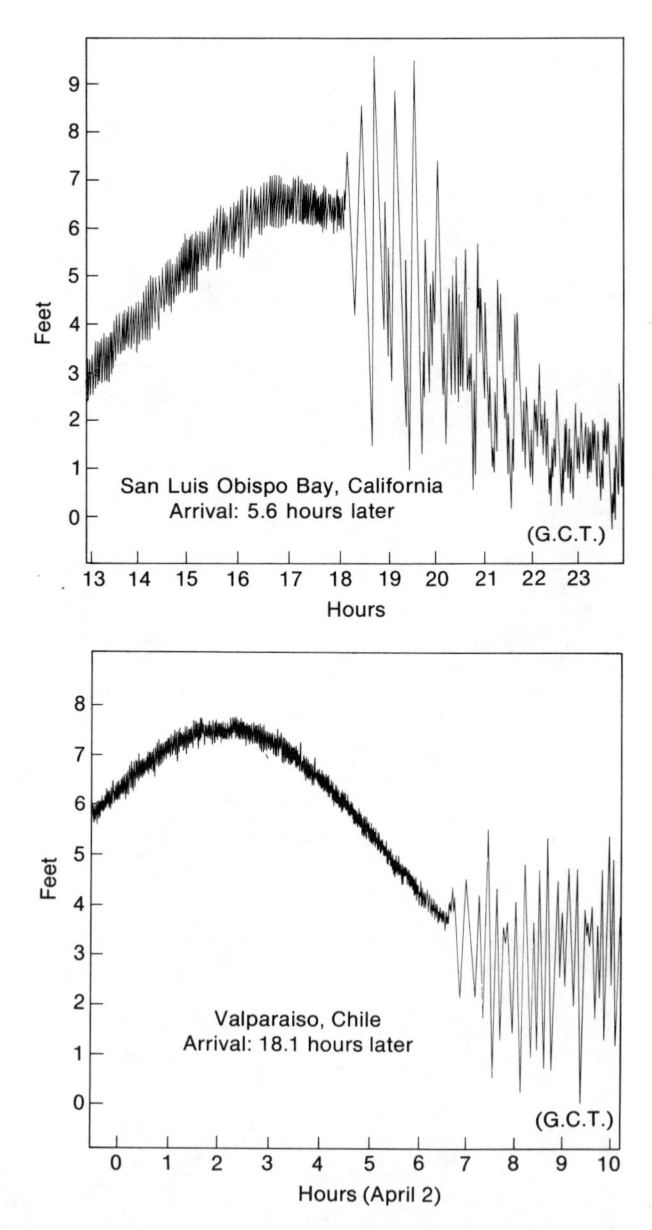

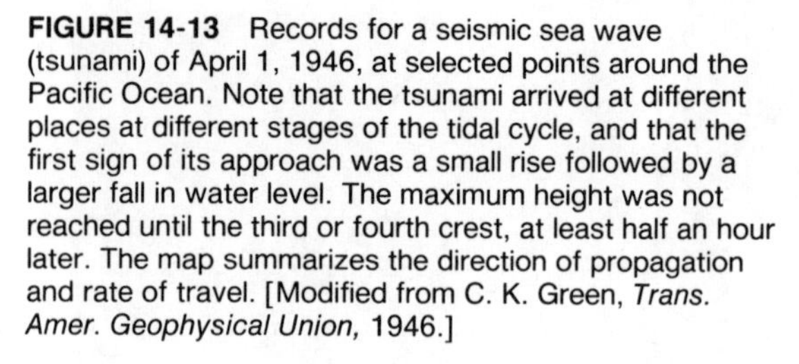
FIGURE 14-13 Records for a seismic sea wave (tsunami) of April 1, 1946, at selected points around the Pacific Ocean. Note that the tsunami arrived at different places at different stages of the tidal cycle, and that the first sign of its approach was a small rise followed by a larger fall in water level. The maximum height was not reached until the third or fourth crest, at least half an hour later. The map summarizes the direction of propagation and rate of travel. [Modified from C. K. Green, *Trans. Amer. Geophysical Union,* 1946.]

EXERCISE 14 REPORT

SHALLOW-WATER WAVES AND COASTAL PROCESSES

NAME

DATE

INSTRUCTOR

Refer to the aerial photographs in the discussion for your answers to questions 1–7.

1. What is the direction of longshore transport in Figure 14-7? State your answer in terms of right- and left-hand sides of the photo.

   How do you know?

2. What is the ultimate fate of Santa Barbara Harbor (Figure 14-10) if it is not dredged?

3. In Figure 14-11, why has sand accreted behind the parallel breakwater at Santa Monica, California?

4. Figure 14-14 is a map view of the Santa Monica beach and breakwater shown in Figure 14-11. Indicate on it the distribution of wave energy at points A and B, using terms such as *high, low,* or *nil.*

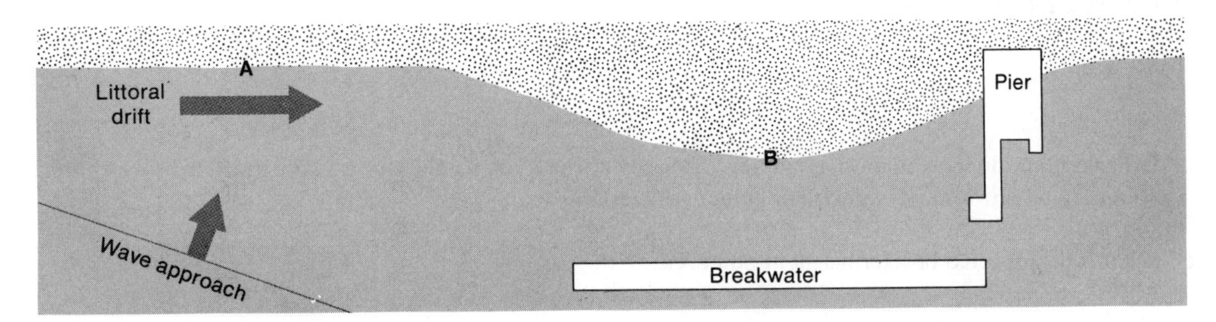

FIGURE 14-14 Diagram of a breakwater and beach at Santa Monica, California.

5. Make a rough sketch of the refraction pattern around the small rocky island at Prouts Neck, Maine (Figure 14-4). Why does the beach extend farther seaward off the island than adjacent areas?

6. In Figure 14-12, what is the direction of wave approach and longshore drift at Sandy Beach, New Jersey?

7. How much has Sandy Beach in Figure 14-12 grown between 1940 and 1963? (Scale of the photos is 1 inch = 2250 feet.)

What is the average annual rate of growth?

8. (a) Complete the wave-crest diagram in Figure 14-15 showing refraction shoreward.

(b) Sketch in the orthogonals with a deep-water spacing as indicated.

(c) Calculate the wave energy coefficients for segments 1–4 in the figure.

(d) Indicate along the shore in the diagram several places at which you would expect a sand beach to be wider because of accretion, or narrower or nonexistent because of erosion.

(e) What is the expected breaker height at point X?

________________ meters.

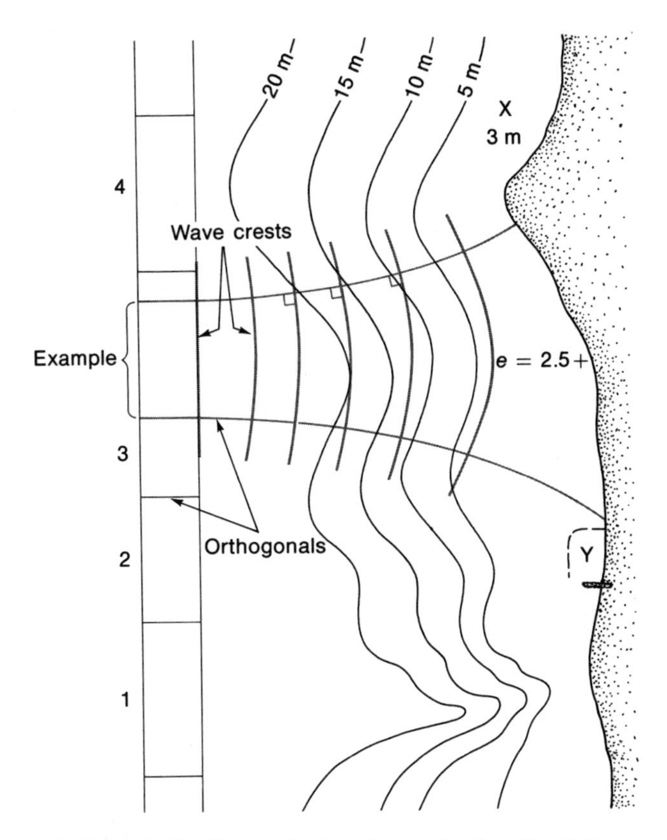

FIGURE 14-15 Wave refraction diagram for Question 2. The numberals along the broken contours indicate the depths in meters. The wave energy coefficient, *e*, is 2.5 or more.

(f) Where would the best board surfing be located and why?

(g) Where would the safest bathing beach be located and why?

(h) Given the predicted wave refraction pattern over the submarine canyon, what would be the expected direction of longshore currents north of the canyon?

How would this affect the proposed breakwater at point Y?

9. (a) Using the travel-time diagram in Figure 14-13, suggest the time required for a tsunami generated in the Aleutian Trench to arrive at the following locations:

Hawaii

San Luis Obispo, California

Valparaiso, Chile

What is the average velocity of the tsunami in its travel to Hawaii?

To Valparaiso, Chile?

Give a rational explanation for this difference in velocity between the two stations.

(b) Estimate the greatest height of the tsunami above still-water level at the four stations shown in Figure 14-13.

Sitka, Alaska

Honolulu, Hawaii

San Luis Obispo, California

Valparaiso, Chile

Why would the wave be almost unnoticeable in Sitka even though it is much closer to the source area of the quake?

(c) What was the first evidence of the arrival of the tsunami on the shoreline at Honolulu—low or high water?

What crest in the series did the greatest damage?

(d) Why do tsunamis occur so much less frequently in the Atlantic Ocean than in the Pacific?

EXERCISE 15

DISTRIBUTION OF LIFE IN THE SEA

In this exercise we will investigate the distribution and abundance of marine organisms. The oceans may be divided into large **biomes,** or living regions. In fact every part of the marine environment is occupied by some form of marine life, even the deepest trenches and most anoxic (low-oxygen) basins.

THE MARINE LIFE ZONES

The following classification of the marine life zones was set up by Joel Hedgpeth* and other biologists and is modified slightly here. It is based on the natural distributions, and breaks in distributions, of marine organisms, as shown in Figure 15-1. The classification is first divided into two major environments—the **pelagic** and **benthic environments.** The pelagic environment is the water column, and it in turn is divided into a *neritic* environment (the waters over the continental shelf) and an *oceanic* environment (the waters beyond the shelf). The oceanic environment is further divided into *epipelagic, mesopelagic, bathypelagic,* and *abyssopelagic* zones. The benthic environment is quite finely divided from the shore seaward into (1) the *supralittoral,* or spray, zone; (2) the *littoral* zone, that between the highest and lowest tides; (3) The *inner sublittoral* zone, from lowest low tides to a depth that can no longer support fixed plants; (4) the *outer sublittoral* zone, from the base of the inner sublittoral to about the shelf edge, or break in slope; (5) the *bathyal* zone, from the shelf break to about 4000 meters in depth; (6) the *abyssal* zone, from about 4000 meters to the deepest sea except for the trenches; and (7) the *hadal* zone, the trenches.

These divisions were established mainly in accord with the natural breaks in the distributions of marine organisms. In other words, if a number of marine biologists have noted changes in the nature of marine life at a certain depth in the oceanic environment—be this a change in species, dominance, or other characteristic—that depth would be a logical place to divide the oceanic environment into biomes or subbiomes. For example, it has been observed that a significant change occurs at depths ranging from about 100 to 200 meters in the open ocean, and therefore the

* After Joel Hedgpeth, "Treatise on Ecology and Paleoecology," *Ecology,* Vol. 1 (Geological Society of America, Memoir 67, 1957; printed by Waverly Press, Baltimore).

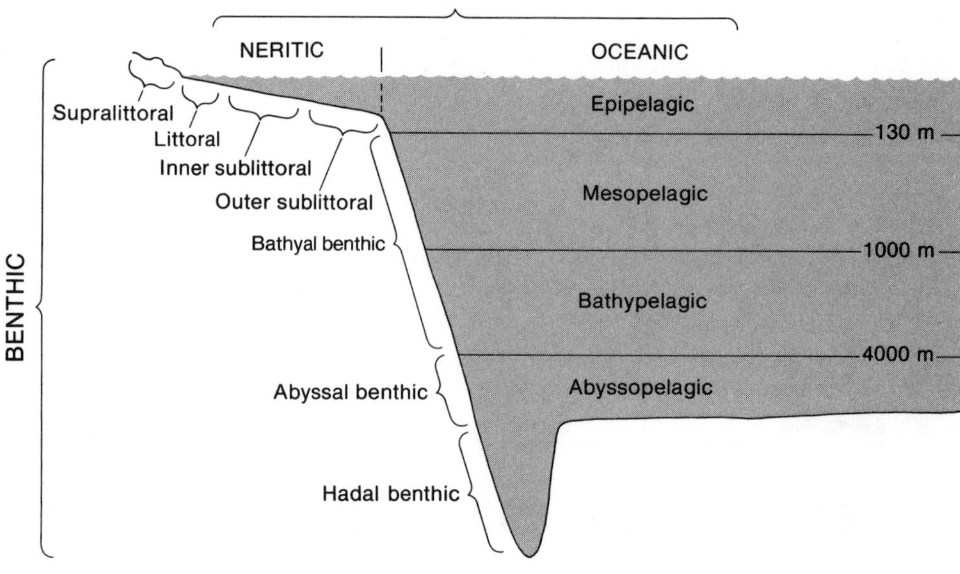

FIGURE 15-1 Modified Hedgpeth classification of life zones in the oceans. The pelagic environment (the neritic and oceanic environments) and the zones of the oceanic environment (approximate depths of these zones are in meters). The benthic, or bottom, is also divided into zones.

zonation into the epipelagic and mesopelagic environments has been established. Although this zonation works reasonably well in the temperate and tropical seas, it is less applicable in the polar regions, where chemical and physical conditions are more uniform.

The organisms that live within these biomes can be divided roughly in accord with the habitat or type of environment they occupy. The two major habitats, and the organisms that live in them, are listed as follows:

Pelagic environment.

- Nektonic. Swimming organisms (for example, fish, seals, whales)
- Planktonic. Floating organisms (for example, diatoms, jellyfish)
  - Holoplanktonic. Organisms that float throughout the life cycle
  - Meroplanktonic. Organisms that float during part of the life cycle and are benthic or nektonic during the rest of it

Benthic environment

- Sessile benthic. Organisms that live in one place (for example, clams, mussels)
  - Infauna or inflora. Animals or plants that grow in the sediments or rocks
  - Epifauna or epiflora. Animals or plants that grow on the sediments or that are attached to rocks, other organisms, and foreign objects (such as bottles)
- Vagrant, or mobile, benthic. Organisms that rove on the ocean bottom (for example, crabs, snails)

FACTORS THAT LIMIT DISTRIBUTION: A HYPOTHETICAL EXAMPLE

Although the Hedgpeth classification is based mainly on the distribution of marine organisms, many of the reasons for the boundaries between its regions are at least in part physical or chemical. To illustrate this point, a hypothetical pattern of distribution for a planktonic organism will be described in terms of hypothetical *limiting* or controlling biological, physical, and chemical factors. The three-dimensional distribution pattern of this hypothetical holozooplankter is illustrated for a specific parcel of ocean water in Figure 15-2. The hypothetical factors that control or limit its distribution are designated by specific temperature, salinity, and biological boundaries, numbered 1–8 in the figure and in the following paragraphs.

Boundaries 1 and 2 are temperature boundaries. The organisms living within these boundaries are *stenothermal,* that is, they can live only within narrow

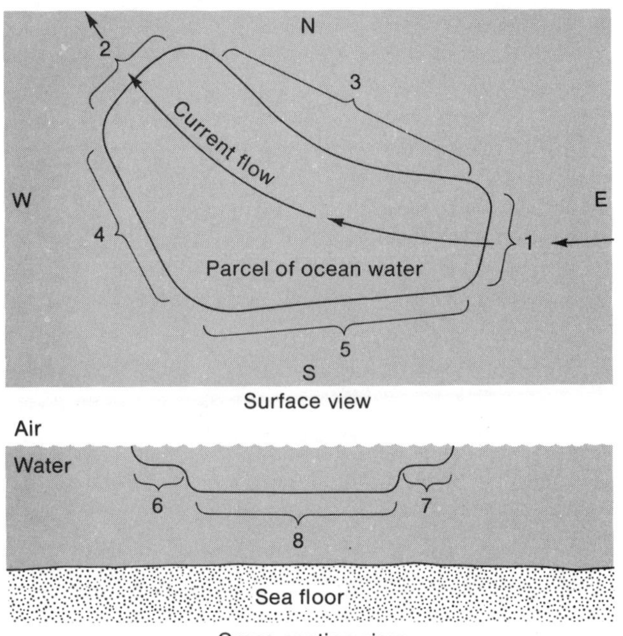

FIGURE 15-2 Diagram of the habitat of a hypothetical holozooplankter, and the boundaries controlling its distribution.

temperature ranges (as opposed to *eurythermal* organisms, which can exist under a wide range of temperatures). A current is flowing through the area, as indicated by the arrow. Within boundary 1 the water temperature is mild enough to support the organism (outside this boundary the water is too warm), but at boundary 2 the water has cooled to a point at which the organism cannot exist.

Boundary 3 is biological. This hypothetical holozooplankter is a copepod (a small crustacean) that feeds exclusively on a certain phytoplankton (a single-celled floating plant), specifically a diatom. Because of the low population of diatoms to the north, the copepod can't get enough to survive and is therefore excluded from that area. Thus, boundary 3 is determined by the sparse supply of the necessary phytoplankton species.

Boundary 4, also biological, is established by the easternmost boundary of the habitat of a predator that feeds on the copepod so efficiently that is depletes the population to essentially zero.

Boundary 5 is determined by an abrupt increase in salinity that is too great for the copepod. The copepod is *stenohaline,* which means that it can tolerate only a narrow range in salinity. (*Euryhaline* organisms are able to tolerate a wide salinity range).

Boundaries 6 and 7, like 1 and 2, are determined by temperature. Below them, the waters are too cold for the copepods to exist even though their food, the diatoms, can.

Boundary 8 is a biological boundary. It is the deepest level at which the diatom can exist, so the copepods do not range below this depth.

These hypothetical controls determine the range of the hypothetical copepod. Sometimes the ranges at which an organism can exist may be split into two or more distinct regions. You will be considering the reasons for such **disjunct distributions** in the report form.

DEFINITIONS

Benthic environment. The bottom of the ocean. An organism that lives in the benthic environment would live on or in the bottom.

Biomes. Large natural assemblages of organisms. On land such assemblages might be a grasslands biome or a tropical rain forest biome.

Disjunct distributions. Occurrences that are separated geographically. An organism that exhibits a disjunct distribution may live at both polar regions but nowhere else.

Pelagic environment. The region of the water column. An organism that lives in the pelagic environment would live in the water column.

EXERCISE 15

REPORT

DISTRIBUTION OF LIFE IN THE SEA

NAME

DATE

INSTRUCTOR

1. Answer the following questions by writing the physical or chemical factor that might explain, or partially explain, various divisions in the Hedgpeth classification. Be specific and scientific in your answer; for example, "the bottom of the seasonal thermocline" or "the bottom of the mixed layer" (Exercise 8), and "the bottom of the photic zone" (Exercise 16) would be appropriate answers. If you don't remember some of the physical or chemical oceanographic divisions or barriers, review previous exercises (for example, Exercises 8–11), or consult your lecture notes or textbook.

   (a) The division between the epipelagic and mesopelagic zones. You should be able to give five or six good answers here.

   (b) The division between the mesopelagic and bathypelagic zones. This boundary is about 1000 meters deep, and it is a level at which many interesting physical and chemical changes occur. Try to suggest four good reasons here.

   (c) The division between the bathypelagic and abyssopelagic zones. Because the differences are subtle, this is a difficult question, but try to give two reasons.

   (d) The division between the abyssal benthic and hadal benthic zones. It will be hard to think of any change in water physics or chemistry at this boundary because the same water mass exists above and below the boundary; therefore, think of a biological answer here. (Hint: Look at the distribution of trenches and think in terms of evolution in the course of geologic time.)

(e) The division between the supralittoral and littoral zones.

(f) The divisions within the littoral environment, the so-called *intertidal zonation.* Many rocky littoral or intertidal environments exhibit an upper zone of snails (periwinkles) and associated organisms followed by a zone of barnacles and associated organisms followed by a zone of algal mat and associated organisms. What might cause these fine breaks, some so discrete you can almost draw a line along them?

(g) The division between the inner and outer sublittoral zones.

(h) The division between the outer sublittoral and bathyal zones. Cite factors different from those you have already given for the division between the epipelagic and mesopelagic zones in part (a) of this question.

(i) Why is there no boundary at about 1000 meters in the benthic environment but a significant one at about the same depth in the pelagic environment? Think of what the main control on the benthic organisms might be that the pelagic organisms would not have to contend with, and vice versa.

(j) The division between the bathyal benthic and abyssal benthic zones. Refer to Exercise 6 if you don't remember what change occurs in sediments at about this depth.

OPTIONAL QUESTIONS

2. Now consider individual species distributions. One kind of disjunct distribution in the marine environment is called *bipolar distribution;* in it the same species (or form) lives in shallow waters around the North and South poles, but not in the shallow or deeper water in between. State a few biological or physical reasons to explain this type of distribution.

3. Another kind of distribution is tropical submergence (or tropical avoidance), where a species lives in shallow polar waters but at depth in temperate and tropical regions. State a few biological or physical reasons to explain this type of distribution.

EXERCISE 16

PRIMARY AND SECONDARY PRODUCTIVITY

Primary productivity (autotrophs) is the conversion of inorganic compounds into organic ones. The most important kind of primary productivity is the process of **photosynthesis,** in which light energy is used to convert carbon dioxide and water into carbohydrates. The basic overall chemical equation for this process is

$$6CO_2 + 6H_2O \xrightarrow{\text{sunlight}} C_6H_{12}O_6 + 6O_2$$
carbon dioxide + water = carbohydrate + oxygen

In reality this process is very complicated, although it can be divided into two sets of reactions, light reactions and dark reactions. In the light reactions, light excites photosynthetic pigments (mainly chlorophylls), which pass that energy on in a way that makes it available to the dark reactions. In this process, water molecules are split and oxygen evolved as a by-product. The dark reactions use the chemical energy converted from light to fix the carbon of CO_2 into carbohydrate. (The dark reactions are so called because, unlike the light reactions, they *may* occur in the dark.)

What takes place is the formation of an organic compound (a carbohydrate, such as a sugar or a starch) that can be used as a building block for other organic compounds or as an energy source. Also, oxygen is given off; essentially all the oxygen in the atmosphere is derived from photosynthesis, and at least half of that photosynthesis takes place in the oceans. When the process of photosynthesis is reversed, the high-energy bonds that were formed during construction of the carbohydrate are broken and the energy of that bond is used as energy to run the organism: this reverse process is oxidative **respiration,** and all organisms respire.

Besides photosynthesis, other types of primary productivity take place: one is chemosynthesis, in which iron or sulfur bacteria break down iron and sulfur compounds and use the energy derived from the breaking of the bonds involved to build carbohydrate from carbon dioxide and water. Although the other processes are not as important as photosynthesis, some biologists believe that chemosynthesis may be important in the deep sea, where no photosynthesis can occur. This appears to be the case around hydrothermal vents on the spreading ridge in the eastern Pacific, where bacteria apparently oxidize the hydrogen sulfide that comes out of these vents to sulfur and sulfate to fix carbon. In doing so the bacteria support a rich vent fauna at 2000 meters. In fact, it appears that a large tube worm (3 meters long) in that environment uses the same chemosynthetic process to at least partially sustain itself.

Secondary productivity (heterotrophs) is the transfer of organic compounds from one **tropic level** to another. In other words, it is the rate of transfer from prey to predator.

THE MEASUREMENT OF PRODUCTIVITY

Productivity, be it primary or secondary, is usually measured as the rate of the process of fixing of carbon, and it is expressed in grams of carbon (dry weight) per square meter of ocean surface per year. There are a number of procedures for measuring this rate, the most common ones being the following.

1. *The light and dark bottle technique.* Seawater is collected along with its natural component of phytoplankton, although sometimes the water sample is passed through a screen to eliminate the larger zooplankton. For example, suppose that we measure the primary productivity from one sample that has been collected from a depth of 10 meters. The water from the sample is placed in two bottles, one of transparent glass (the light bottle) and the other a bottle that has been taped or painted so that no light can penetrate (the dark bottle). The original amount of oxygen is measured and then the bottles are either placed on deck (in a water bath) or resubmerged in the ocean to their original depth of 10 meters. After a period of time (usually at least a few hours) the bottles are retrieved and the oxygen content is again measured. Since we know the amount of oxygen from each bottle, we can note the rate of photosynthesis in two ways—the *gross,* or total, photosynthesis that has occurred; and the net photosynthesis, the amount of photosynthesis (the total amount of oxygen produced) minus the respiration (the amount of oxygen used in respiration).
2. *The hot carbon or ^{14}C method.* A known amount of "hot," or radioactive, carbon in the form of bicarbonate is added to the sample and the phytoplankton are allowed to photosynthesize for a time (see Exercise 24, "The Use of Radioisotopes in Oceanography"). The amount of ^{14}C that has been fixed by photosynthesis in the phytoplankton is measured by counting the hot carbon in phytoplankton. This method can be used to measure secondary productivity also by letting herbivores eat hot phytoplankton. Even if you use a light and dark bottle in this measurement, you may still have difficulty making an accurate measurement of gross and net photosynthesis.

The preceding methods are the two most important and common ways to measure primary productivity; however, other techniques have also been tried, such as the following.

3. *Standing crop of phytoplankton.* A **standing crop** refers to either the number of phytoplankton per volume or unit area (number under 1 square meter throughout the **photic zone**), or the **biomass** (living weight) of phytoplankton. Determination of either the density (number of individuals per volume) or the biomass is an adequate means of getting an approximate measure of primary productivity, and this technique can be used to estimate secondary productivity as well. Of course, because you are measuring at an instant in time you are not measuring a rate, so that a significant problem arises. The method has other drawbacks, and you will be asked to enumerate some of them in the report form.
4. *Amount of chlorophyll method.* For this method, it is assumed that if one sample has more chlorophyll than another, the higher content signifies greater productivity. However, this technique poses a number of problems, and you should be able to list some in the report form after you have read the rest of this exercise (see also Exercise 18).

THE DISTRIBUTION OF PRIMARY PRODUCTIVITY

Dominant factors that affect primary productivity and the standing crop of phytoplankton at any particular time include (1) the availability of light, which varies with latitude, time of the year, time of day, turbidity of water, amount of organisms in the water (biological overshadowing), depth, and so on; (2) the availability of limiting, or *nonconservative,* nutrients, such as nitrates, phosphates, and silicates (see Exercise 17); (3) the rate of grazing by herbivores, such as copepods.

Of these three factors, the principal ones are the amount of light and the amount of available nutrients. For the oceans in the world as a whole, light can be considered a principal limiting factor only at extremely high latitudes. The availability of nutrients is a much more significant factor for a much larger portion of the ocean. Generalized curves for dissolved

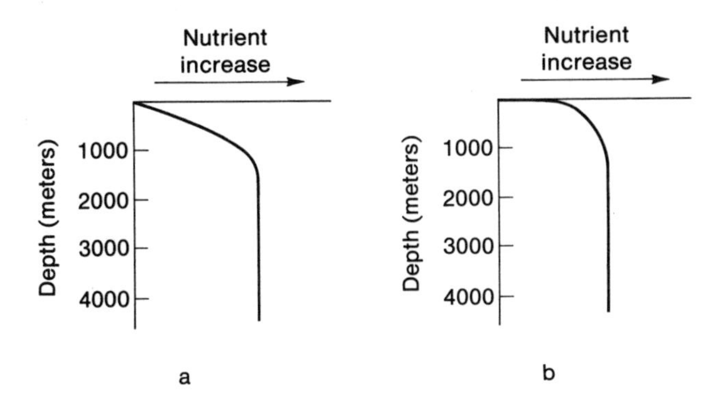

FIGURE 16-1 Dissolved nutrient curves for various conditions at a hypothetical geographic location in the world ocean: (a) curve for tropical and temperate open-ocean nonupwelling regions; (b) curve for high-latitude and coastal upwelling regions.

nutrients under various conditions at various geographic regions are presented in Figure 16-1.

Most of the world's oceans exhibit a curve similar to that shown in Figure 16-1a, in which the nutrients in shallow water are depleted. This depletion is due to the consumption of nutrients by organisms and the removal of the nutrients from shallow water by the passage of fecal pellets (excrement), molts (sheddings) from organisms, and the dead organisms themselves to greater depths. The shallow water is replenished with inorganic nutrients by (1) **upwelling** of nutrient-rich deep water, (2) **runoff** and **mixing** from land (rivers), and (3) **additions** from the atmosphere. Of these sources of replenishment, upwelling is the most important. Although the runoff from land is important in the course of geologic time, more than 90 percent of the nutrients present in shallow water have been brought into the surface layer through the process of upwelling. Upwelled waters bring with them the nutrients that were originally produced in surface waters but that have fallen through the water column into deeper water. Thus the distribution of productivity is determined mainly by the distribution of upwelling, and, in general, regions of strong upwelling are also regions of high productivity. Upwelling is discussed in Exercise 8, and the main kinds of upwelling are illustrated in Figure 8-4. Figure 16-2 (on the next two pages) shows the global distribution of regions of high, medium, and low productivity. Although photosynthesis is the most important kind of primary productivity, there are other kinds of primary productivity, such as chemosynthesis by sulfur bacteria at the newly discovered hydrothermal vents, which supports an entire ecosystem (Exercise 4).

EUTROPHISM

Regions in which productivity is high are sometimes designated as *eutrophic,* meaning well fed, and regions of low productivity as *oligotrophic.* In the report form you will see that some of the important oceanic phenomena that you have already studied—such as upwelling, nutrient concentration, and so on—vary significantly between eutrophic and oligotrophic regions. Indeed, you might think that a eutrophic condition would, by its very definition, be a desirable one. However, the term *eutrophism* is now widely used to refer to the pollution of streams, lakes, and the ocean. It can occur where raw sewage, industrial wastes, or the like are dumped into the water, which thereby gets "overnourished." A few ramifications of eutrophism are the killing off of the natural organisms in the area, the proliferation of unwanted ones, and the reduction of the amount of dissolved oxygen, which in turn can be lethal or upset the balance of the environment.

DEFINITIONS

Addition. Incorporation from another source, such as the addition of nutrients from the atmosphere.

Biomass. Total weight of organisms in a habitat (such as grams per cubic meter), or of a species or group of species (for example, the phytoplankton biomass).

Mixing. Combining of waters with different properties, such as stream runoff and ocean water.

Photic zone. The layer of a body of water that receives ample sunlight for the photosynthetic processes of plants.

Photosynthesis. The process of converting carbon dioxide and water into cabohydrates by the use of chlorophyll.

Primary productivity. The conversion by some life form of inorganic compounds into organic compounds. Photosynthesis is one type of primary productivity.

Respiration. The process by which organisms convert organic compounds into usable energy. Oxidative respiration (where carbohydrates are burned biologically in the presence of oxygen) is essentially the reverse of photosynthesis.

Runoff. Water and nutrients added to the ocean from streams.

Secondary productivity. The conversion by some life form of organic compounds into other organic compounds. Primary productivity takes place at the first tropic level, and secondary productivity takes place at all the other trophic levels.

Standing crop. The number of organisms or biomass per volume or unit area at any instant in time.

Trophic level. The general position in the ecosystem; for example, one such level would be the plant eaters.

Upwelling. The physical oceanographic process of deep water moving into shallow water.

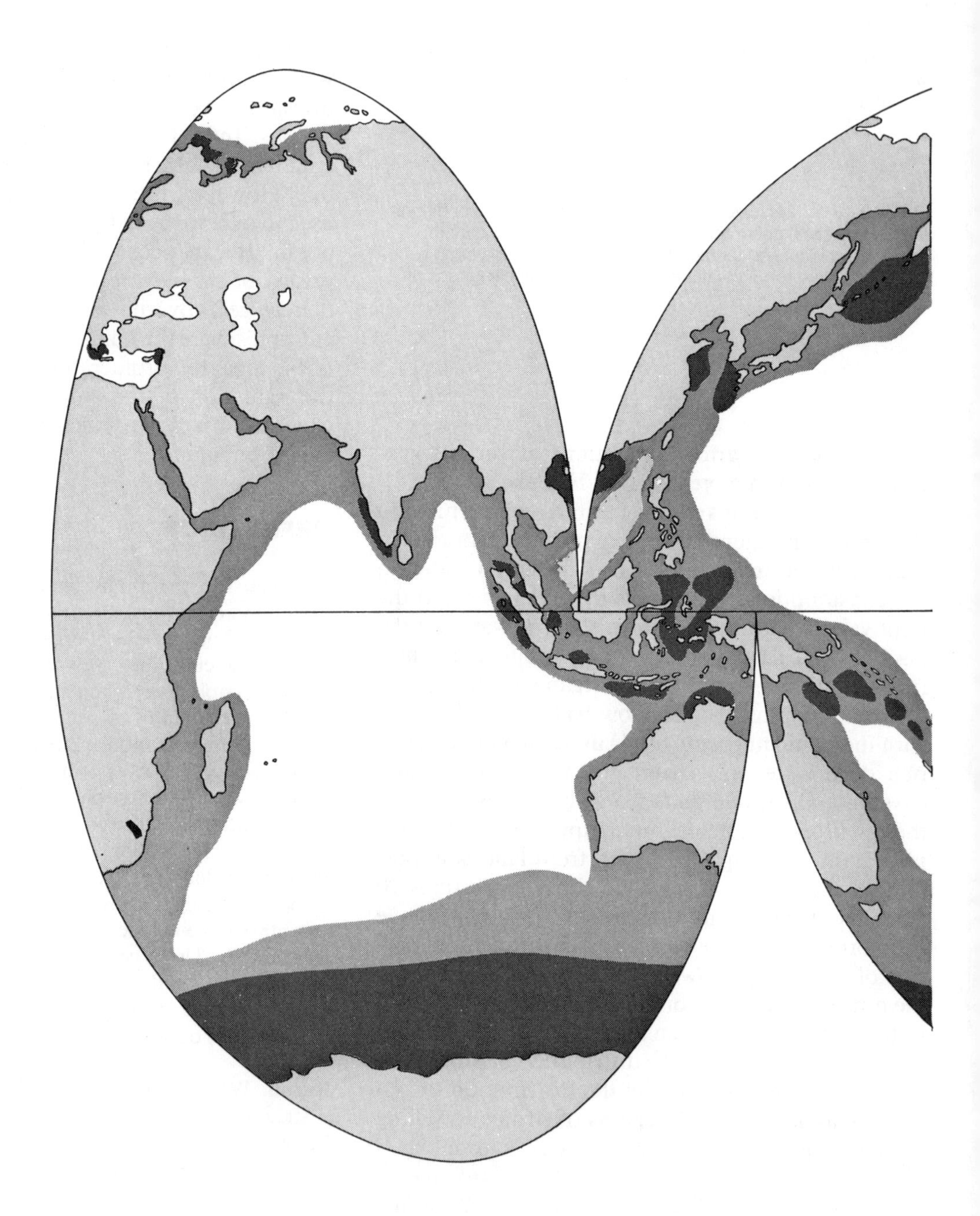

FIGURE 16-2 Map showing global distribution of waters in which upwelling and mixing occur. These waters are nutrient rich and the productivity of marine life would be expected to be very high (dark gray) or moderately high (medium gray). [After John Isaacs, "The Nature of Oceanic Life." Copyright © 1969 by Scientific American, Inc. All rights reserved.]

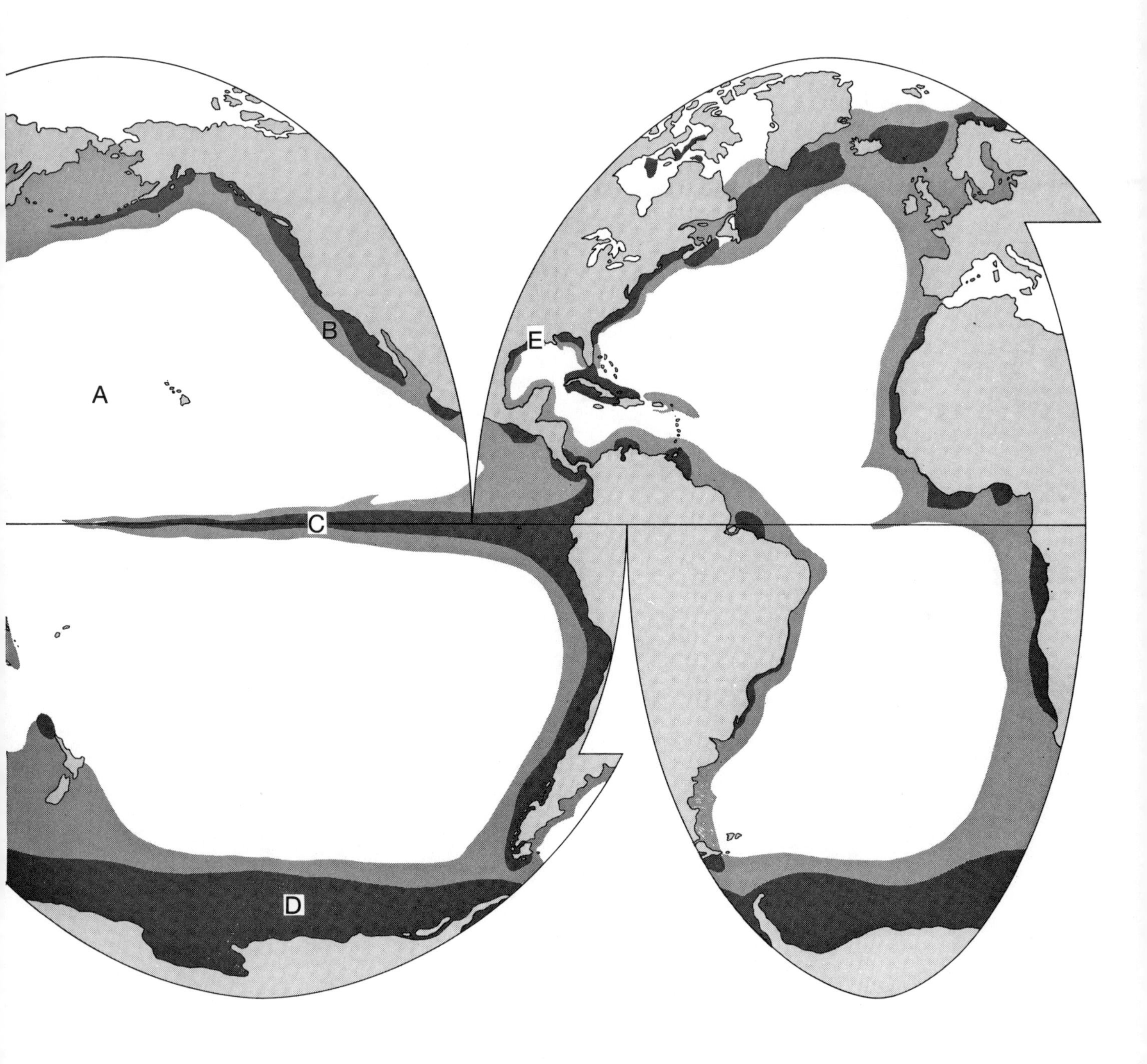
A
B
C
D
E

EXERCISE 16

REPORT

PRIMARY AND SECONDARY PRODUCTIVITY

NAME ____________________

DATE ____________________

INSTRUCTOR ____________________

1. The following questions refer to the light and dark bottle technique for measuring productivity.

(a) After the second oxygen measurement, which bottle do you think would generally have more or about the same amount of oxygen in it, and which bottle would generally have less oxygen in it than the original amount? Why?

(b) How would you calculate the gross photosynthesis and net photosynthesis from data generated in this manner? Describe the mathematical technique, not the actual numbers.

(c) Suggest some of the problems you think would arise with this procedure. One clue is the removal of larger zooplankton by some investigators before the experiment begins. You should be able to think of three serious problems.

2. Refer to the photosynthetic equation and the discussion at the beginning of this exercise for assistance in answering the following questions.

(a) How do you explain the fact that gross productivity can take place at greater depths than positive net productivity?

(b) The depth at which gross productivity and respiration are equal (where net productivity is zero) is usually termed the photosynthetic compensation depth. A photosynthetic plant (phytoplankter) must be above its photosynthetic compensation depth at least a significant part of the time. Why?

(c) As suggested in part (b) of this question, different species may have different photosynthetic compensation depths. Why might this be so? (Hint: Think of the morphologic and color differences between house plants and garden plants.)

(d) Different members of the same species, or the same individual during its lifetime, may also have different photosynthetic compensation depths. Why? (Think of differences in food intake between a growing child and an elderly person.)

(e) In fact, different geographic regions may have different photosynthetic compensation depths. For example, consider the following regions:

(i) A tropical open-ocean region far from land
(ii) A polar open-ocean region far from land
(iii) A nearshore region off the mouth of a river coming from a large continent
(iv) A nearshore region just off a small volcanic island

Determine the relative photosynthetic compensation depths of each, and list each type of region in order from that with the shallowest depth to that with the deepest. State why each region is shallower than that below it and/or deeper than that above it. If necessary, refer to the discussion on the distribution of productivity.

(f) The photosynthetic compensation depth will vary temporally (with time); state why this will happen seasonally and daily. (Hint: Remember that plants require sunlight for photosynthesis.)

3. Sample A has a higher primary productivity (measured by both the light and dark bottle method and the ^{14}C method) than sample B from the same area. However, sample B has a higher standing crop of photoplankton than sample A. List two reasons why this situation (higher standing crop but lower primary productivity) might occur.

4. Give at least three good reasons why the chlorophyll method may not give good results. The reasons should be different from those given for the preceding question on the standing crop of phytoplankton method.

5. Refer to Figure 16-2 and answer the following questions on the distribution of primary productivity. In your answers do not forget to list such phenomena as the Coriolis effect or Ekman spiral (Exercise 11), upwelling (Excercise 8), and others. Also remember that density-driven upwelling occurs all around the Antarctic, just as it does in the South Atlantic.

(a) Why does region A (the North Pacific) have low productivity?

(b) Why does region B (the California Current) have high productivity?

(c) Why does region C (at the equator) have high productivity?

(d) Why does region D (at the Antarctic) have high productivity?

(e) Why does region E (which designates the estuary systems along the Gulf Coast of the United States) have high productivity? (Consult a more detailed map of the area, such as one for the Galveston, Texas, region.)

6. (a) In the Antarctic the nutrient supply is ample, and the limiting factor is light. Think about how seasonal primary productivity in the Antartic might operate. Within the coordinates below, draw a curve showing the seasonal primary productivity peak or peaks, and in the space at the side state why your curve looks as it does.

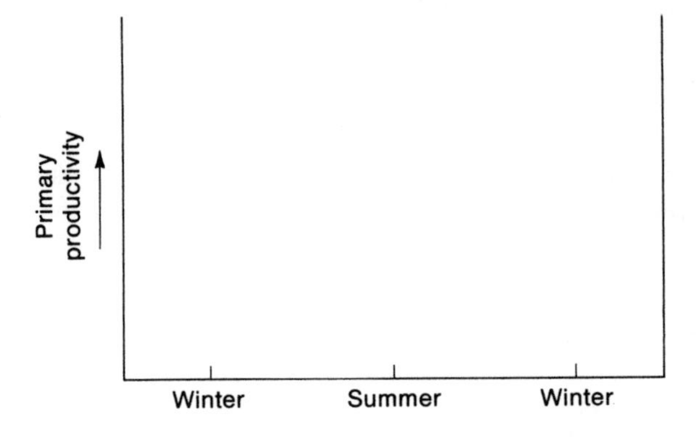

(b) Keeping in mind the seasonality in productivity, draw in below the peak or peaks of primary productivity for regions B and C shown on the map in Figure 16-2, and explain the occurrence of these peaks.

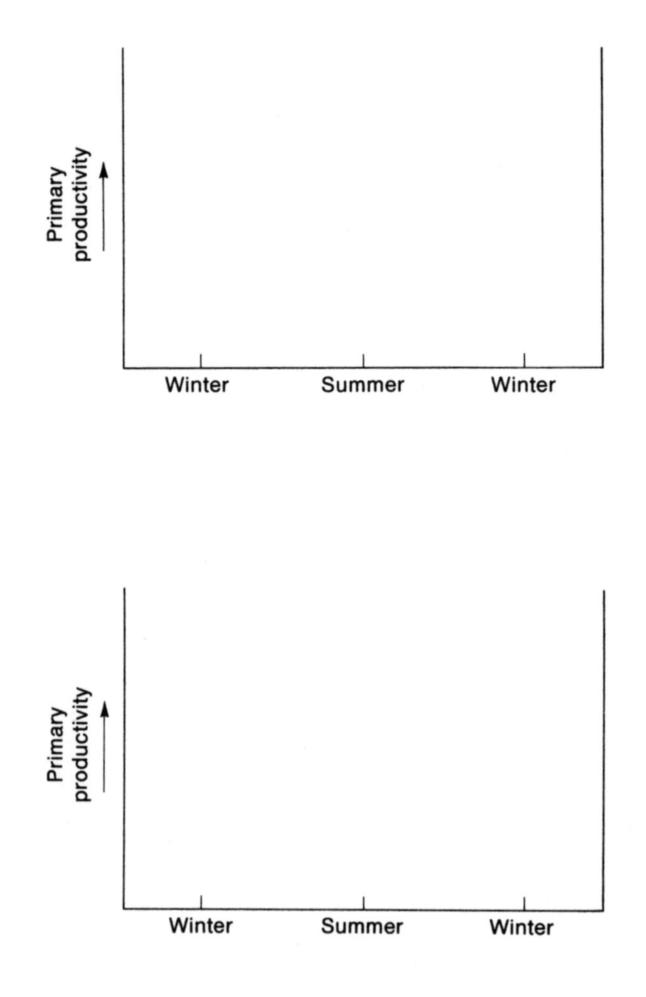

OPTIONAL QUESTIONS

7. You probably had trouble drawing the curve for region B in part (b) of the preceding question. The curve, with other related factors, is shown on the chart in Figure 16-3. Answer the following questions about the relationships shown in this figure. Since there is more than one answer for each question, list as many as you can.

   (a) Why do the dissolved nutrients drop in the spring?

   (b) Why does the spring phytoplankton bloom start in the spring and die out in the early summer?

   (c) Why is there a difference in the steepness of the zooplankton and phytoplankton biomass curves during the spring bloom?

   (d) What are some possible reasons for a fall phytoplankton bloom?

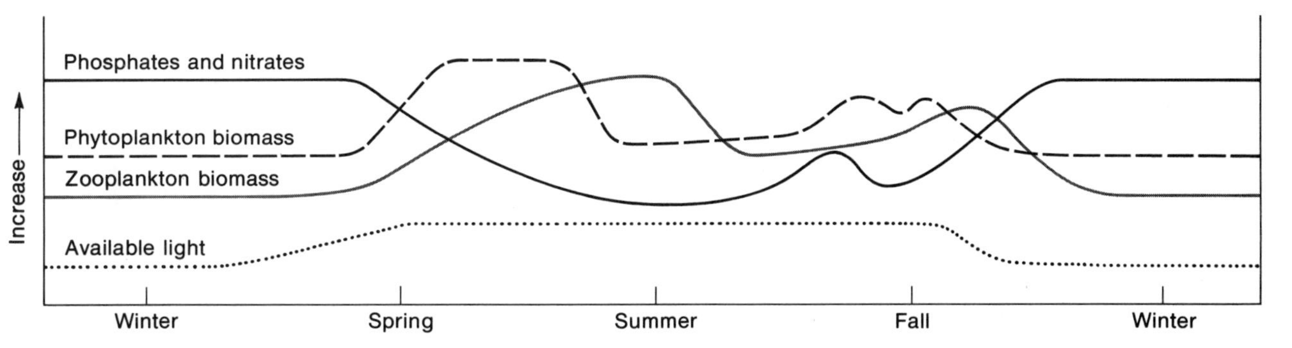

FIGURE 16-3 Seasonal productivity and variability of nutrients and light. The increase along the vertical axis refers to the increases of all curves—phytoplankton biomass, zooplankton biomass, phosphates, nitrates, and available light.

8. Certain important characteristics are directly related to the eutrophism or oligotrophism of a region. Some of the phenomena that vary between the two region types are given below. Compare them for each type of region by stating whether they would be high or low, great or small, and so on.

	Eutrophic	Oligotrophic
Primary productivity	______________	______________
Biomass	______________	______________
Density (number of individuals per volume of water)	______________	______________
Upwelling	______________	______________
Nutrient concentration (nitrate, phosphate, silica; see also Exercise 17)	______________	______________
Stability of food web	______________	______________

If you have had trouble in deciding how the final item, the stability of the food web, is affected by each region, consult Exercise 18 on marine food chains and nutrient cycles. Now, explain your answer on stability below. Why is one condition more stable than the other?

In general, fewer species exist in eutrophic regions than in oligotrophic regions. Why? (See Exercise 18 if necessary.)

EXERCISE 17

PRIMARY NUTRIENTS

Primary nutrients are those substances consumed in phytoplankton growth (through **photosynthesis**) and released by phytoplankton decay (through **respiration**). Although many elements are necessary for phytoplankton growth, the more important ones that are easily measured in the ocean are phosphorus, nitrogen, carbon, and silicon. Tissues, or soft parts, of living organisms are composed largely of carbon, hydrogen, and oxygen. The next most abundant elements are nitrogen, sulfur, and phosphorus. Many organisms have hard skeletons composed of silica, SiO_2, or calcium carbonate, $CaCO_3$, in addition to their tissue.

If a nutrient is scarce in comparison to demand for it, it is a **limiting nutrient.** Thus, if phytoplankton were grown in a closed system, the limiting nutrient would be the first one to be completely consumed. The factor that makes a nutrient limiting, then, is not the supply alone, but the supply in relation to the demand. In the surface ocean, phytoplankton growth is limited by the available quantities of nitrogen and/or phosphorus. Available nitrogen is found primarily in the form of nitrate, and phosphorus is found primarily in the form of phosphate (PO_4^{-3}).

Addition of a limiting nutrient to a system will stimulate phytoplankton growth. This effect will be amplified throughout the food chain, increasing productivity at all trophic levels. In the ocean, this stimulation is provided primarily by **upwelling** of nutrient-rich deep water into the photic zone (see the discussion of upwelling in Exercise 16).

CYCLING OF NUTRIENTS

Photosynthesis and plant growth can take place only in the **photic zone,** whereas respiration occurs throughout the ocean. The rate of addition of nutrients through river runoff is much lower than the rate of nutrient uptake by phytoplankton throughout the ocean, and so the nutritional demands of phytoplankton have to be satisfied primarily by the recycling of nutrients through respiration. The rate of respiration must nearly balance that of photosynthesis in the ocean as a whole.

The cycling of phosphorus by means of photosynthesis, respiration, mixing, and runoff is diagrammed in Figure 17-1. The numerical values in parentheses

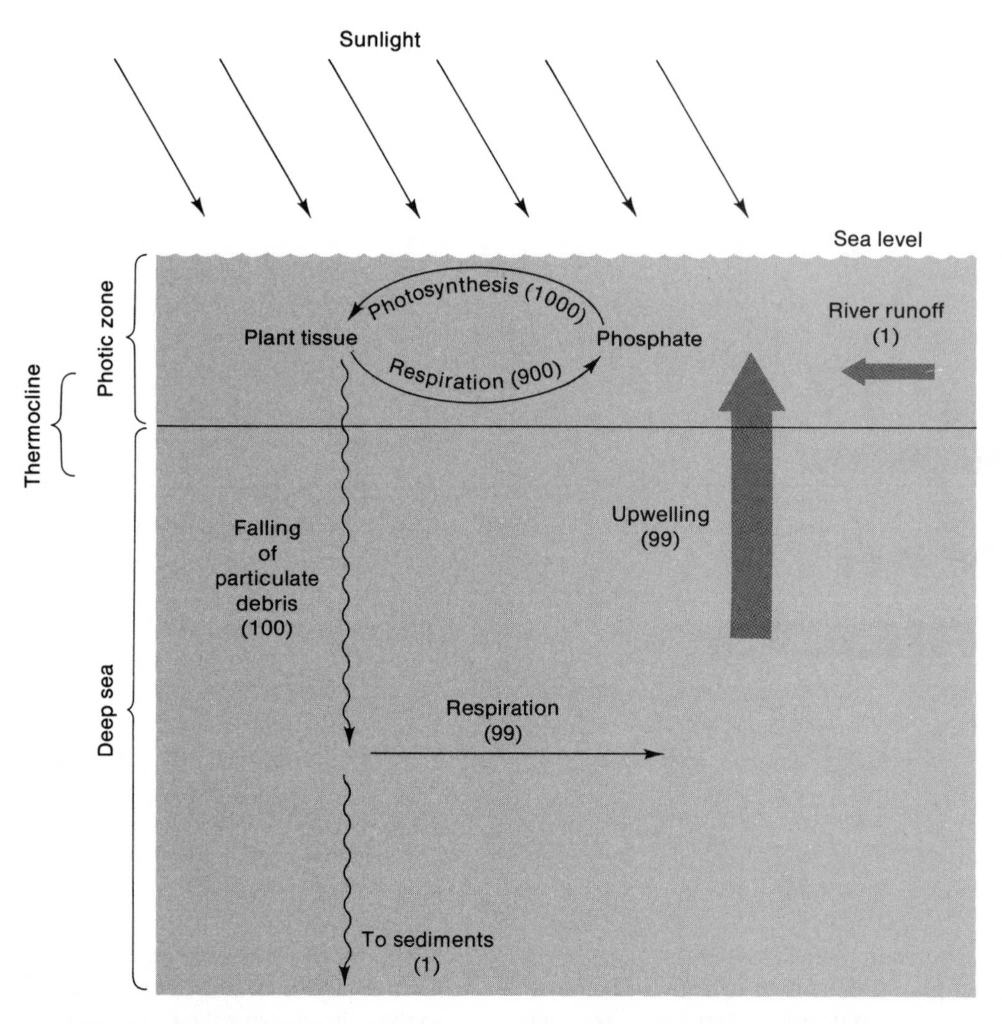

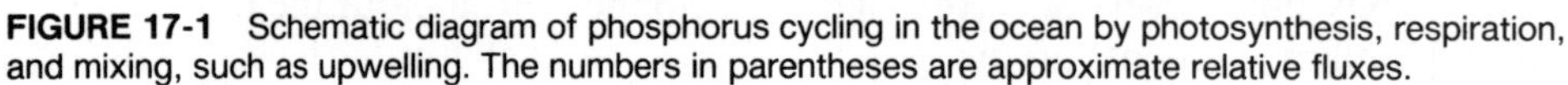
FIGURE 17-1 Schematic diagram of phosphorus cycling in the ocean by photosynthesis, respiration, and mixing, such as upwelling. The numbers in parentheses are approximate relative fluxes.

show the proportions of phosphorus cycled by each process. As the figure shows, for every atom of phosphorus added to the surface ocean by river runoff, approximately 99 atoms are upwelled. These atoms are cycled in surface waters, but respiration there cannot keep pace with photosynthesis, and approximately once every ten cycles particulate plant or animal debris will sink out of the photic zone, carrying phosphorus into the deep sea. Ninety-nine out of 100 times these atoms later will be released from particulates by deep-sea respiration, but one time in 100 the atoms will be buried in the sediment. Thus the ocean remains in steady state, because phosphorus input by rivers is balanced by loss to sediments. Note that the flux of particulate debris removing nutrients from surface waters must be balanced by upwelling of nutrient-rich deep water, or productivity in surface waters would cease. Similarly, respiration in the deep sea consumes oxygen. All oxygen in the deep sea would soon be consumed if respiration were not balanced by the sinking of cold, oxygen-rich water at high latitudes (see Exercise 10 on water masses and temperature–salinity diagrams). A similar cycling exists for other nutrients, although the details of their fluxes differ from those of the phosphorus cycle.

CONSERVATIVE AND NONCONSERVATIVE PROPERTIES

As we have seen, nutrient distribution in the ocean is controlled by two kinds of processes, biological reactions (photosynthesis and respiration) and mixing of ocean water. In order to distinguish the effects of these two processes on phosphorus distribution in the ocean, it is necessary that we understand the concept of **conservative** and **nonconservative** properties. A conservative property is one that is changed only by mixing of different types of water. Recall that temperature and salinity are used to identify different types of water (see Exercises 8 and 9). These two parameters are found to be conservative properties in the deep sea, because they can be changed only if the waters are mixed with those having different temperature–salinity characteristics. (But note that in the surface ocean, temperature and salinity are *not* conservative because the ocean can exchange heat and moisture with the atmosphere.) If two **water types** are mixed in various proportions, the resulting **water mass** will exhibit a linear relationship between temperature and salinity. This linear relation can be seen when any two conservative properties are plotted on a graph. In some regions of the ocean, nutrients are conservative (because the rate of photosynthesis or respiration is negligible in comparison with the rate of mixing) and nutrients are useful for identifying water types, just as temperature and salinity are. In other regions, biological reactions are more rapid than the mixing rate, so that nutrients exhibit nonconservative behavior when plotted against salinity. Any plot of a nonconservative property versus a conservative property (such as temperature or salinity) will be nonlinear. This technique is useful for identifying regions of the ocean in which biological processes, particularly respiration, are occurring. A large deviation from a linear mixing line indicates that the rate of production or consumption is much faster than the rate of mixing. The direction of the deviation will indicate whether the nonconservative property is being produced or consumed.

DEFINITIONS

Conservative property. A property of seawater that changes only when one water type is mixed with another water type. Examples are temperature and salinity in the deep ocean.

Limiting nutrient. The ingredient necessary for plant growth that is least abundant in proportion to demand.

Nonconservative property. One whose value may be changed as a result of chemical reactions as well as the mixing of different water types. An example is oxygen, which is consumed by animals in respiration.

Photic zone. The zone extending downward from the ocean surface within which light is sufficient to sustain photosynthesis. The depth of this layer varies with water clarity, time of the year, and cloud cover, but is about 100 meters in the open ocean.

Photosynthesis. Process by which plants manufacture carbohydrates from carbon dioxide, water, and sunlight. It results in carbon dioxide uptake and oxygen release.

Respiration. Process whereby food (sugar, for example) is "burned" by animals and energy is released. It results in oxygen uptake and carbon dioxide release.

Upwelling. The upward movement of cold deep water to the ocean surface.

Water mass. Ocean waters produced by the mixing of water types. A water mass occupies a region of ocean and may have a variation in temperature, salinity, and nutrient composition.

Water type. A water having well-defined temperature, salinity, and nutrient characteristics.

TABLE 17-1
Depth data on temperature, salinity, and nutrients collected at a station in the central North Pacific

Depth (meters)	Potential temperature, T (°C)	Salinity, S (‰)	Oxygen, O_2, concentration (millimoles per kilogram)*	Nutrient concentration: Carbon dioxide, $\Sigma\ CO_2$† (millimoles per kilogram)	Nutrient concentration: Phosphate, PO_4 (micromoles per kilogram)	Nutrient concentration: Silica, SiO_2 (micromoles per kilogram)
20	26.53	34.658	0.203	–	0.18	1.8
49	26.53	34.663	0.204	1.946	0.18	1.8
99	25.21	34.811	0.209	1.973	0.18	1.6
149	21.32	35.007	0.205	2.041	0.21	1.8
248	13.97	34.403	0.169	2.093	0.85	10.3
544	6.10	34.235	0.053	2.303	2.61	71.7
742	5.27	34.429	0.044	2.299	2.83	86.8
988	4.18	34.513	0.050	2.331	2.88	107.8
1185	3.57	34.542	0.058	2.357	2.90	121.6
1534	2.72	34.583	0.073	2.365	2.90	139.2
2126	2.15	34.622	0.102	2.353	2.78	156.4
2530	1.82	34.640	0.117	2.366	2.70	162.2
3130	1.29	34.671	0.135	2.375	2.61	164.4
3735	1.16	34.680	0.148	2.341	2.48	162.5
4385	1.07	34.687	0.157	2.323	2.44	152.8
4953	0.91	34.698	0.182	2.411	2.33	136.9

Source: Unpublished data collected primarily by the GEOSECS program.
* One mole = 6×10^{23} molecules; 1 millimole = 10^{-3} mole; 1 micromole = 10^{-6} mole.
† Σ stands for summation or total; it includes CO_2 as well as the bicarbonate ion, HCO_3^-, and the carbonate ion, CO_3^{2-}.

EXERCISE 17

REPORT

PRIMARY NUTRIENTS

NAME ______________________

DATE ______________________

INSTRUCTOR ______________________

Refer to Table 17-1 to answer the following questions.

1. On the grid below plot the following from the table: (a) O_2 versus depth; and (b) PO_4^{-3} versus depth. It is suggested that you plot O_2 versus depth with circles and PO_4^{-3} with X's. Note that when you have plotted (a), it is easy to plot (b) by matching depths. Connect the O_2 points with a dashed line and the PO_4^{-3} values with a solid line.

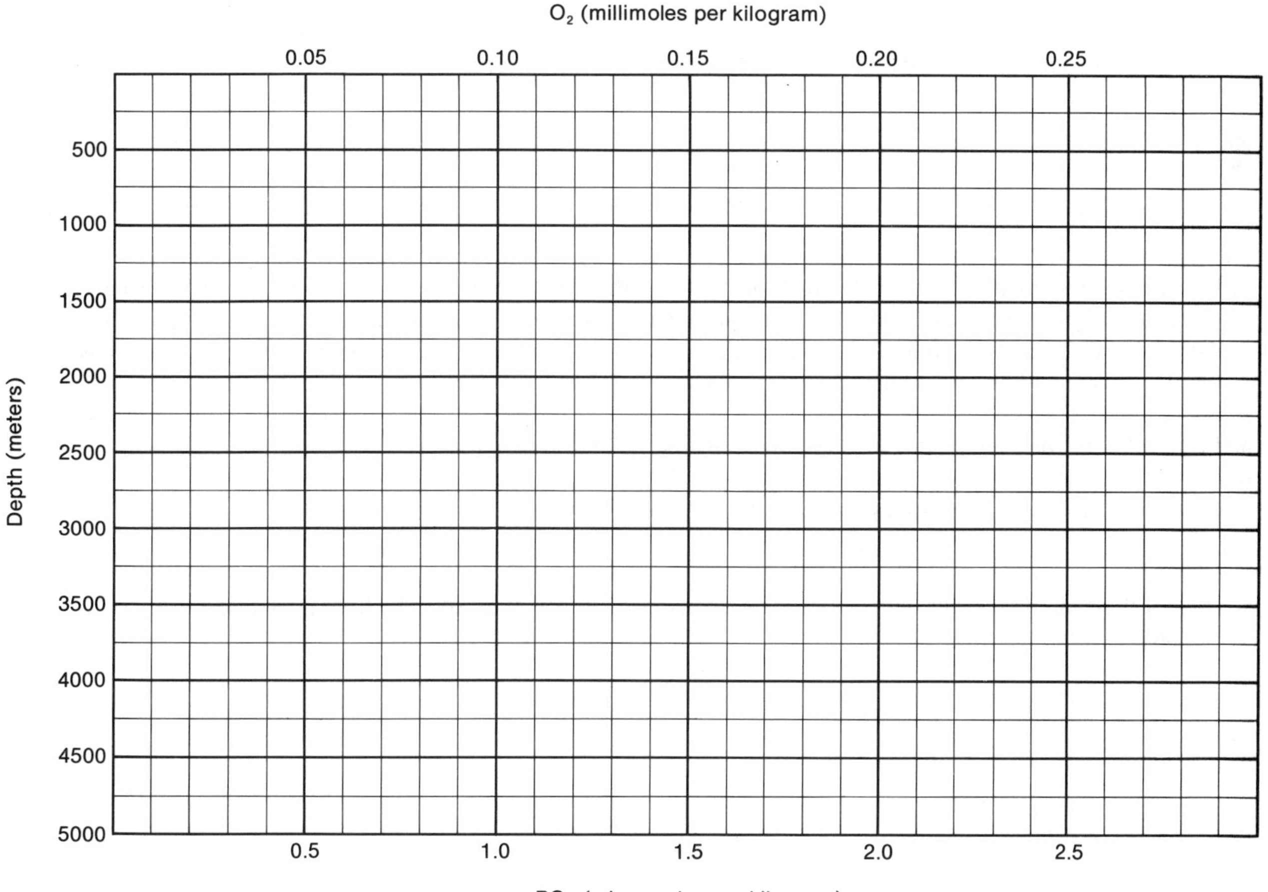

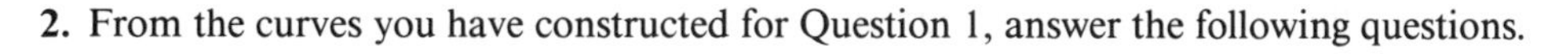

2. From the curves you have constructed for Question 1, answer the following questions.

(a) In what portion of the ocean does photosynthesis exceed respiration?

(b) In what portion of the ocean does respiration exceed photosynthesis?

(c) What effect should this have on the CO_2 content?

Is this effect observed?

(d) Why are phosphorus and silica contents so low in surface waters?

(e) Why does the amount of dissolved oxygen increase near the bottom?

3. For samples below depths of 900 meters in Table 17-1, plot the following in the grids provided: (a) T versus S; (b) O_2 versus S; and (c) SiO_2 versus S. As in Question 1, once you have plotted T versus A, the rest is easy. Just line up the O_2 and SiO_2 data with the plotted S points for T. Use a line of alternate dashes and dots for T, a dotted line for SiO_2, and a dashed line for O_2 to distinguish the properties. Different colors may also be used.

4. From the data you have plotted in Question 3, answer the following questions.

(a) How many water masses are present in samples below 900 meters?

(b) For these samples, which of the following are conservative properties, and which are nonconservative: T, S, O_2, and SiO_2?

(c) Of the nonconservative properties, which are consumed and which are released in the deep ocean? How do you know? (Hint: Look at the shapes of the curves you plotted in Question 3. Note that the SiO_2 axis increases downward.)

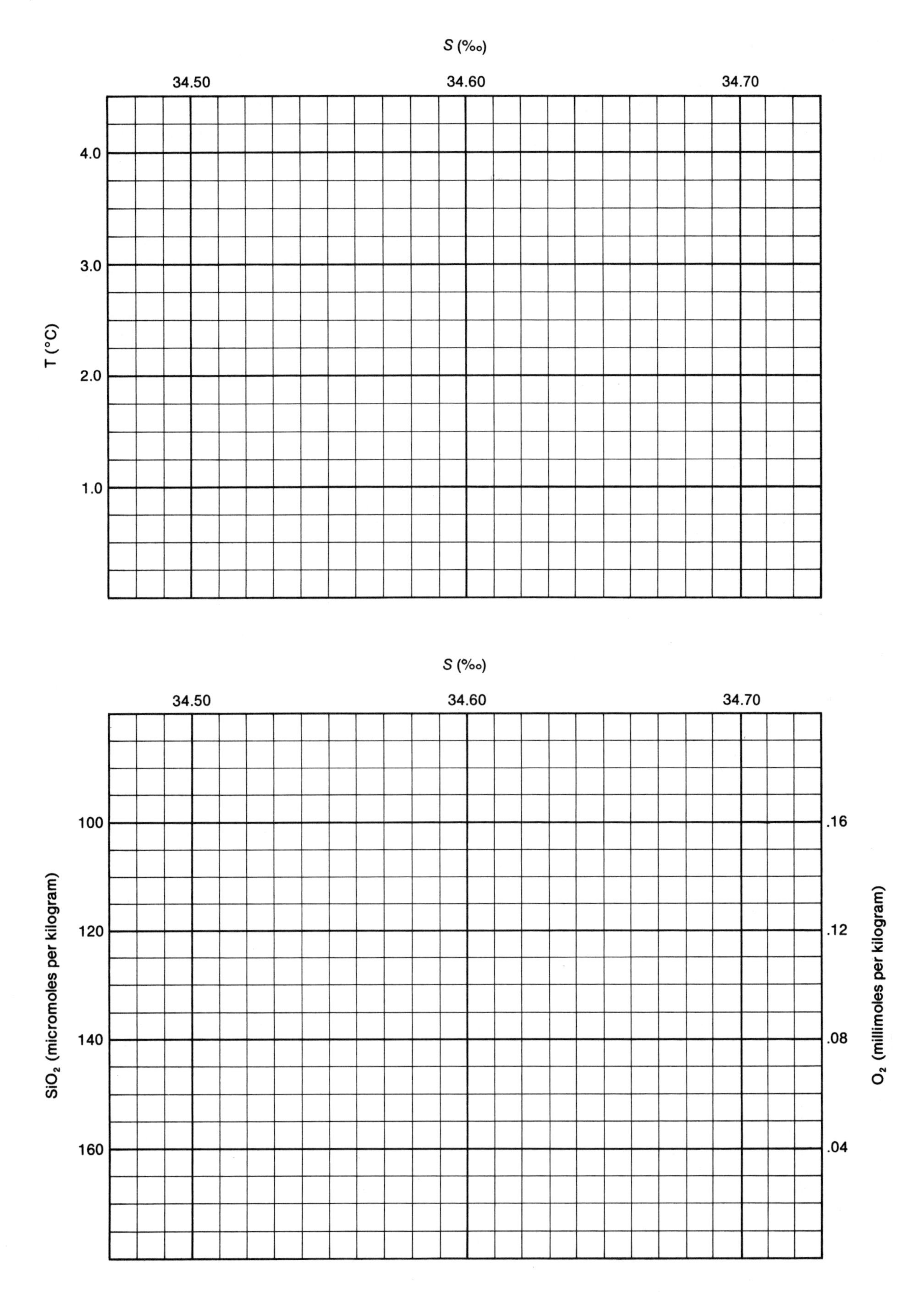

S (‰)
34.50
34.60
34.70
4.0
3.0
2.0
1.0
T (°C)
S (‰)
34.50
34.60
34.70
100
120
140
160
SiO2 (micromoles per kilogram)
.16
.12
.08
.04
O2 (millimoles per kilogram)

OPTIONAL QUESTION

5. As stated on page 165 of this exercise, "The rate of respiration must nearly balance that of photosynthesis in the ocean as a whole." However, it appears that over geologic time there has been a slight excess of photosynthesis over respiration. How would this affect (a) the composition of the atmosphere, and (b) the quantity of hydrocarbons (such as oil) in sedimentary rocks?

EXERCISE 18

MARINE FOOD CHAINS AND NUTRIENT CYCLES

In this exercise we will study marine food chains and nutrient cycles. As we saw in Exercise 16, nutrients and nutrient cycles are closely associated with primary productivity. Here we will study "complete" nutrient cycles and relate them to marine food chains.

TROPIC PYRAMIDS AND FOOD CHAINS

The primary producers make up the first **tropic level** in a typical trophic pyramid, or food chain; a very simple trophic pyramid is illustrated in Figure 18-1. The first trophic level in this example consists of seven different species of diatoms, or marine phytoplankton, representing the **primary producers.** They are *primary* producers because they have the ability, with the aid of light energy, to convert inorganic compounds in seawater into simple organic compounds (such as carbohydrates) and give off oxygen. This process, termed *photosynthesis,* is the most important kind of primary productivity (see Exercise 16). Indeed, all life in the sea (or anywhere on earth) is dependent on primary productivity, whether it originates from marine phytoplankton in the open ocean, algae in shallow bottom areas, or grass on land. The next trophic, or feeding, level shown in Figure 18-1 is composed of shrimplike organisms known as *krill* (euphausiids). They are termed either herbivores (because they eat plants) or **primary consumers** (because they do not produce food by primary productivity but are the first in line to consume the organic matter produced by the first trophic level). In this example whales make up the third trophic level, making a very short food chain. In this particular food chain whales are the **secondary consumers** (they eat the primary consumers), or first-stage carnivores (they are also the first to eat other animals and are therefore carnivores), or top carnivores (since they are on top of the pyramid). This three-level trophic pyramid is a simplification of that actually existing in the Antarctic waters.

A slightly more complicated and more realistic diagram of the Antarctic open-ocean trophic pyramid is shown in Figure 18-2. Although a complete description of Antarctic biology has not been developed, portions of the food chain, such as that shown in

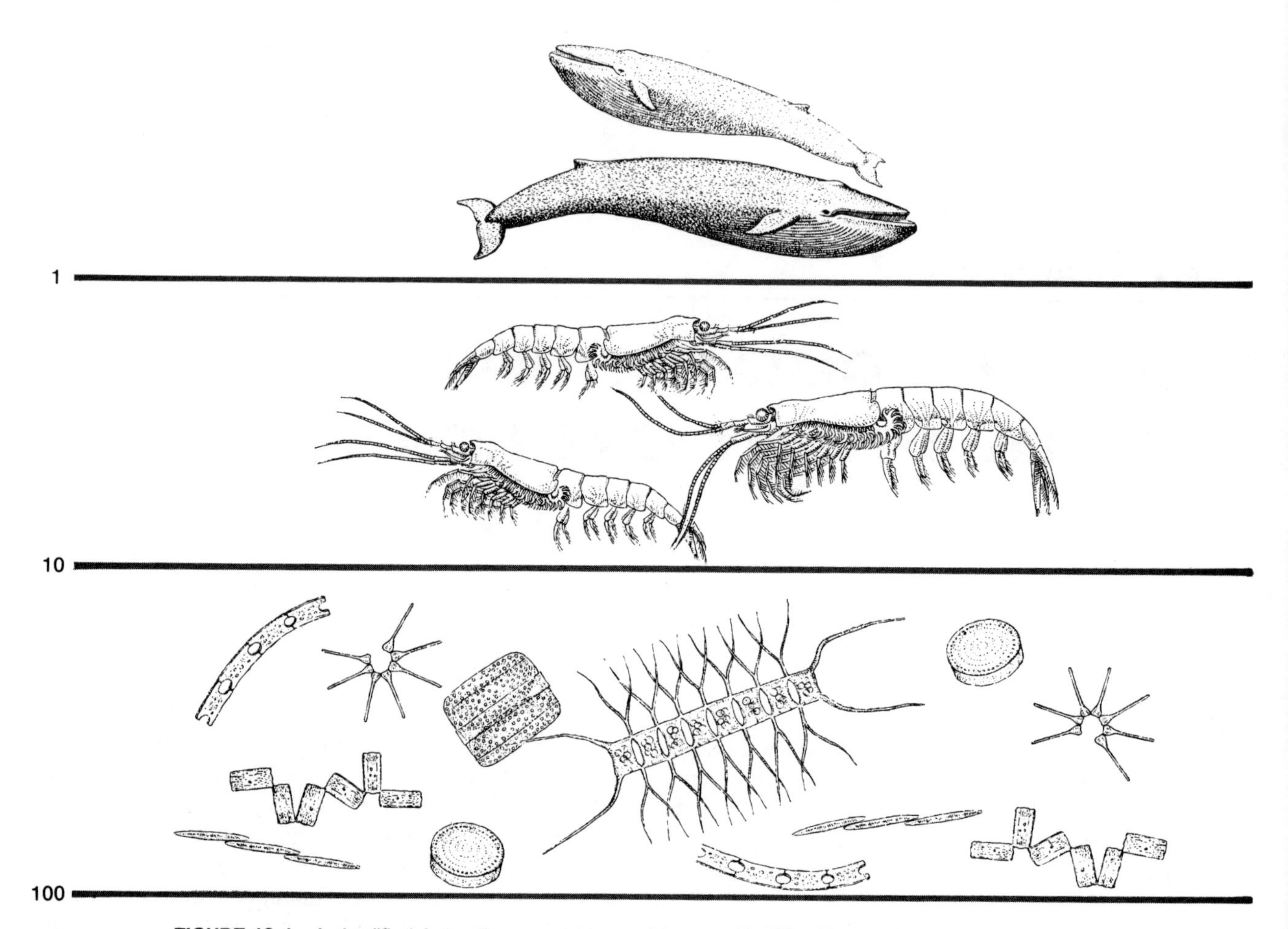

FIGURE 18-1 A simplified Antarctic open-ocean trophic pyramid. [After Robert C. Murphy, "The Oceanic Life of the Antarctic." Copyright © 1962 by Scientific American, Inc. All rights reserved.]

Figure 18-3, have been well studied. In this example, krill (the second trophic level) are fed upon by members of the third trophic level, including the blue whale, the crabeater seal, certain birds, and small fish and squid. These third-level organisms are in turn fed upon by the emperor penguin, large fish, and carnivorous seals. The fourth trophic level is occupied by the leopard seal. It becomes increasingly more difficult to place the relatively sophisticated animals operating at the top of the food chain or trophic pyramid into their "correct" trophic levels; many occupy different levels. For example, the killer whale is the top carnivore and functions within the fifth trophic level when it eats the leopard seal, or within the third when it eats the crabeater seal or blue whale (it usually prefers small blue whales). The same species may also change its trophic levels in the course of a life cycle, or seasonally; for example, as some small fish grow into larger fish, they shift from tropic level 2 to trophic level 3.

Although third-trophic-level organisms may eat other organisms, their staple is the krill. In fact, about 13 species are dependent, either directly or indirectly, on the krill for their existence in Antarctic waters. If

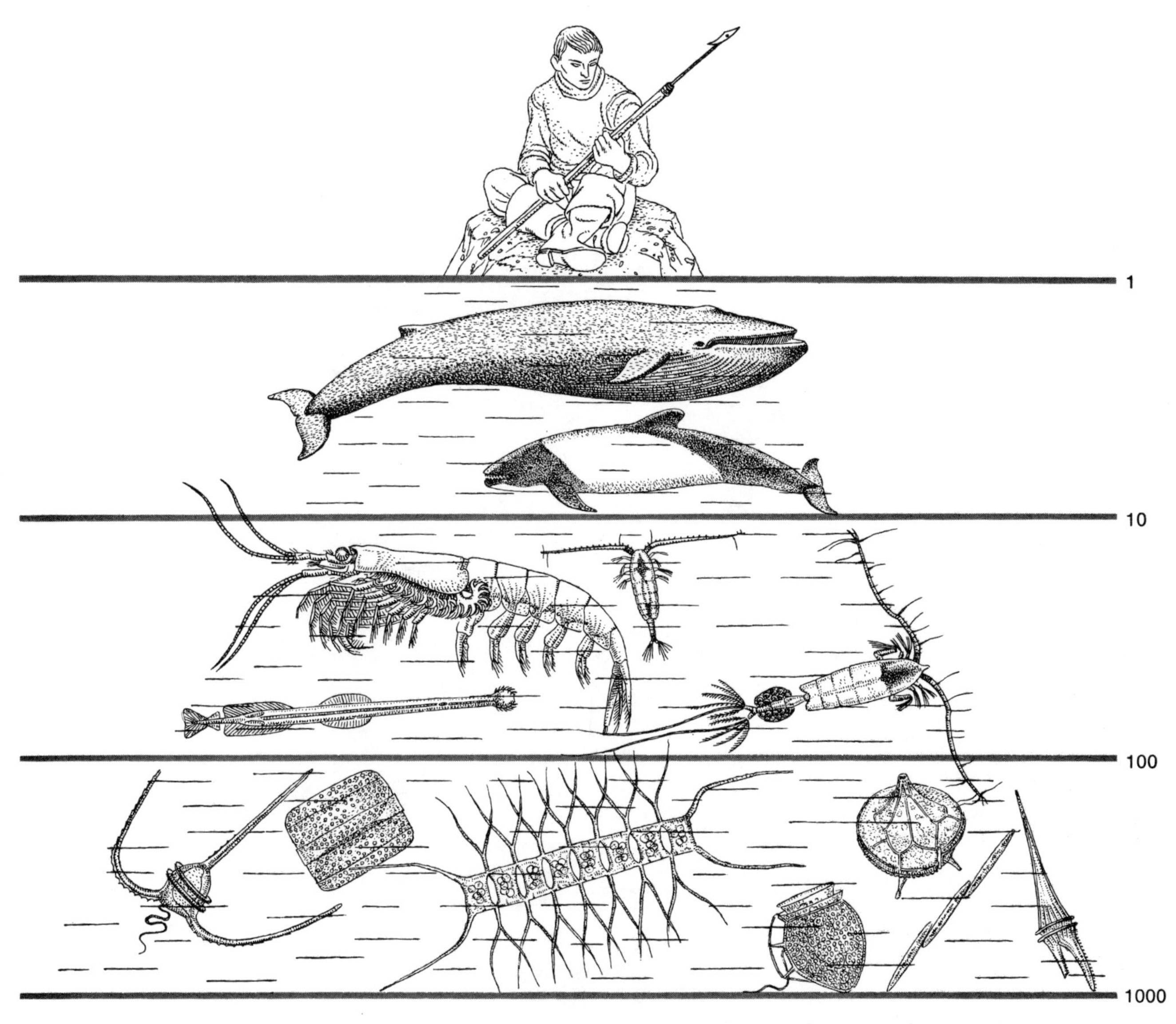

FIGURE 18-2 A more complicated Antarctic open-ocean trophic pyramid. [After Willis E. Pequegnat, "Whales, Plankton and Man." Copyright © 1958 by Scientific American, Inc. All rights reserved.]

something should happen to the krill, the effect on this part of the Antarctic food cycle would be disastrous. When you answer Questions 1 – 3 in the report, remember that they may have more than one answer, owing to the dependence of one trophic level upon another; disruptions of one level thus have a potential for affecting many other levels as well.

Having examined examples of *open-ocean* trophic pyramids and food chains, let us now turn our attention to a *nearshore* trophic pyramid and food chain — one that most of us have a much greater likelihood of observing at some time, if we have not already done so. An example illustrated in Figure 18-4, which shows a portion of a fcod chain or web studied in a Long Island estuary.* It manifests many of the important characteristics found in most food webs. One of the principal characteristics of marine food webs is their complexity (many species occupy more than

* This particular food web has been thoroughly investigated by George M. Woodwell, Charles Wurster, and Peter Isaacson. For a detailed commentary on their work, see George M. Woodwell, "Toxic Substances and Ecological Cycles," *Scientific American,* March 1967.

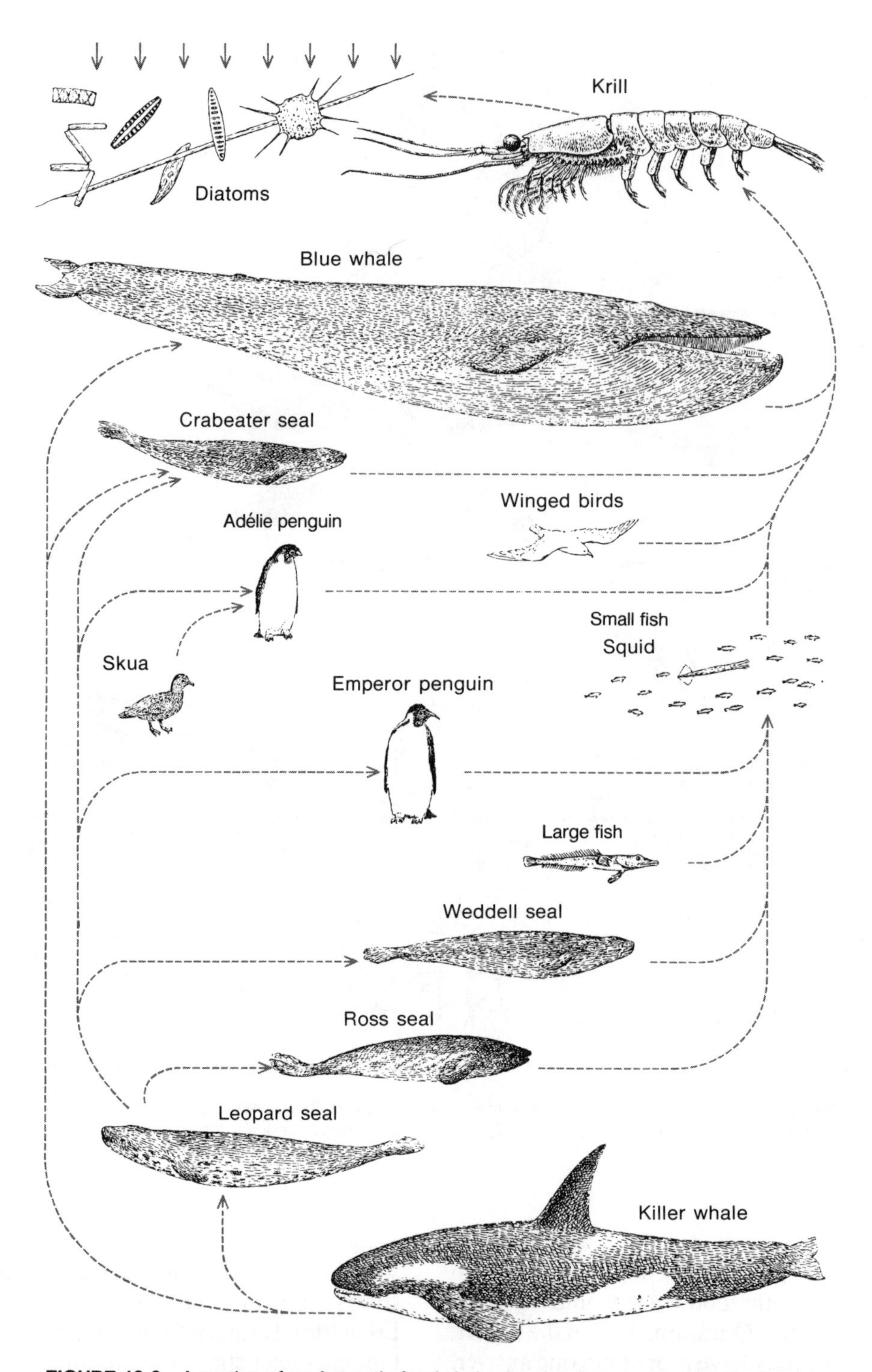

FIGURE 18-3 A portion of an Antarctic food chain showing the way in which the food cycle of marine animals is based on the krill. [After Robert C. Murphy, "The Oceanic Life of the Antarctic." Copyright © 1962 by Scientific American, Inc. All rights reserved.]

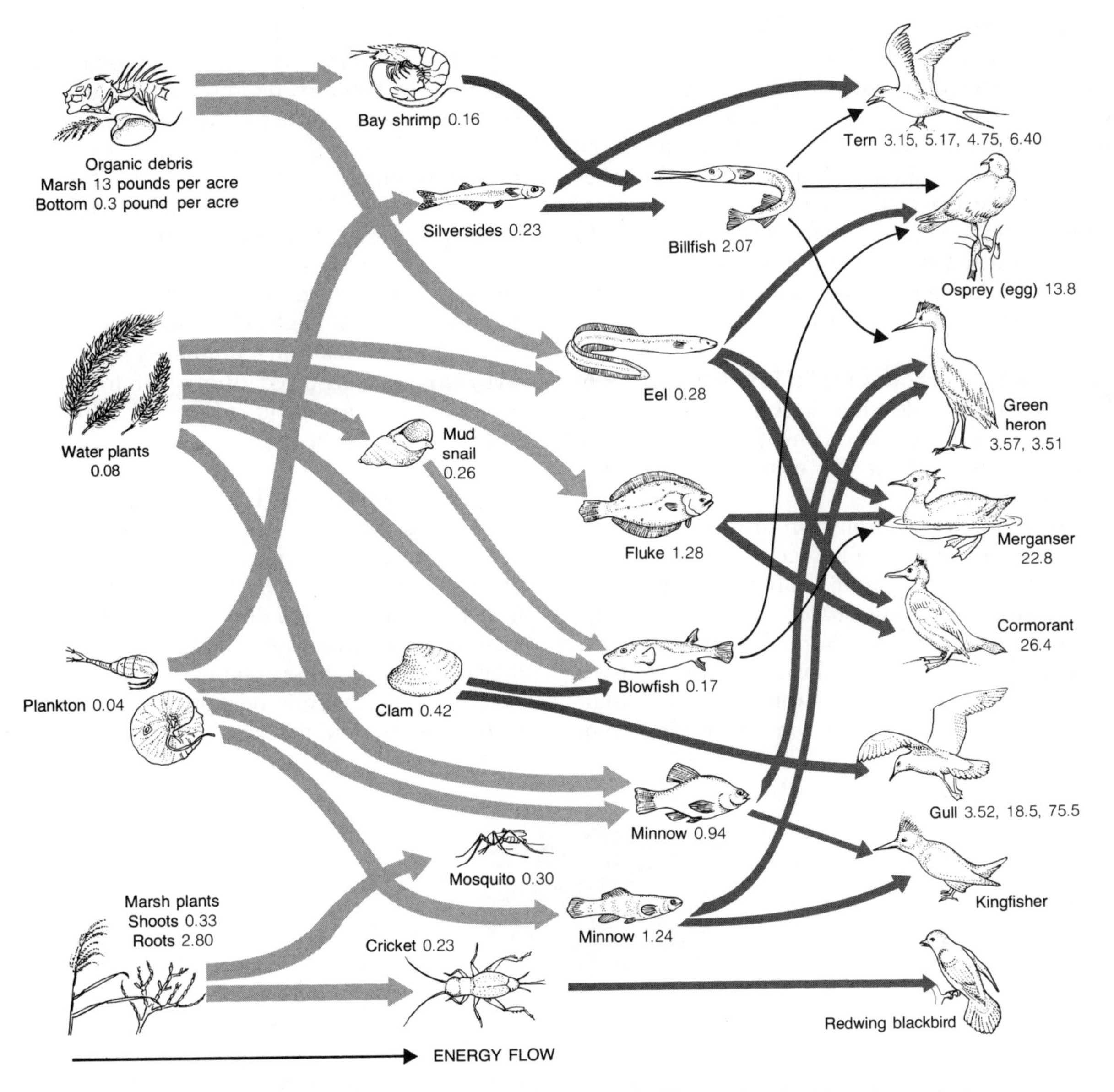

FIGURE 18-4 A portion of a food web in the Long Island estuary. The numbers beside each organism's name refer to the DDT content, in parts per million, found in each organism. [After George M. Woodwell, "Toxic Substances and Ecological Cycles." Copyright © 1967 by Scientific American, Inc. All rights reserved.]

one trophic level), so that multiple cross-connecting links and pathways result. Most ecologists believe that the more biological pathways there are leading from the primary producers to the upper portions of the food web, the greater the *stability* of a given ecosystem, stability being the ability of a system to rebound biologically should something upset it. Thus, stable ecosystems are usually composed of many trophic levels, many species, and many species occupying different trophic levels at the same time, all resulting in many pathways to the top of the web.

BIOGEOCHEMICAL CYCLES

Energy from the sun is needed for photosynthesis. This energy is harnesed as described in Exercise 16 and is passed through the earth's ecosystem via food chains. Light is a limiting factor in photosynthesis, and affects not only the presence but also the abundance of life. At high latitudes light limits productivity at certain seasons, but the loss of light with increasing depth in the ocean creates a phytoplankton compensation depth, at which photosynthesis equals plant respiration. However, the main limiting factors in well-lighted marine waters are the principal nutrients of nitrates, phosphates, and carbon (though carbon is not limiting in most areas). We will study the carbon, nitrogen, and phosphorus cycles on land and in the sea, and then look at the way in which their distribution affects the food chains. Keep in mind that microorganisms (especially bacteria and fungi) are vital to all these cycles and to food chains in that they decompose organic compounds, thus making them available again to the cycles and chains.

Carbon Cycle. Carbon is the basic building block of all the larger molecules characteristic of life on earth. Thus life as we know it on this planet is "carbon based" because without this element it could not function or exist.

The carbon cycle — its entry from the carbon reservoir into the living part of the ecosystem and its passage through the system until its ultimate return to the reservoir — on earth is illustrated in Figure 18-5. The main reservoir of carbon is the pool of carbon dioxide in the atmosphere. Dissolved carbon dioxide enters the sea by mixing at the air-water interface. Additional carbon dioxide comes from the respiration of plants and animals and is also provided in the seawater by the sea's carbonate-bicarbonate system. Dissolved carbon dioxide in seawater is available to plants for photosynthesis; carbon that has been "fixed" (converted to organic carbon) by photosynthesis can then move through the food web. Carbon is also present in sediments in the carbonates and limestones (composed of the lime skeletons of marine organisms) or as peat, coal, or oil and gas. The abundance of carbon in these sorts of deposits is being stored and is not immediately available for photosynthesis. The stored carbon is ultimately returned to the sea by erosion in dissolved forms or is burned (by man or other agents) and returned to the carbon dioxide pool in the atmosphere.

Nitrogen Cycle. Nitrogen is another element required by all living organisms. It is especially important as a building block of proteins. Nitrogen moves through the ecosystem as illustrated in Figure 18-6. Gaseous nitrogen cannot be used directly by most organisms, but some bacteria and algae are capable of converting the gas into compounds that can be used. Through these *nitrogen-fixing* microorganisms, nitrogen enters the living portion of the ecosystem directly from the

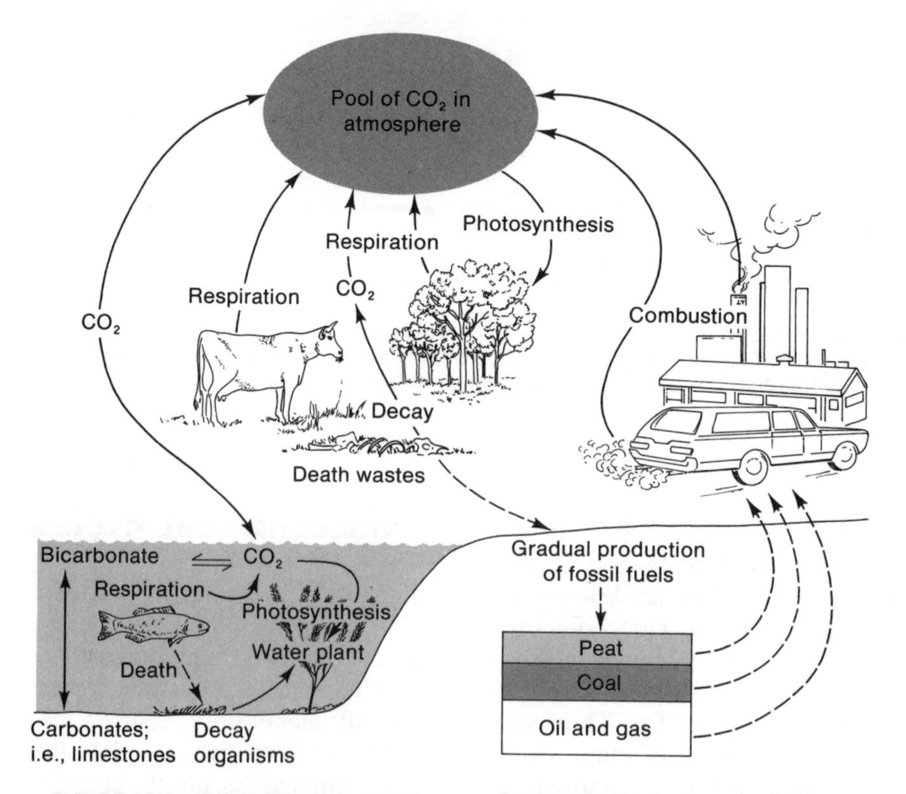

FIGURE 18-5 The carbon cycle. Solid arrows represent the flow of carbon dioxide, CO_2. [From Paul R. Ehrlich and Anne H. Ehrlich, *Population, Resources, Environment: Issues in Human Ecology,* 2d ed. W. H. Freeman and Company. Copyright © 1972.]

atmosphere: it comes into the terrestrial environment via nitrogen-fixing bacteria in soil and roots, and into the marine environment via mainly nitrogen-fixing blue-green algae; or it is taken up by plants directly through their consumption of water. It is used primarily to make proteins and is released by organisms in their excretory products or by the decay of dead organisms. Another kind of bacteria, denitrifying bacteria, returns nitrogen to the atmosphere or water by breaking down the organic nitrogen (such as ammonia, NH_3). Not only is it important that nitrogen be returned to a form that can again be used by organisms; it is also important that these waste products break down, since some (like ammonia) are toxic to many forms of life.

Phosphorus Cycle. Phosphorus is an essential element in the composition of the genetic material (DNA and RNA) and in the phosphorus compounds (such as ATP and ADP) that living things use as the prime energy-manipulating devices. The phosphorus cycle is illustrated in Figure 18-7. The main phosphorus reservoirs are phosphate rocks, which are deposits of fossilized organisms. Phosphorus is released from these reservoirs by the processes of erosion and leaching. It can enter the living part of the ecosystem as phosphate compounds in the sea or soil, and can pass through several levels of organisms before being returned to the soil by the decay of organisms, excretory products, and so on.

DISTRIBUTION OF NUTRIENTS IN THE SEA

The amount of carbon dioxide available in the seawater or in shallow marine waters does not appear to be a factor limiting to photosynthesis. However, the amounts of nitrogen, phosphorus, and sometimes silicon compounds (nitrate, phosphate, and silicate) are often limiting, especially in the open ocean. As shown in Table 18-1, 90 percent of the ocean is termed *open ocean;* this is the region removed not only from the shelf but also from the boundary-current and open-ocean upwelling areas (for a diagram of these types of upwelling, see Figure 8-4a). Open-ocean nonupwelling regions are essentially biological deserts (this means that about 90 percent of the sea, or about three-fourths of the earth, is nonproductive) owing to the lack of dissolved nutrients in the shallow parts of the water column (the **photic zone**), where the pri-

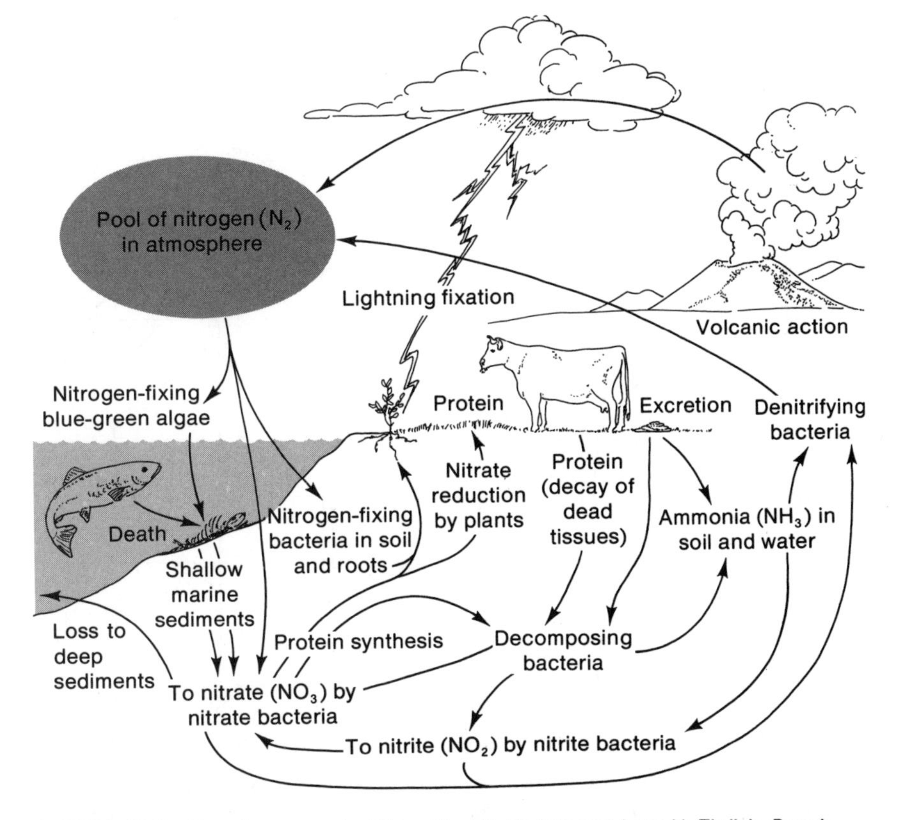

FIGURE 18-6 The nitrogen cycle. [From Paul R. Ehrlich and Anne H. Ehrlich, *Population, Resources, Environment: Issues in Human Ecology,* 2d ed. W. H. Freeman and Company. Copyright © 1972.]

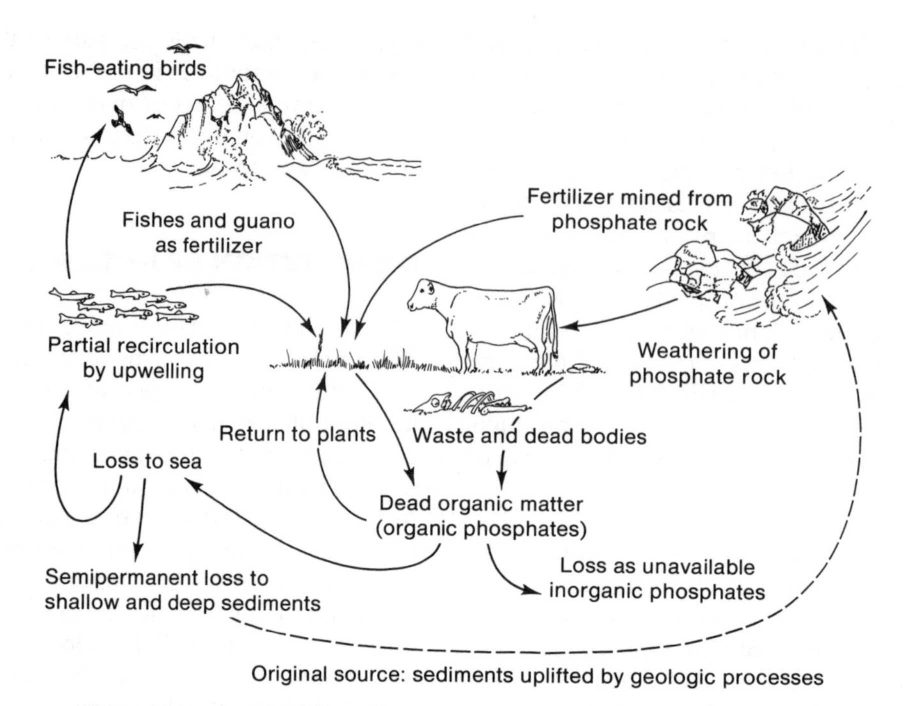

FIGURE 18-7 The phosphorus cycle. [From Paul R. Ehrlich and Anne H. Ehrlich, *Population, Resources, Environment: Issues in Human Ecology,* 2d ed. W. H. Freeman and Company. Copyright © 1972.]

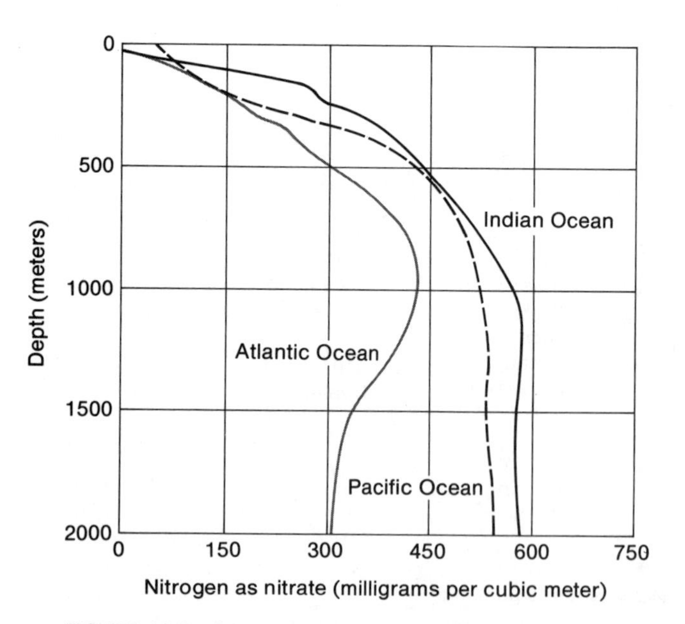

FIGURE 18-8 Curves showing the distribution of dissolved nitrate in seawater (of three ocean basins) from an open-ocean nonupwelling region. [After Gifford B. Pinchot, "Marine Farming." Copyright © 1970 by Scientific American, Inc. All rights reserved.]

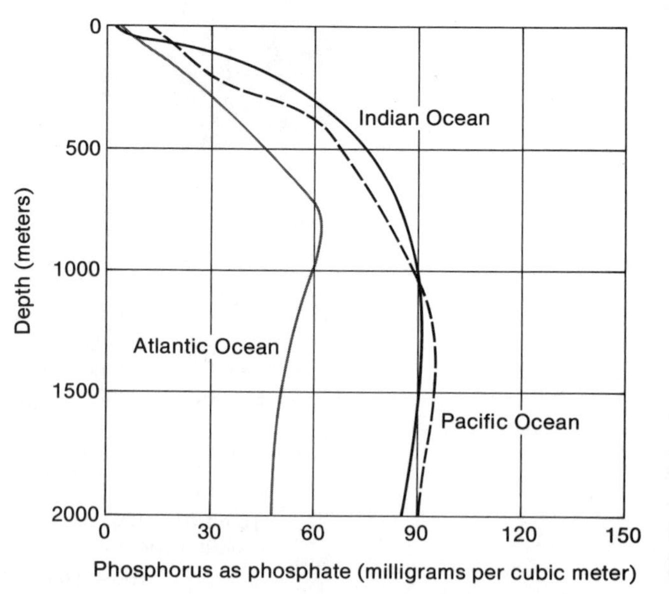

FIGURE 18-9 Curves showing the distribution of dissolved phosphate in seawater (of three ocean basins) from an open-ocean nonupwelling region. [After Gifford B. Pinchot, "Marine Farming." Copyright © 1970 by Scientific American, Inc. All rights reserved.]

TABLE 18-1
Productivity and fish production of the ocean.

Area	Percentage of ocean	Area (square kilometers)	Average productivity (grams of carbon per square meter per year)	Average number of trophic levels (approximate)	Annual fish production (metric tons)
Open ocean	90	326,000,000	50	5	160,000
Boundary-current and open-ocean upwelling areas*	9.9	36,000,000	100	3	120,000,000
Coastal upwelling areas	0.1	360,000	300	1.5	120,000,000
Total annual fish production					240,160,000
Amount available for sustained harvesting†					100,000,000

SOURCE: After Ryther, *Science,* 1969.
* Including certain offshore areas where hydrographic features bring nutrients to the surface.
† Not all the fish can be taken; many must be left to reproduce or the fishery will be overexploited. Other predators, such as seabirds, also compete with us for the yield.

mary producers need them. The ways in which the distribution of dissolved nitrate and of dissolved phosphate varies with depth in these marine "desert" areas are shown in Figures 18-8 and 18-9. Note that the nutrient curves (see Exercises 16 and 17) from all three oceans are very similar in both figures.

Essentially very few nutrients are available in the photic zone, where they could be used by the phytoplankton for primary productivity. The reason for their scant supply in shallow water and their increase in deep water is that many of the nutrients in shallow water are fixed in the tissues of living organisms. Moreover, many of them are removed from the water column by an organic "rain" of fecal pellets, dead carcasses, molts of crustaceans, and so on, and are concentrated at the lower depths. Thus, if vigorous primary productivity is to be stimulated, this deeper nutrient-rich water must somehow get to the surface: a fertilization of surface water must occur. Vertical circulation in enhanced in areas of upwelling, as described in Exercises 16 and 17.

As Table 18-1 shows, only 10 percent of the ocean has a reasonable amount of primary productivity and a significant fish production. In fact, only 0.1 percent of the ocean produces about one-half of the fish currently available from harvesting. These numbers are a function of the average productivity of the various areas and the number of trophic levels presented in different regions. As demonstrated in the table, the open ocean is about one-sixth as productive as the boundary-current and open-ocean upwelling areas, and these are about one-third as productive as the coastal upwelling areas. However, the coastal upwelling areas produce as much fish in metric tons as do the boundary-current and open-ocean upwelling areas, and both types of upwelling areas produce 1500 times as much fish as does the remaining 90 percent of the world ocean. The main reason for these differences is the number of trophic levels present in each area. The coastal upwelling areas have an average of 2.5 trophic levels only (1.5 if you don't count the phytoplankton as first level, as is done in Table 18-1), so that the food chain in these waters is very short and simple, as shown in Figure 18-10a. In these areas the phytoplankton at the bottom of the food chain (trophic level 1) are usually aggregates of colonial diatoms that are large enough to feed fish of a size large enough to be exploited by man.

In a boundary-current or open-ocean upwelling area, represented in Figure 18-10b, the solitary diatoms are first eaten by copepods and the copepods then are eaten by the exploitable fish. Figure 18-10c illustrates the general condition prevailing on the face of the earth, whereby the solitary diatoms (or other phytoplankters) are eaten by microplankton (such as radiolarians), which are in turn eaten by mesoplankton (such as the copepods and chaetognaths shown in the figure), which are in turn eaten by small generally nonexploitable fish, which are finally consumed by an exploitable fish such as the tuna.

Why does the number of trophic levels influence the ultimate fish production? The reason is that there is roughly a 10 percent **ecological efficiency** (conversion efficiency) from one trophic level to the next up the food web. This means that of the radiant energy changed into chemical energy by photosynthesis, only about 10 percent of the phytoplankton biomass (energy) on the average is finally converted into animal tissue by primary consumers (the second trophic level). Then at each subsequent link in the food chain,

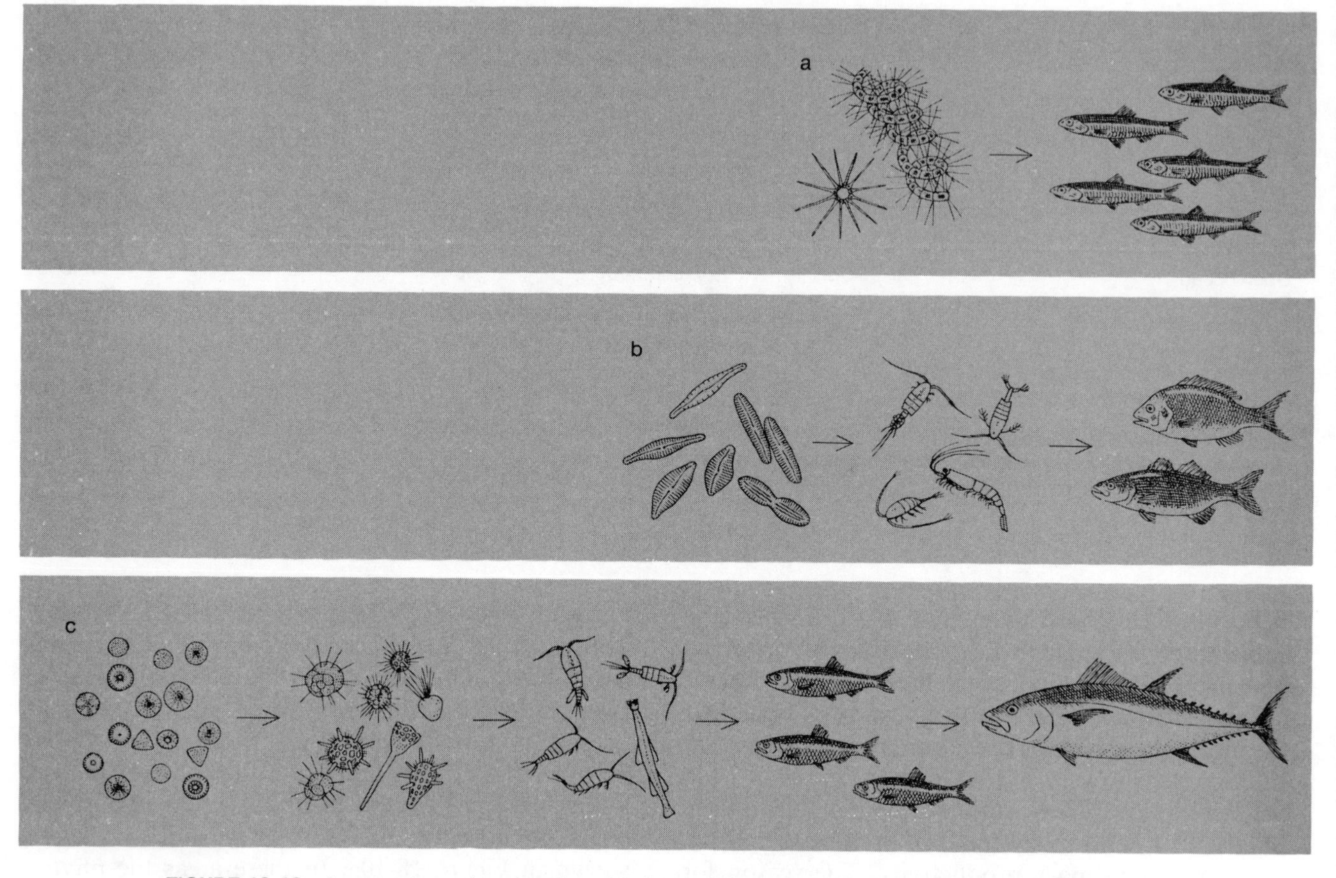

FIGURE 18-10 A comparison of the length and makeup of food chains from the following areas: (a) high-productivity coastal waters; (b) a boundary-current upwelling area; (c) low-productivity open-ocean waters. [After Gifford B. Pinchot, "Marine Farming." Copyright © 1970 by Scientific American, Inc. All rights reserved.]

another 90 percent of the energy that has been ingested is lost by respiration, other metabolic activities, and so on. If you refer back to Figure 18-1 you will understand the meaning of the numbers beside each trophic level: in this simple food chain 100 grams of phytoplankton (primary producers) are necessary to feed 10 grams of krill (primary consumers), and 10 grams of krill are required to feed 1 gram of whale. Therefore the more trophic levels there are, the less efficient the food chain is in converting radiant energy into edible fish. Figure 18-2 illustrates four levels, and the ecological efficiency is shown on the side of that simplified diagram.

Pyramids of biomass, standing crop, or numbers can be inverted as a result of differences in turnover rates between trophic levels (differences in rates of reproduction or predation and the like), but energy pyramids can never be inverted. You can't make something out of nothing.

MAN-MADE CRISES IN THE MARINE ECOSYSTEM

There are of course natural upsets or crises in marine ecosystems. A well-known example is the red tide, or red water, produced by dinoflagellates that bloom to such an extent that fish kills occur. But in this discussion we are concerned with upsets precipitated by human beings. One such obvious ecological crisis is that induced by overfishing of certain species to the extent that they cannot maintain their normal reproductive cycle. The total annual fish production is approximately 240,160,000 metric tons and, as suggested in Table 18-1, the amount available for sustained harvesting (so as not to deplete the stock) is about 100 million metric tons annually. It had been predicted that by 1980 the world harvest would be as great as 70 million tons, and that to surpass 100 million metric tons would require the harvesting of lower trophic levels in the food chain. Since the collapse of the Peruvian anchovy fishery (over 25 percent of the world's harvest) due to overfishing and El Niño conditions, it appears that lower trophic levels will be exploited to a greater degree.

Another problem is pollution of the marine environment. For example, people have been dumping the insecticide DDT into the sea in massive amounts since the 1940s (although its use in the United States has been banned since 1972). DDT is taken up by marine phytoplankton and has been shown to cause a reduction in their ability to photosynthesize, an especially serious effect at the base of the marine food web. In fact it was once suggested that if the oil tanker *Torrey Canyon,* which broke up off the coast of England in 1967, had been carrying insecticide instead of crude oil and the insecticide had spread throughout the North Atlantic, it would have seriously reduced photosynthesis in the North Atlantic for years.

An event of this magnitude would have had far-reaching effects on the marine ecosystem: for example, it was suggested that the oxygen production of the earth would have been reduced by 10 percent. Not only does DDT reduce photosynthesis, but it cannot be broken down or physically or chemically converted in any way by the phytoplankton and other organisms (this is one characteristic that makes it a good insecticide). Consequently it is merely stored at each level. Thus, when the DDT-enriched phytoplankton are eaten by the primary consumers, the 10 percent ecological efficiency mechanism goes into effect for the organic compounds but not for the DDT. At this level, too, it is simply stored. As DDT is transferred up the food chain, it occurs in increasingly higher concentrations as a result of this effect, which is called *biological magnification,* or biological amplification (illustrated in Figure 18-11). What happens is that, although small amounts of DDT may be given off through respiration and excretion, the bulk of it remains in each organism and is thus magnified up the food chain to become highly concentrated in the top carnivores, such as the diving predatory birds. For example, in the final few years before the 1972 ban, the brown pelican, which nested on the islands off the coast of southern California, had acquired so much DDT that this concentration interfered with the female's ability to deposit a well-calcified egg. Thus many of the eggs broke and the pelican population greatly declined. It is only since the ban that the pelican population is again building up.

It should be noted that DDT has also been found in the Adélie penguin and the skua bird, signifying that the pesticide has found its way into the Antarctic ecosystems as well. Consequently it is no surprise that DDT is also found in the fatty tissues of man throughout the world. Although DDT is currently banned in the United States, it is extensively used in many other parts of the world and is therefore well represented today in the marine environment and its organisms.

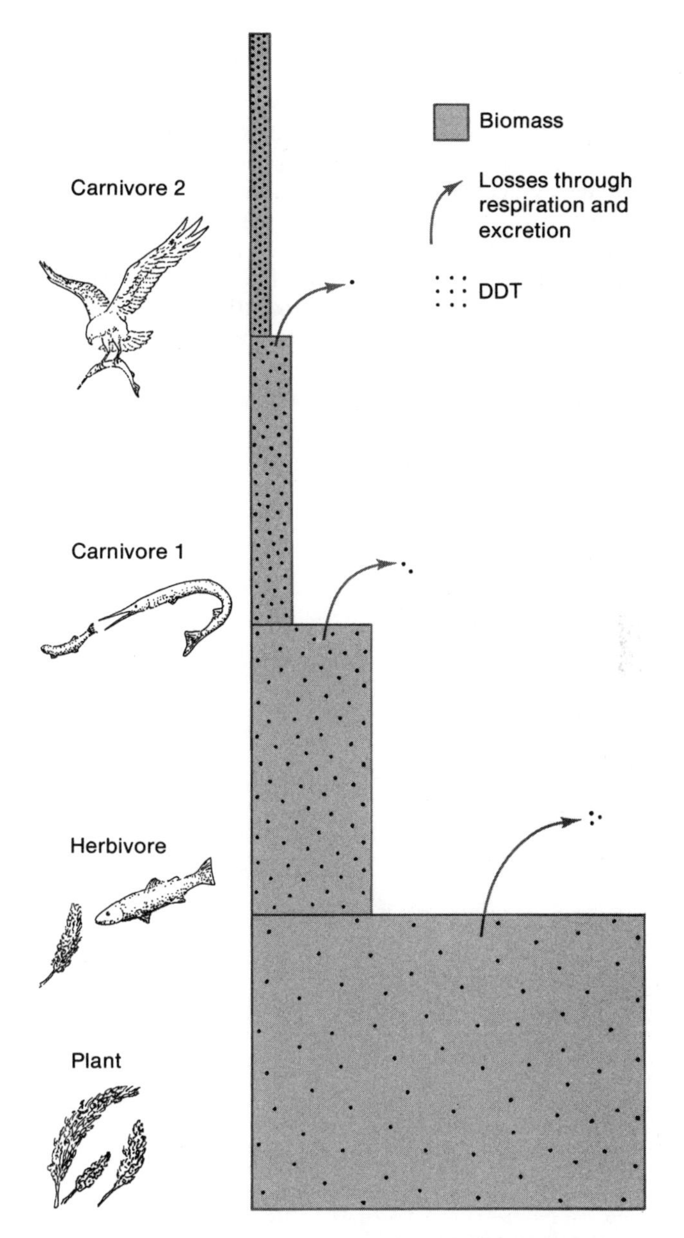

FIGURE 18-11 Biological magnification of DDT in a marine food chain. [After George M. Woodwell, "Toxic Substances and Ecological Cycles." Copyright © 1967 by Scientific American, Inc. All rights reserved.]

DEFINITIONS

Ecological efficiency. The efficiency with which energy is transferred from one trophic level to the next. Usually expressed as a ratio or percentage, it is the amount of living matter (biomass) added to a trophic level in comparison with the amount of living matter (food) required to produce it.

Photic zone. The depth zone within the sea (from surface to some depth) within which photosynthesis can occur. It may be considered the depth to which all light is filtered out except for about 1 percent.

Primary consumers. Organisms (herbivores) that eat the primary producers and occupy the second trophic level.

Primary producers. Organisms (herbivores) capable of converting enough inorganic matter into organic matter so that they are autotrophic (or self-nourishing) and occupy the first trophic level.

Secondary consumers. Organisms (carnivores) that eat the primary consumers and occupy the third trophic level.

Trophic level. A nourishment level in a food chain.

EXERCISE 18

REPORT

MARINE FOOD CHAINS AND NUTRIENT CYCLES

NAME ____________________

DATE ____________________

INSTRUCTOR ____________________

1. Referring to the discussion on open-ocean food chains and trophic pyramids in the first part of the exercise, suggest what would happen if for some reason the crop of diatoms was poor for a few years.

2. What might possibly cause a poor crop of phytoplankton (other than overgrazing by the krill)? Suggest a couple of physical oceanographic phenomena that could cause a lower standing crop (that is, a lower amount at any one time in a specific volume of water).

3. The skua bird preys on Adélie penguin eggs and juveniles. What would be the main effects of an overpopulation of skua birds on the following three populations?

   (a) The Adélie penguin.

   (b) The crabeater seal.

   (c) The leopard seal.

4. (a) Suppose that the plant–cricket–redwing blackbird pathway of the marsh (Figure 18-4) represents an isolated ecosystem. How might someone start a cricket plague in the Long Island estuary area?

Such a plague would effectively wipe out the ecosystem, but it would probably be reestablished with time by migration of the component parts.

(b) Now suppose the cormorants were removed from the larger coastal ecosystem (Figure 18-4). What would happen to the flukes and eels?

What would this removal do to the population of water plants?

What would happen to populations of osprey and merganser?

How might the water plants recover even if the cormorants did not?

(c) Keeping in mind your answers to part (b), how would you modify the Long Island estuary ecosystem in Figure 18-4 so that the following population effects would occur?

(i) The blowfish decrease in abundance but the fluke do not.

(ii) The terns decline in number, but the ospreys do not.

5. Considering the two food webs we have investigated (the open-ocean and nearshore systems), answer the following general questions about ecosystems.

(a) Do you think ecological upsets are more disturbing to the ecosystem if they occur at the lower trophic levels or the higher trophic levels, and why?

(b) Why is a more complex ecosystem (that is, one that has more potential pathways to the top of a given food web) probably more stable than a simple ecosystem?

6. Answer the following questions on ecological efficiency, relying on Figure 18-4 for *general concepts.* However, for purposes of this particular question, assume that the organic debris and the plankton depicted in the figure (one is a copepod and the other a dinoflagellate that eats other phytoplankton) are second-trophic-level creatures; the water plant and the marsh plants still belong to the first level.

(a) Rank all the birds according to their ecological efficiency. Thus one will be listed as the most ecologically efficient; it should gain its food by the shortest pathway (the fewest trophic levels) from the first trophic level. To determine how directly a species obtains its food, assume that each broad line represents one unit, and each thin line a half unit. For example, the tern gets one unit from the silversides and a half unit from the billfish. You will find that some birds will share a ranking.

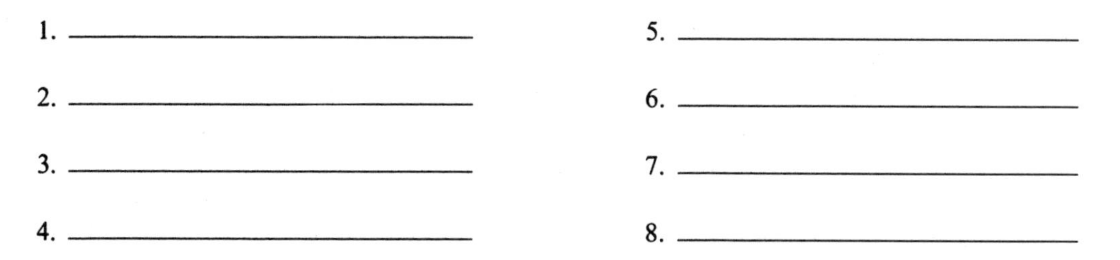

(b) All other factors (rate of reproduction, predation on each species, and so on), being equal, which would you consider to be the most abundant and the least abundant bird species in the Long Island estuary area, and why?

7. The study of Long Island estuarine ecology (Figure 18-4) was focused on the DDT problem. The numbers beside each species or species group in the figure indicate the parts per million of DDT found in each organism.

(a) What is the amount of biological magnification of DDT in going from the first trophic level to the second trophic level? (Remember that in this example the water plants and the marsh plants are the only first-trophic-level organisms for which we have DDT data, so that you should average these two magnifications.)

(b) Using the DDT data, state whether you think the blowfish eats more mud snails, water plants, or clams, and why.

(c) The tern and the merganser are birds of the fourth trophic level, but they contain quite different amounts of DDT. Can you think of two possible reasons for this difference? (Hint: The average age for a tern is about one-half the average age of a merganser.)

(d) Can you see any correlation between the ranking of Long Island estuary birds by ecological efficiency (which you did in Question 6) and the amounts of DDT in their tissues? If so, why would you think ecological efficiency and concentration of DDT were linked?

EXERCISE 19

PLANKTON

Plankton are floating organisms. In general they are **pelagic** organisms (organisms in the water column) that are unable to swim against a current of, say, 1 knot for an extended period of time. Pelagic animals capable of swimming are called **nekton.** Plankton may be classified in a number of ways. They can be grouped into **phytoplankton** (plant organisms) or **zooplankton** (animal organisms). They may also be classifed by habitat, such as **oceanic** (offshore) or **neritic** (nearshore) plankton, or **epipelagic** (surface to 100–200 meters) or **mesopelagic** (200 to about 1000 meters) plankton. Sometimes special terms are used for plankton occupying specific habitats: for example, the name *neuston* refers to organisms that live at the surface film of the water. Also, some plankton, the **holoplankton,** exist as such throughout life, whereas **meroplankton** are planktic for only a portion of their life cycle, as are clams at the juvenile stage or fish at the larval stage. Finally, plankton can be classified according to size, as in the following list.

megaplankton	larger than 2000 micrometers
mesoplankton	200–2000 micrometers
microplankton	20–200 micrometers
nannoplankton	2–20 micrometers
ultrananoplankton	smaller than 2 micrometers

In this exercise you will attempt to identify plankton and try to put them into a rough planktonic ecosystem diagram. First you will need some plankton. You should try to collect some marine plankton if possible. If you can't, preserved plankton may be purchased from biological supply houses. Should you purchase plankton from a supply house, we suggest that you purchase different types of samples so that you can see a variety of forms. If you can't get any marine plankton, freshwater plankton will do. When collecting freshwater plankton you should take tows in the largest and oldest natural body of water you can find. Streams are generally poor sources of plankton, most having only a few species of meroplankton, such as insect larvae. New bodies of water, like a cow pond, just haven't had enough time to acquire a good crop of plankton.

OBSERVATION OF LIVING PLANKTON

Observing living plankton is a fascinating experience. First try to observe living specimens by shaking the bottle of plankton (to mix and suspend the organisms) and pouring small portions into a few petri dishes or the like. Place the petri dish containing plankton on the stage of a dissecting microscope, with light coming from under the glass stage, and observe the plankton at low magnification (about 20 power). After focusing up and down you will probably find that most of the plankton are on the bottom of the dish, although some (those with more fat) may be on the surface at the air–water interface. Try focusing at different powers, but at high magnifications be prepared for rapid movement of the plankton to cause problems. Study the activity—for example, try to determine which organisms are eating which. This information will help you when you construct your plankton ecosystem diagram.

The rapid movement of the plankton can make it difficult to observe them. You will probably notice that many of them become more active the longer you look at them. The reason is that the microscope light is heating the water, causing them to move faster. You can slow them down by adding to the water cotton fibers or something similar that will trap them, or you can add gelatin or Protoslo to increase the viscosity of the water and impede their motion. Other chemical methods for slowing plankton movement should be avoided because they tend to kill the organism. Another interesting experiment is to drop in some ink and observe the currents made by the planktic **crustaceans** (shrimplike organisms) around their bodies. Some of these currents are feeding currents the crustaceans generate to bring food to their mandibles.

You may notice that some of the plant organisms, or phytoplankton, move as well. For example, many of the **flagellates** (phytoplankton with whiplike extensions) exhibit motion, and if your sample contains dinoflagellates (meaning "horrible whips") you may see them spiraling through the water. Even some of the **diatoms** move! Generally speaking there are two main types of diatoms; the round ones are called *centric* diatoms, and the elongated ones *pennate* or *raphate* diatoms. The elongated diatoms seem to glide along the bottom of the dish. They do so by creating waves of their cell membrane in a groove (the raphe) that runs between their two valves. Of the zooplankton, many of the ciliate **protozoans** (microorganisms with hairlike processes, or cilia, extending from the cell surface) tend to glide along, and the tintinids (ciliates with a vaselike shell) may dart around. Other small animal organisms that you may see are rotifers, or wheel animalcules; ostracods, which are crustaceans that appear to have a clamlike shell around them; cladocerans, another crustacean; and numerous meroplanktic forms of **benthic** and **nektonic** organisms that are illustrated in the plankton guide (Figure 19-1) that follows the discussion in this exercise.

Some larger zooplankton that you might see include copepods and the naupliar stages of copepods, mysids, shrimp, chaetognaths (glass or arrow worms), salps, and medusae of **coelenterates.** The copepods are common in most seawater samples, and it is interesting to note how they use their antennae to swim: in fact, the word copepod means oar foot. If chaetognaths are present you will probably find that they are eating everything else.

You should also attempt to see some micro- and nannoplankton, with the aid of a more highly powered light transmission microscope if possible. With an eye dropper, remove some of the water from the bottom of the sample jar. Place a few drops on a depression slide if available, or on a regular glass slide. Use a cover slip and observe under the microscope. If smaller plankton are present, they will appear to be moving rapidly owing to the small field of view, so that you will have to add something (gelatin or Protoslo) to slow them down.

OBSERVATION AND IDENTIFICATION OF PRESERVED PLANKTON

Be careful when working with preserved plankton samples, because some people are allergic to the liquid preservative (formalin). You may prefer to view the preserved plankton in a petri dish with a cover on it. In any case, be sure that the room is well ventilated. A covered petri dish has a tendency to fog, but this can be alleviated to some extent by rubbing mineral oil on the inside of the top cover. If you can't tolerate the smell, pour the sample through the same net you used to collect it and wash with a little fresh water. To observe nannoplankton from a water-bottle sample, allow the sample to stand for a few hours and then remove a few drops from the bottom of the bottle. Observe the plankton, looking for the various characteristics described in the discussion on live samples. Refer to Figure 19-1, or to a field guide.

DENSITY CALCULATION

For quantitative work—determining the quantity of plankton in a volume of water—shake the preserved sample vigorously, remove a known volume (10–50 milliliters), and allow to stand for an hour. Carefully decant off the liquid except for a few milliliters; pour these into a small petri dish and observe under high magnification (100 power or more). If you know the

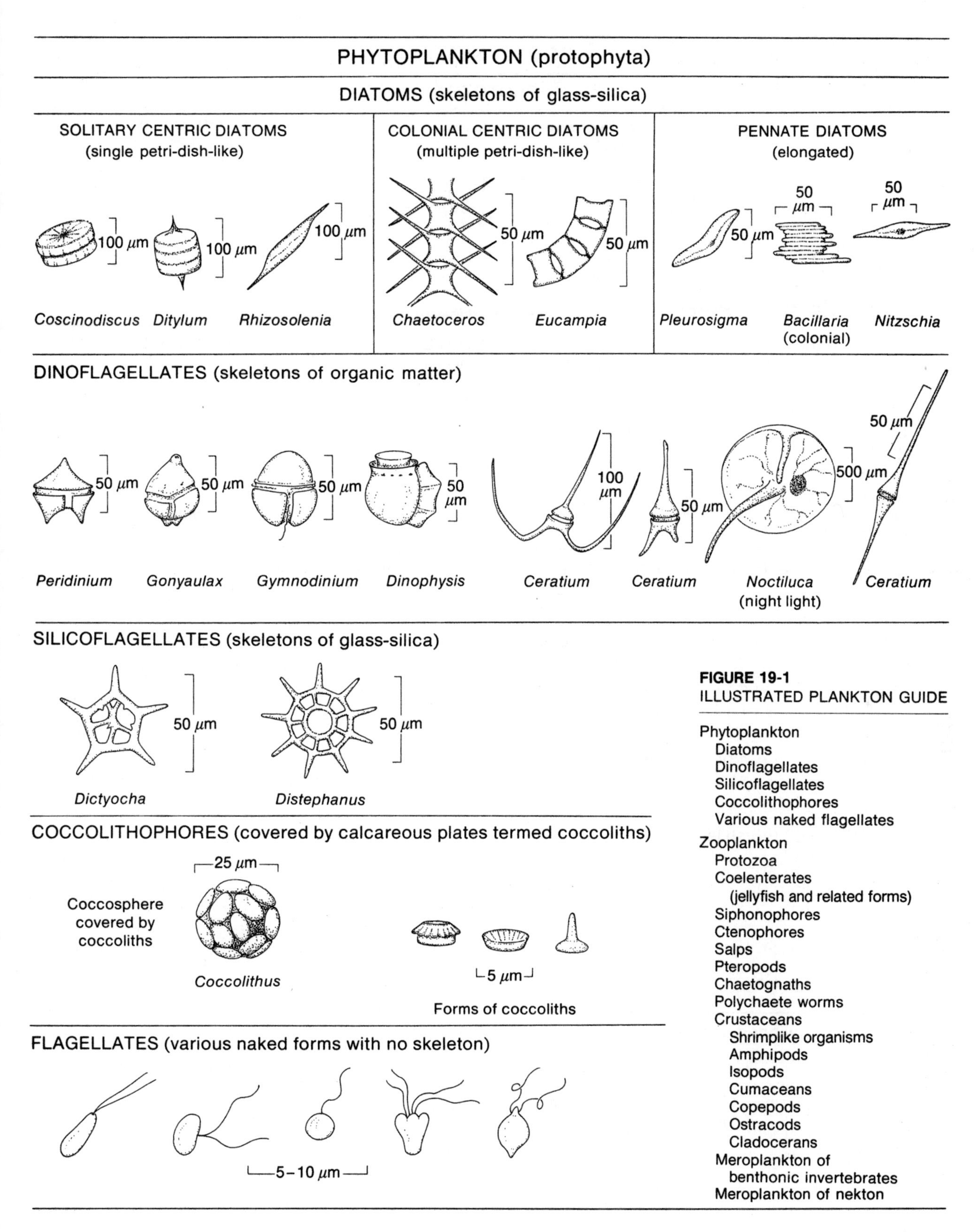

FIGURE 19-1
ILLUSTRATED PLANKTON GUIDE

Phytoplankton
- Diatoms
- Dinoflagellates
- Silicoflagellates
- Coccolithophores
- Various naked flagellates

Zooplankton
- Protozoa
- Coelenterates (jellyfish and related forms)
- Siphonophores
- Ctenophores
- Salps
- Pteropods
- Chaetognaths
- Polychaete worms
- Crustaceans
  - Shrimplike organisms
  - Amphipods
  - Isopods
  - Cumaceans
  - Copepods
  - Ostracods
  - Cladocerans
- Meroplankton of benthonic invertebrates
- Meroplankton of nekton

(continued)

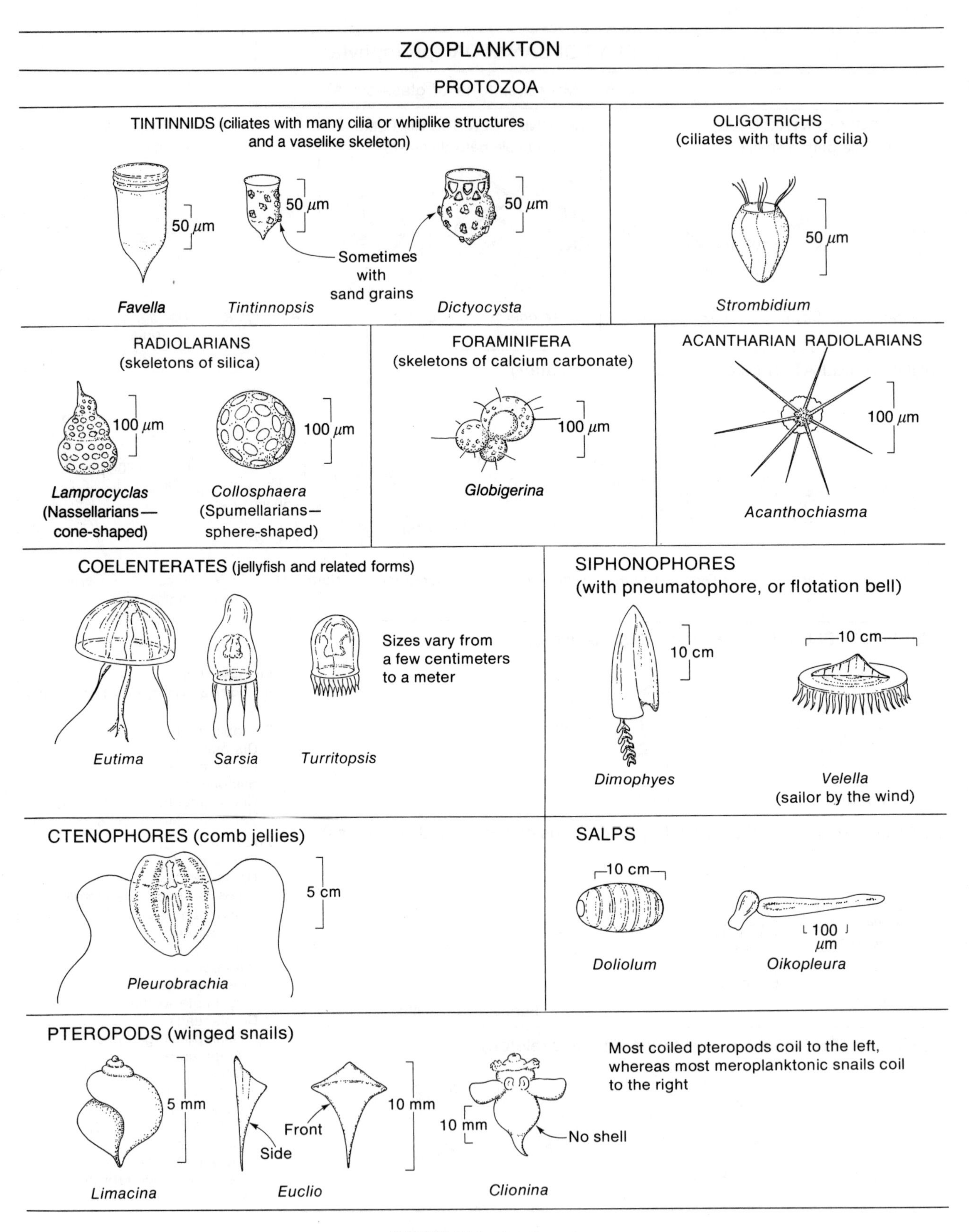

FIGURE 19-1 *(continued)*

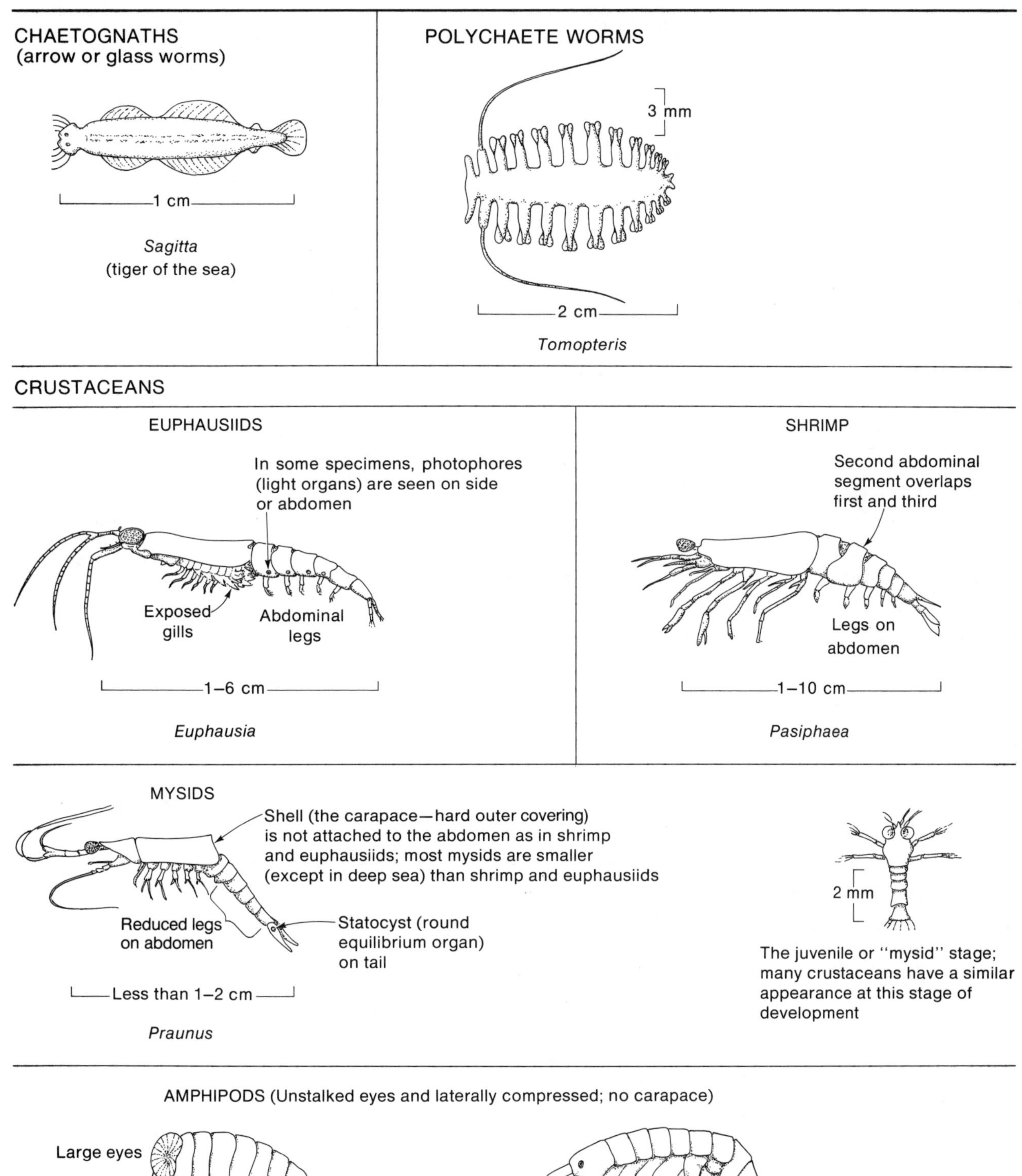
CHAETOGNATHS
(arrow or glass worms)
1 cm
Sagitta
(tiger of the sea)
POLYCHAETE WORMS
3 mm
2 cm
Tomopteris
CRUSTACEANS
EUPHAUSIIDS
In some specimens, photophores (light organs) are seen on side or abdomen
Exposed gills
Abdominal legs
1–6 cm
Euphausia
SHRIMP
Second abdominal segment overlaps first and third
Legs on abdomen
1–10 cm
Pasiphaea
MYSIDS
Shell (the carapace—hard outer covering) is not attached to the abdomen as in shrimp and euphausiids; most mysids are smaller (except in deep sea) than shrimp and euphausiids
Reduced legs on abdomen
Statocyst (round equilibrium organ) on tail
Less than 1–2 cm
Praunus
2 mm
The juvenile or "mysid" stage; many crustaceans have a similar appearance at this stage of development
AMPHIPODS (Unstalked eyes and laterally compressed; no carapace)

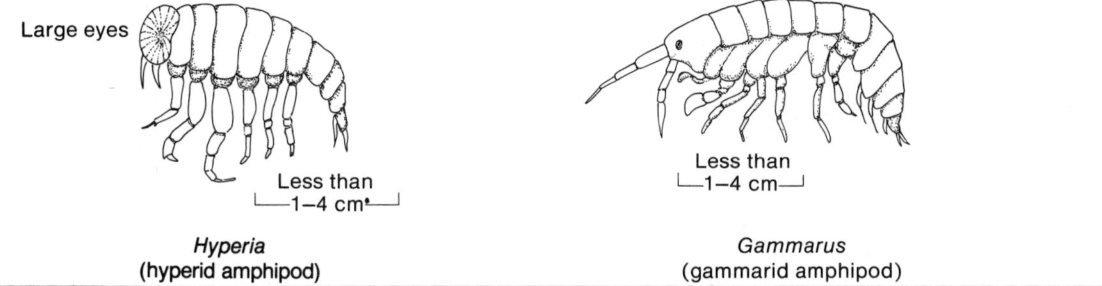
Large eyes
Less than 1–4 cm
Hyperia
(hyperid amphipod)
Less than 1–4 cm
Gammarus
(gammarid amphipod)

(continued)

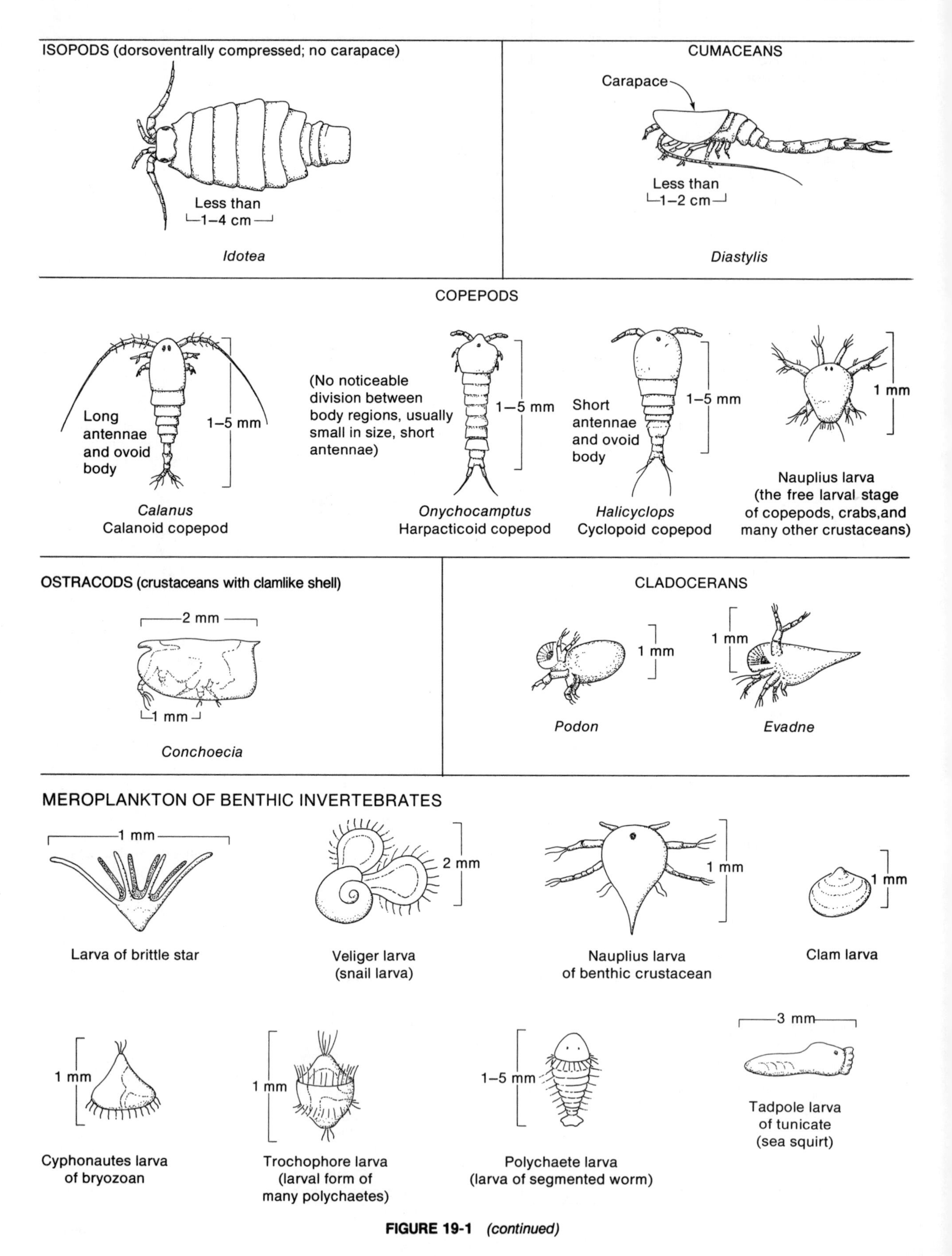

FIGURE 19-1 *(continued)*

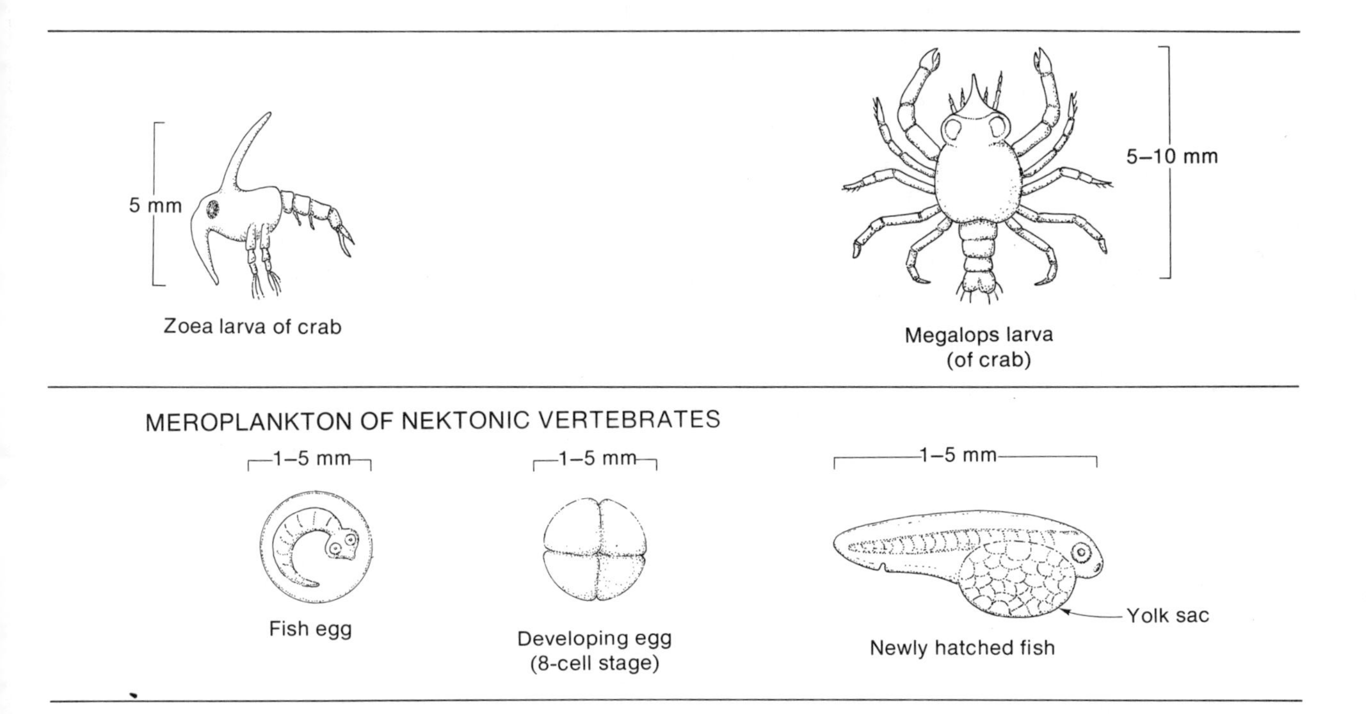

volume of water originally sampled, you can calculate densities in number of plankton per liter or per milliliter of water. To obtain truly quantitative information on plankton abundance (if you want to have this information on your plankton counting sheet in the report), you need either to split the sample into smaller portions (so that you examine a percentage of the original sample) or to observe the entire sample.

COUNTING THE PLANKTON

Now try to identify and count the different plankters from your preserved sample (quantitatively, if you wish, as described in the preceding paragraph). If you do not have field keys or other literature, use the illustrated plankton guide provided in Figure 19-1. Note your observations on the counting sheet provided in Question 1 in the report. Identify by species if possible; if not, describe as precisely as you can, to the lowest taxonomic unit you can determine (such as centric diatom #1), and draw a rough sketch of each one. Don't worry if you can't identify everything, neither can your instructor! Try also to count at the species level: for example, "23 individuals of centric diatom #1," "13 of centric diatom #2," and so on.

At the top of the counting sheet is space for information on where, when, and how the sample was taken; for example, you should specify what kind of net was used, the mesh size, how much water was filtered, and the exact locale and depth. However, if you are using material from a biological supply house, you probably do not have these data, and so you should merely note as much information as you can on the species observed and the numbers of each. If you *have* taken planktonic tows, and have taken more than one (for example, at different locations, different depths, or different times of day), note the results from each on separate counting sheets, compare them, and see if you can explain any variations. For example, a nearshore sample may have more meroplankton than an offshore sample, or the biomass near a sewer outfall may be greater and the diversity of life less than will be found for a sample farther away from the outfall. Try to specify the standing crops of the different groups by determining the volume of water filtered and the number of each group per cubic meter, or kiloliter (if you have split the sample, remember to take this into account by multiplying by the appropriate number). To do so (without a current meter), assume you have had 100 percent filtering efficiency (there was no head of water at the mouth of the net) and determine the volume filtered by multiplying the surface area of the mouth of the net by the distance over which the net was towed.

DIAGRAMMING A PLANKTON ECOSYSTEM

You have already studied marine ecosystems or food webs in Exercise 18. In Question 2 of the report you will diagram a planktonic ecosystem, or planktonic food web, by placing your species in what you would consider their right niche (that is, the organism's place, or feeding level, in the ecosystem). For example, the base of the ecosystem would consist of the primary producers, such as diatoms, dinoflagellates and so on; thus if you have 10 phytoplankton species, put them down along the base line. This is your first trophic level. (For information on marine food webs and trophic levels, see Exercise 18.) For the second trophic level, the primary consumers, put down those organisms you think should belong there, such as copepods. Construct as many trophic levels as you think exist in your sample.

Remember that your sample is probably not a *closed* ecosystem, one that is self-sufficient and reasonably uninfluenced by other organisms. Rather, you probably have an *open* ecosystem, in which the real top carnivore is a fish, or other animal that wasn't in the collection; alternatively you may not have primary producers if a coarse net was used. In your diagram you should add what other organisms you would expect to be present if the ecosystem were indeed closed. Remember also that any ecosystem consists of more than the living organisms observed within it. Incorporate into your diagram, with a colored pencil or in parentheses, those inorganic ingredients necessary for the primary producers (such as light and nutrients).

DEFINITIONS

Benthic organisms. Organisms that live on or in the bottom of an aquatic environment.

Coelenterates. Multicellular simple animals such as jellyfish.

Crustaceans. Animals with an exoskeleton (outside skeleton) and jointed appendages with antennae, head appendages for grasping and chewing, and compound eyes (many eyes collected into one or more structures or eyes).

Diatoms. Single-celled plants with a siliceous (glass) skeleton.

Epipelagic plankton. Floating organisms in the epipelagic region (surface to 100 to 200 meters).

Flagellates. Single-celled plants or animals possessing whiplike locomotory appendages.

Holoplankton. Organisms that are floating throughout life.

Meroplankton. Organisms that are floating for only part of life.

Mesopelagic plankton. Floating organisms in the mesopelagic region (below the epipelagic region, 100 or 200 meters, to about 1000 meters).

Nektonic organisms. Pelagic organisms that swim in the water.

Neritic plankton. Floating organisms living in the waters over the continental shelf.

Oceanic plankton. Floating organisms living in the waters beyond the continental shelf (all other waters).

Pelagic organisms. Organisms living in the water column.

Phytoplankton. Floating plant life.

Protophyta. Single-celled plants.

Protozoans. Single-celled animals.

Standing crop. The number of organisms, or the biomass, per volume or unit area at any instant in time.

Zooplankton. Floating animal life.

EXERCISE 19

REPORT

PLANKTON

NAME ______________________

DATE ______________________

INSTRUCTOR ______________________

1. In the plankton counting sheet on page 199, give the species name under the appropriate category, if possible, and/or draw and count each plankter. *Try to work on the species level:* even if you can't name the species, you can designate by numbers (for example, centric diatom #1, centric diatom #2, and so on).

2. (a) Construct a diagram of a planktonic ecosystem by placing the name or a drawing of each species in your sample in what you think is its correct trophic level. Diagram for as many trophic levels as you think exist in your sample. Space is provided for six levels; however, you may find either that you have to add others or that you may not be able to diagram for as many as six; you may find you have no more than three levels.

______________________________ Top carnivore or carnivores (trophic level 6)

______________________________ Third-level carnivores (trophic level 5)

______________________________ Second-level carnivores (trophic level 4)

______________________________ First-level carnivores (trophic level 3—carnivores)

______________________________ Herbivores (trophic level 2—primary consumers)

______________________________ Primary producers (trophic level 1—plants)

(b) Now draw in energy, or feeding, pathways from one species to another, showing the various predator–prey relationships. If you are sure of a particular pathway—for example, if you saw the copepod eat a centric diatom, or see a copepod inside a coelenterate—indicate the connection between the two with a *solid* arrow, the head showing the direction of energy flow. If you are not sure of a particular pathway but have good reason to think that it exists, draw a *dashed* arrow to indicate high probability, and a *wiggly* arrow to indicate possibility only.

(c) Assuming that your ecosystem is an open one, add the other organisms that you think might be present if it were closed. Use a pencil of a different color.

(d) Complete your diagram of this ecosystem by indicating below trophic level 1 those inorganic ingredients necessary for the primary producers (such as light and nutrients). Note, in the space below the diagram or with an arrow at the side, what happens within the various trophic levels as the organisms at each die (uneaten) and are returned as nutrients by the decomposers (bacteria and other organisms that return the carcasses and fecal material as basic inorganic compounds that can be used by the first trophic level).

3. (a) In the course of your observations you will have obtained some information on the standing crops of the various trophic levels: that is, you will have a general idea of the quantities of organisms at each level, even if you don't know the densities in numbers per liter or milliliter. Assuming that you have efficiently collected a true sample of the community, make a general comparison of the standing crops of each level. (For example, are there more primary producers than herbivores, more herbivores than primary carnivores, and so on?) If you have quantitative data, calculate densities—for example, 200 primary producers per cubic meter, or kiloliter. Your instructor will tell you how to register this information.

(b) In general only about 10 percent of the potential energy contained in the original food is passed on to the consumer at the next trophic level. Is this what your diagram of the ecosystem shows? If not, give some reason.

(c) Planktonic organisms have special adaptations that enable them to cope with their environment, such as appendages to aid in flotation. List as many specific examples of planktonic adaptations as you can from your own observations of the plankton. (Be specific—describe the organ itself and specify its function.)

PLANKTON COUNTING SHEET

STATION__________ LOCATION (locale and depth)__________

EQUIPMENT USED__________ DATE AND TIME OF COLLECTION__________

AMOUNT OF TOTAL SAMPLE COUNTED (in density calculation)__________

PHYTOPLANKTON		
Diatoms	Dinoflagellates	Other phytoplankton

ZOOPLANKTON		
Protozoa	Coelenterates	Crustaceans (adults)
Crustaceans (larvae)	Other holoplankton	Meroplankton

EXERCISE 20

MARINE ADAPTATIONS AND THE DEEP-SEA ENVIRONMENT

The processes by which an organism apparently becomes better suited to its environment or for particular functions within that environment are termed adaptations. These adaptations may be structural, biochemical, behavioral, or other, and they develop in the course of geologic time in response to physical and biological stresses. We will be investigating mainly those structures or activities of marine organisms that tend to equip them better for life in the marine environment.

An **adaptation** is a modification of a previous structure, biochemical process, or behavioral pattern that occurs in the course of evolution. Commonly known examples of adaptations in terrestrial animals are hooves, which equipped horses and cattle for roaming and feeding on grasslands, the prehensile tail enabling some species of monkey to live in an arboreal or tree habitat, and the long neck that allows the giraffe to feed on leaves from the scattered trees of the savanna region. However, most of us are less familiar with adaptations of marine organisms, especially some of the most interesting—those of the deep sea.

Some organisms have traits that are used one way in one environment but are suitable for use in another way in another environment: these are termed **preadaptations,** because they are adaptations that have already equipped the organism to exist in one environment and tend to make it preadapted for another environment. A marine example is the brittle star. Although those existing in the deep sea look much the same as their shallow-water relatives, they have been preadapted to the deep-sea environment. Their major preadaptations are the long appendages and the ability to feed on detritus, or disintegrated matter (including mud). In the report section you will be asked to explain why these characteristics are good preadaptations for life in the deep-sea benthic environment.

THE OCEAN ENVIRONMENTS

For many species, both marine and terrestrial, it is possible to compare the ancestral stock with the descendants, observe what the adaptations are, and determine their function if we know the environmental conditions of the ancestral stock and the environmental conditions of the adapted descendants. In this discussion, we will be concerned mainly with the adaptations of deep-sea organisms that exist in the different marine environments that prevail at different ocean

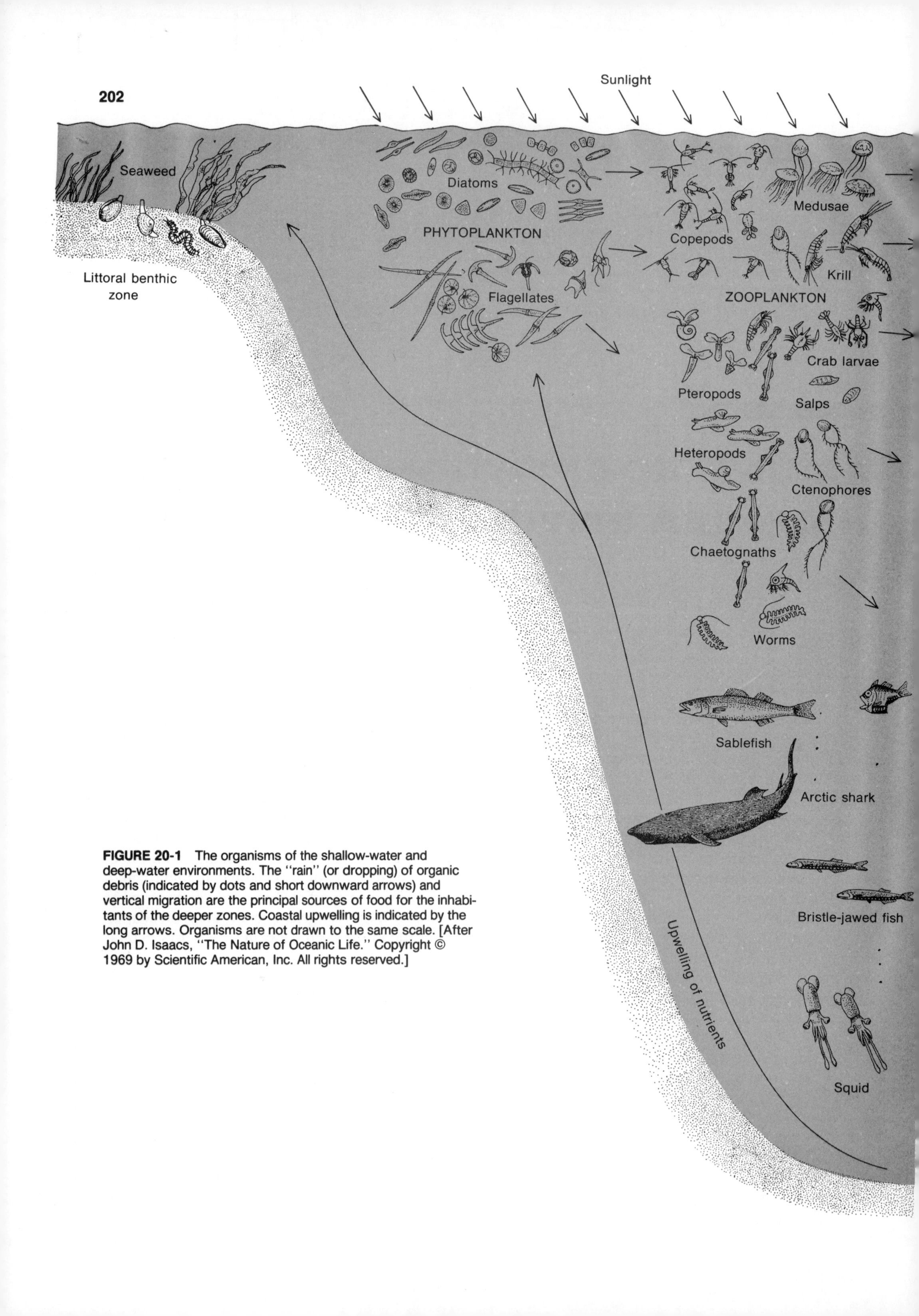

FIGURE 20-1 The organisms of the shallow-water and deep-water environments. The "rain" (or dropping) of organic debris (indicated by dots and short downward arrows) and vertical migration are the principal sources of food for the inhabitants of the deeper zones. Coastal upwelling is indicated by the long arrows. Organisms are not drawn to the same scale. [After John D. Isaacs, "The Nature of Oceanic Life." Copyright © 1969 by Scientific American, Inc. All rights reserved.]

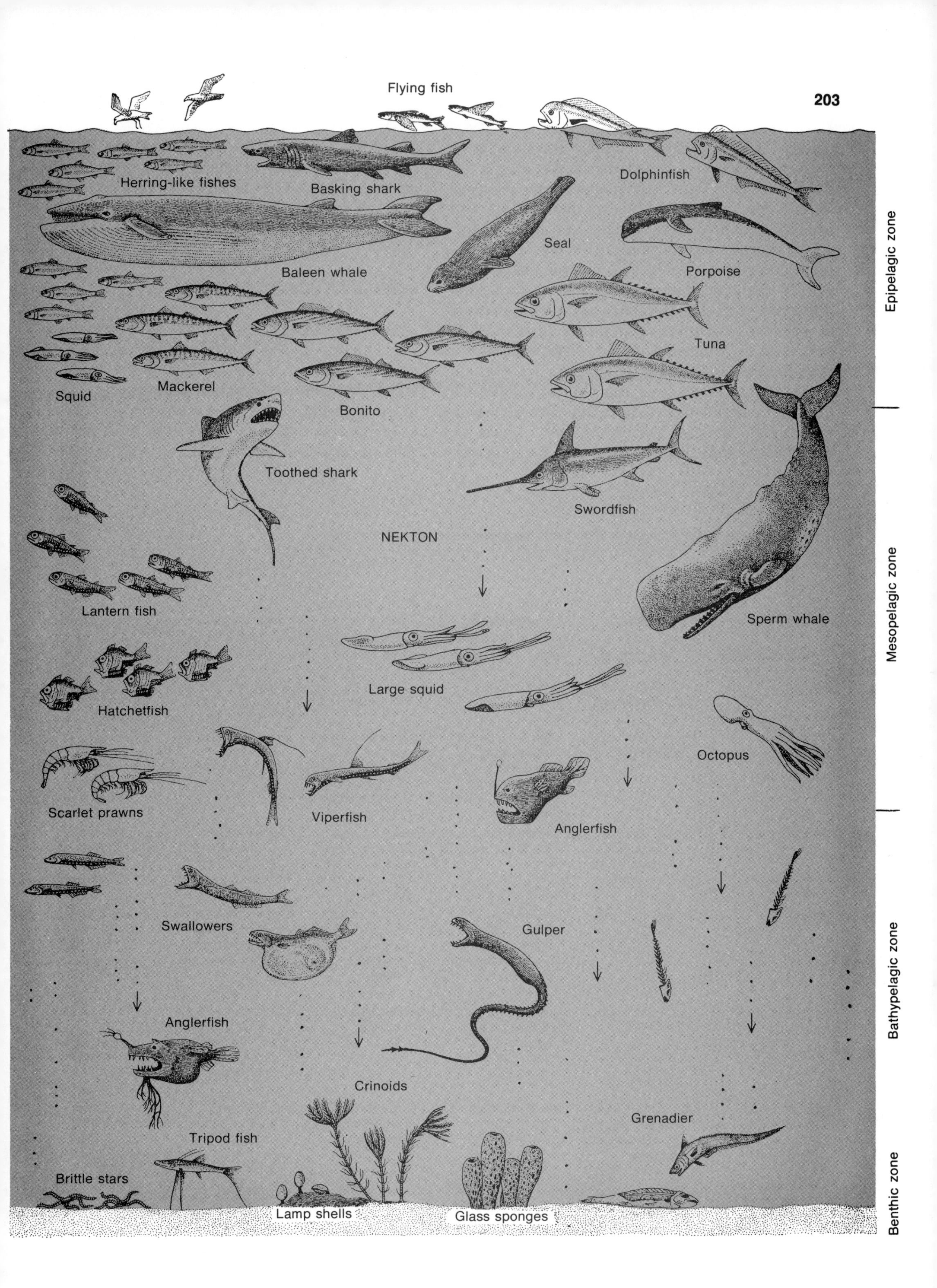
Flying fish
Herring-like fishes
Basking shark
Dolphinfish
Seal
Baleen whale
Porpoise
Tuna
Squid
Mackerel
Bonito
Toothed shark
Swordfish
NEKTON
Lantern fish
Sperm whale
Large squid
Hatchetfish
Octopus
Scarlet prawns
Viperfish
Anglerfish
Swallowers
Gulper
Anglerfish
Crinoids
Grenadier
Tripod fish
Brittle stars
Lamp shells
Glass sponges
Epipelagic zone
Mesopelagic zone
Bathypelagic zone
Benthic zone

depths. These organisms (or their ancestors) have evolved from shallow-water marine organisms. By comparing them with their shallow-water relatives, we can determine what these adaptations are and what they are used for in the deep-sea environment.

The characteristics of the various ocean environments are given in Table 20-1, and the information will aid you in determining some of the reasons that different species evolved in the way they did in order to survive in a particular environment. First, the shallow-water environment is the epipelagic zone shown in Figure 20-1 and listed in the table (for a description of the marine life zones, see also Exercise 15). We will assume this shallow area to be the ancestral zone containing the "standard" organisms from which the deep-water species evolved. But even in the epipelagic zone itself, we can find examples of adaptations from the "standard" fish type: one is the flying fish with its adaptive modification of fins to "wings" and its behavioral modification of "flying"; plankton, too, display adaptations to their environment.

DEEP-SEA ADAPTATIONS

Among the adaptations for the deep-sea environment, an important one is the ability to **bioluminesce** (to create and use biological light). Some organisms such as squid squirt an ink that luminesces as it comes in contact with the dissolved oxygen in the seawater. Other marine organisms have even more sophisticated mechanisms for bioluminescence, the **photo-**

TABLE 20-1
The characteristics of the marine environments (temperate and tropical regions)

	Zones				
Characteristics	Epipelagic (0 to 100 or 200 meters)	Mesopelagic (100 or 200 to about 1000 meters)	Bathypelagic and deeper (about 1000 meters to bottom)	Shallow benthic (water over the shelf)	Deep benthic (water beyond the shelf)
Degree of illumination	Enough for photosynthesis	Twilight zone	Essentially no illumination	Many portions lighted	Essentially no illumination from above
Food supply	Primary productivity occurring	Little or no primary productivity; organisms migrate up to food or wait for it to fall	Little or no primary productivity; organisms migrate up to food or wait for it to fall	Primary productivity occurring	No primary productivity except for chemosynthesis; organisms wait for food to fall
Temperature	Usually about 28°C to about 10°C; sometimes near 0°C in winter	Usually from about 15°C to about 5°C	Usually between 5°C and −2°C; usually down to 1°C or less below 4000 meters	Usually about 30°C to about 10°C, and down to freezing at times	Usually between 15°C and −2°C; usually down to 1°C or less below 4000 meters
Salinity	Usually from about 37 to 32‰	Usually from about 35–34.5‰; intermediate waters from high latitudes less saline	Usually from about 35–34.5‰ and about 34.52‰ below 4000 meters	Usually between about 40‰ and 30‰ with some freshwater runoff	Usually from about 35–34.5‰ and about 34.52‰ below 4000 meters
Oxygen content	Usually from about 7 to 3.5‰	Usually from about 5 to 4‰, with values of less than 1 in oxygen minima	Usually from about 6 to 5‰	Usually from about 7 to 3.5‰, with some supersaturation and some anoxic regions	Usually from about 6 to 4‰, with near anoxic conditions in oxygen minima and in isolated basins
Nutrient content (phosphate given for pelagic environments, and general organic carbon for the benthic environments)	Usually from about 0 to 30 milligrams per cubic meter; higher in upwelling regions	Usually from about 30 to 90 milligrams per cubic meter higher in upwelling regions	Usually about 90 milligrams per cubic meter	Usually high in shallow benthic sediments	Usually low in deep benthic sediments, but high under upwelling regions

FIGURE 20-2 Deep-scattering-layer organisms include crustaceans and fishes. Among them are (top to bottom) a euphausiid, a sergestid, and two forms of myctophid, or lantern fish. They range in size from about an inch for the euphausiid to as long as 3 inches for the lantern fish. The four spots on the euphausiid and those on the myctophids are photophores. [After Robert S. Dietz, "The Sea's Deep Scattering Layers." Copyright © 1962 by Scientific American. All rights reserved.]

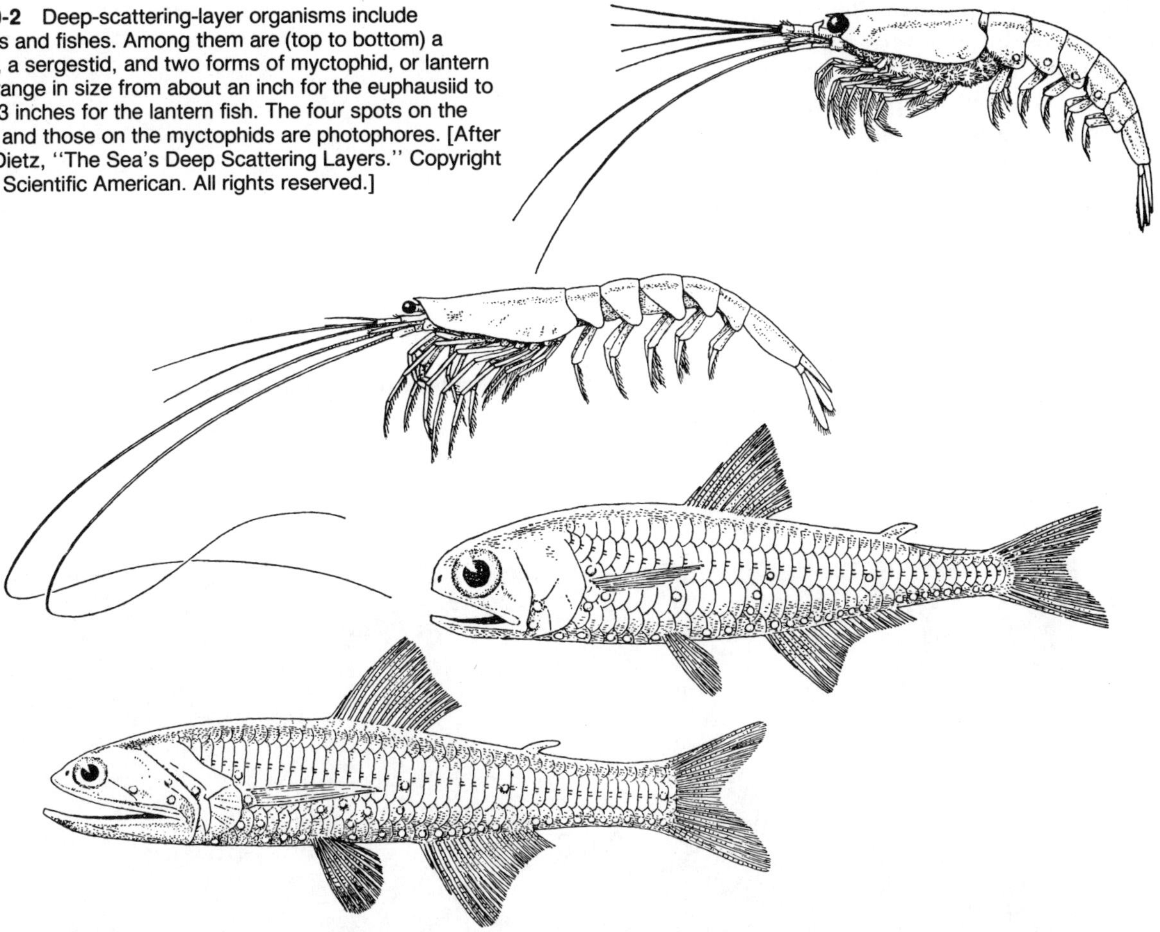

phores, or light organs. Some of these photophores are really nothing more than collections of luminescent bacteria but others are highly evolved organs in which the light can be turned off and on. Many deep-sea fish use photophores for a variety of tasks. For example, the myctophid, or lantern fish, illustrated in Figure 20-2, swims in schools that migrate vertically through the water column, usually a few hundred meters every day. As the figure shows, these fish have large numbers of photophores. On this one species, the photophores probably have a number of different functions, and you will be asked to suggest some in Question 5 of the report section.

Numerous other important structural adaptations can be seen in various deep-sea fish. One is the lack of scales on many species (scales assist the organism in swimming). Another is poorly developed musculature. Also, some of the deep-sea fish have developed a pattern whereby the male becomes totally dependent on the female for its existence, and many of its systems, except the reproductive system and a few others, degenerate.

Adaptations can also be biochemical or behavioral. A good example of a behavioral adaptation is that of **vertical migration.** Many kinds of marine organisms migrate vertically. Even some of the phytoplankton make short vertical migrations in the water column; an example is shown in Figure 20-3. You will notice that the phytoplankton cluster during the day (in response to the light) and disperse, or "wander," at night (when there is no light to which they can orient themselves).

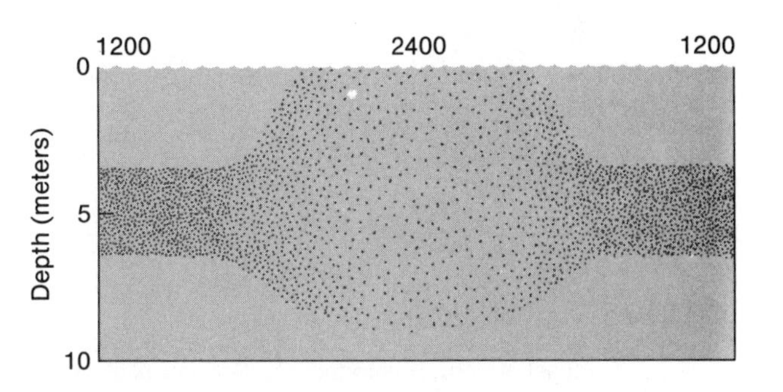

FIGURE 20-3 Phytoplankton migration in the course of a 24-hour period. Note that the plankton are spread out during the nighttime hours and packed together during the day. Hours of the day are noted in accord with the 24-hour maritime system.

One migration pattern that has been well studied is the vertical migration of the copepod *Calanus finmarchicus,* a herbivore and member of the second trophic level. The diatoms it eats are in the upper 50 meters of the water column. In turn it is eaten by herring and similar species of fish that also occupy the upper 50 meters of the water. Figure 20-4 illustrates the migration pattern of *Calanus finmarchicus* at one of its life stages. Notice that the population clusters at noon of one day, migrates toward the surface at night, disperses throughout the water column at midnight, starts to reassemble at dawn except for some of the individuals at deeper levels, and return to its so-called daytime depth at noon.

The phytoplankton migration and the *Calanus fin-*

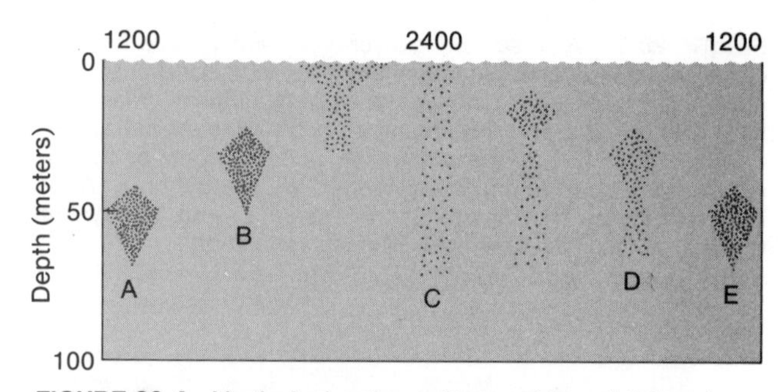

FIGURE 20-4 Vertical migration patterns of the adult female copepod of the *Calanus finmarchicus.* The example demonstrates a pattern of migration whereby the population of copepods (A) clusters at midday, (B) migrates toward the surface at night, (C) disperses throughout the water column at midnight, (D) starts to reassemble at dawn, (E) moves to its daytime depth at noon.

FIGURE 20-5 Deep scattering layers are well developed in this echogram made in the deep Pacific off the coast of Peru. [Photograph by Scripps Institution of Oceanography.]

marchicus migration are usually termed *shallow-water vertical migrations.* Another interesting migration is that of an entire community of various organisms from fairly deep water (say, 500 meters) to the surface and back again. The density of these organisms is so high that they may reflect the sound waves sent out by bottom-finding oceanographic equipment, and be erroneously recorded on the receiver as the bottom (as they have been many times). Because these communities reflect the sound waves, occur in more than one layer, and frequently descend to deep levels, they have been given the name *deep scattering layers,* or DSL's. An actual recording of a migration of DSL's off the coast of Peru is shown in Figure 20-5.

We do not know all the organisms that constitute the DSL's, but we do know that euphausiids, sergestid shrimp, and myctophids, or lantern fish (all shown in Figure 20-2) are common components of DSL's throughout much of the world ocean. The myctophids eat copepods (such as *Calanus finmarchicus*), and they in turn are eaten by squid, larger fish, and, in the polar regions, fur seals. This vertical food chain is illustrated in Figure 20-6.

DEFINITIONS

Adaptation. The process by which an organism becomes better suited to its environment.

Bioluminescence. Emission of light from living organisms.

Photophores. Specialized organs capable of bioluminescence.

Preadaptation. An adaptation for one environment or one aspect of that environment that is also suited for another environment or another aspect of that environment.

Vertical migration. An active (not passive) sustained movement up or down.

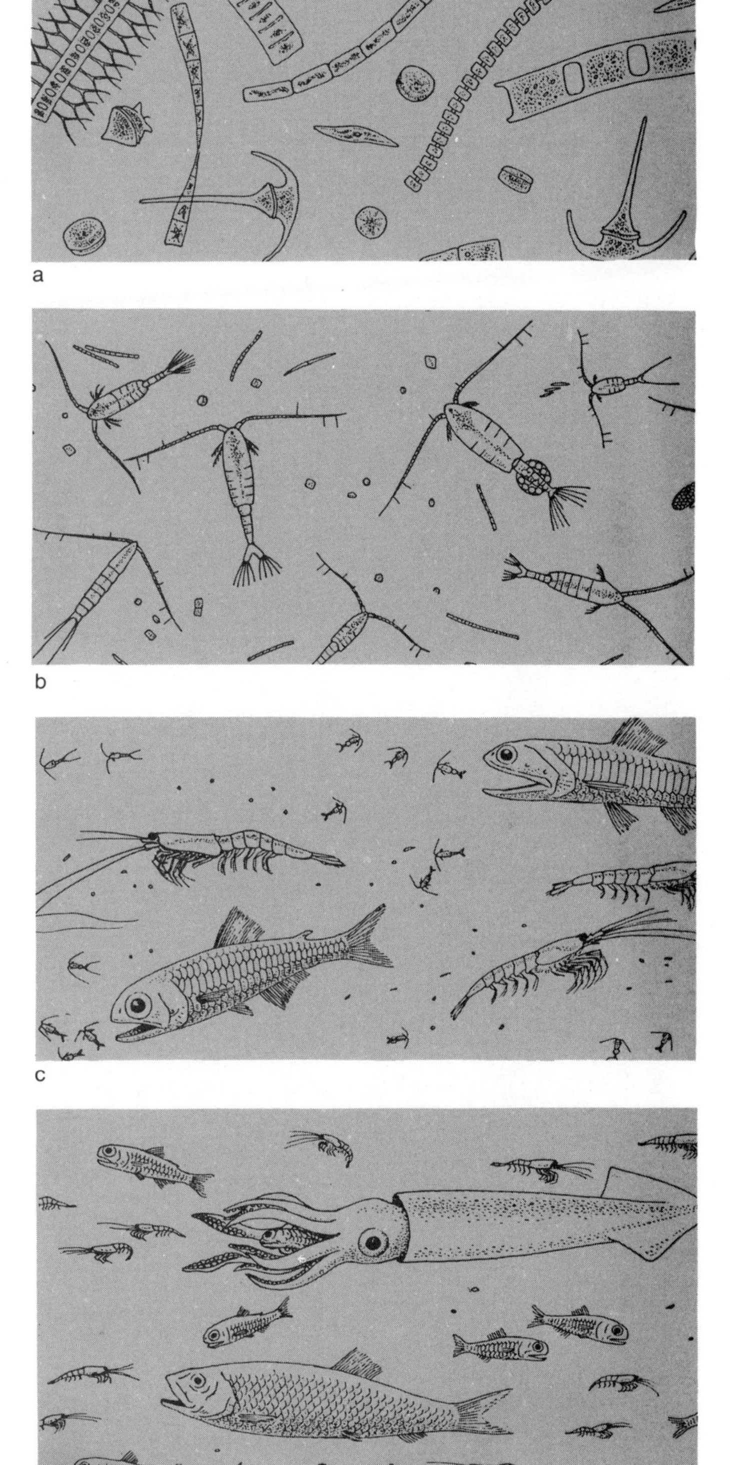

FIGURE 20-6 The deep scattering layer is an important link in this ocean food chain. (a) Phytoplankton are the "grass" of the sea. These are diatoms and dinoflagellates. (b) Copepods feed on phytoplankton and in turn furnish food for larger animals. (c) "Deep scatterers" rise from depths at night to feed in the epipelagic zone. The myctophids (lantern fish) eat copepods; the euphausiids consume the phytoplankton. (d) Larger animals, such as squid and herring, sometimes eat the organisms of the deep scattering layers at night near the surface. Fur seals are known to feed extensively on myctophids. [After Robert S. Dietz, "The Sea's Deep Scattering Layers." Copyright © 1962 by Scientific American, Inc. All rights reserved.]

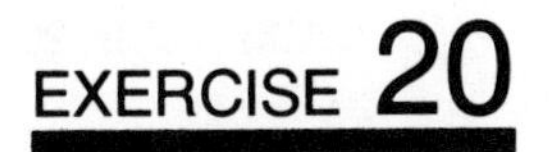

REPORT

MARINE ADAPTATIONS AND THE DEEP-SEA ENVIRONMENT

NAME ______________________

DATE ______________________

INSTRUCTOR ______________________

1. (a) Like other marine organisms, plankton also display adaptations to their environment. Referring to Figure 20-1, list one morphologic adaptation for each of the following plankters that helps it to float or remain suspended.

Diatom

Flagellate

Copepod

(b) Why do the diatoms and flagellates need to stay at the surface or at least in the shallow epipelagic zone?

(c) Why might a certain species of copepod need to stay at the surface or in the epipelagic zone?

2. For the following four species, (a) list two adaptations for depth and (b) give the function of each adaptation, specifying the condition of depth to which the adaptation evolved. Refer to Figure 20-1 and Table 20-1 for your answers.

	Adaptation for depth	Function
Lantern fish	________________	________________
	________________	________________
Anglerfish	________________	________________
	________________	________________
Swallowers	________________	________________
	________________	________________
Hatchetfish	________________	________________
	________________	________________

3. Brittle stars are a dominant organism in the abyssal benthic environments of the world ocean. As mentioned in the text, their main preadaptations are the long appendages and the ability to feed on detritus. Suggest why these two characteristics are especially good preadaptations for life in the deep-sea benthic environment.

4. After an examination of Table 20-1, where would you expect the greatest biomass of deep-sea benthos to be found?

   (a) Consider the three environments listed below, and rank them by benthic biomass from greatest to least.

   Shallow benthic

   Deep benthic under upwelling

   Deep benthic not under upwelling

   (b) Explain your answer.

5. Answer the following questions on bioluminescence:

   (a) Some organisms squirt an ink that luminesces as it comes in contact with the dissolved oxygen in seawater. Suggest the purpose of this biochemical adaptation.

   (b) The photophores on a given species of fish exhibit a species-specific pattern (that is, a pattern unique to that species). In fact, these patterns are so distinctive that fish taxonomists can use them to identify the various species of fish. Of what use might this specific photophore pattern be to the fish?

   (c) Not only does each species have a characteristic photophore pattern, but male and female members of the same species usually have slightly different patterns. Why?

(d) On the lantern fish, light organs are mainly in one region of the body (Figure 21-2). Suggest a reason for this particular concentration. (Note: The hatchetfish shown swimming under the lantern fish in Figure 20-1 are often found in this position, looking for the outlines of lantern fish so that they can attack them.)

(e) There are probably many other reasons for photophores. Provide brief answers to the questions below.

(i) . Some anglerfish have a light organ (the only one they possess) at the end of their lure. Why?

(ii) Some bottom-dwelling fish in the deep sea have a large aggregation of photophores around their heads and also very large eyes. Why?

6. Suggest what the following adaptions might be used for and to what characteristic of the deep sea they are adapted (such as that food is in scant supply).

(a) Lantern fish have scales, but many fish living in the deeper environments do not. Why?

(b) Why do many of the deep-sea fish have poorly developed musculature?

(c) Many of the deep-sea fish pass their juvenile stages in shallow water. What would be the advantages of this?

(d) In many deep-sea species, the male becomes totally dependent on the female, and only its reproductive system continues to function fully. What is this an adaptation for, and why would you think it is a necessary adaptation for the deep-sea environment?

7. In phytoplankton vertical migration (Figure 20-3), the organisms congest during the day in response to the light and disperse at night, in the absence of light. What advantages do you think there might be in packing together during the day?

Why do they avoid the surface waters?

8. Recalling the trophic level of *Calanus finmarchicus*—its prey and predators—answer the following questions about its migratory behavior by referring to Figure 20-4:

(a) Why do the *Calanus finmarchicus* migrate to shallow water?

(b) What is the most logical stimulus for this migration? In other words, how do these organisms know when to migrate upward?

(c) Why are they in shallow water at night instead of in the daytime?

(d) Why do they disperse at midnight, when all the food is in the upper 50 meters?

(e) Why do the deeper forms not migrate up at dawn?

(f) The migration pattern of *Calanus finmarchicus* is a behavioral adaptation that probably evolved in the course of a long geologic period, owing to natural selection. In a few well-reasoned sentences, state how this natural selection might have occurred so that a previous population of *Calanus finmarchicus* having equal numbers of members that (1) didn't migrate at all, (2) migrated up in the daytime, and (3) migrated up at night evolved into the type of population that we see today.

9. Answer the following questions about the deep scattering layers:

(a) In the recording of the DSL migration in Figure 20-5, three distinct layers can be seen (assume they are moving toward the surface as night approaches). There are probably many organisms in each layer, but name at least one organism that you think would be in the uppermost migrating layer.

Why is this organism migrating to the surface?

Why do you suspect it is migrating at night instead of during the day?

What would be a logical reason for the organisms in the layers below to be migrating upward after the top layer? Also, name at least one species that you would expect to find in the next layer.

(b) The migration pattern of most DSL's is diel (that is, they migrate daily), but that of the DSL's in the Arctic is not. What do you think the Arctic temporal migration pattern is, and why?

(c) DSL's are more highly developed (having a greater biomass and more layers) in some geographic areas than in others. Where would you expect to find the most highly developed DSL's and why? Refer to the information in Table 20-1 if necessary, or to Exercise 15 on the distribution of life in the sea.

EXERCISE 21

ESTUARIES

An estuary is a semienclosed basin in which river water mixed with seawater. Estuaries have been formed in a variety of ways.

1. The drowning of river valleys when sea level rose at the end of the Pleistocene epoch (Ice Age). This type is common on the East Coast of the United States and is called a *coastal plain estuary.*
2. The subsidence of fault blocks on tectonically active coastlines. South San Francisco Bay is an example of this type.
3. The carving of fjord-type estuaries by glaciers. Naturally this type of estuary is found in high latitudes. The fjords of Norway, Greenland, and Alaska are examples.
4. The development of sand spits and offshore bars along coastlines to form barrier-beach type estuaries. These are common on the Atlantic and Gulf coasts of the United States. Pamlico Sound in North Carolina is an example of this type. Note that major rivers are not required to develop estuaries formed by processes 2, 3, and 4.

FACTORS IN ESTUARINE CIRCULATION

One of the distinguishing features of an estuary is the unusual circulation pattern, influenced primarily by two factors—density and tides. First let us consider the **density-driven circulation** (see Figure 21-1). As river water flows into the estuary it raises the height of the water surface above the height of the ocean surface and creates a horizontal pressure gradient. Water will flow in the direction of this pressure gradient, resulting in a seaward flow of surface water in the estuary (the water flows downhill). The pressure gradient and surface flow are maintained by the continuous input of river water. However, seawater is more dense than river water. Since pressure at depth depends on the product of water density times water-column height, at some depth (shown by the dashed line in Figure 21-1a) the pressure created by a column of low-density river water will be equal to the pressure created by a shorter column of high-density seawater. At any level below this depth, the pressure in the seawater column will exceed that in the river-water column and the pressure gradient will reverse. Because flow must follow pressure gradients, as a consequence of the density contrast between river water and seawater, a seaward flow of water will occur in the surface waters of the estuary and an inland flow will occur in the deep waters of the estuary. The effect of this circulation on salinity is shown schematically in Figure 21-1b. Note that the net seaward flow must be greater than the inland flow to remove the river water that is continuously entering.

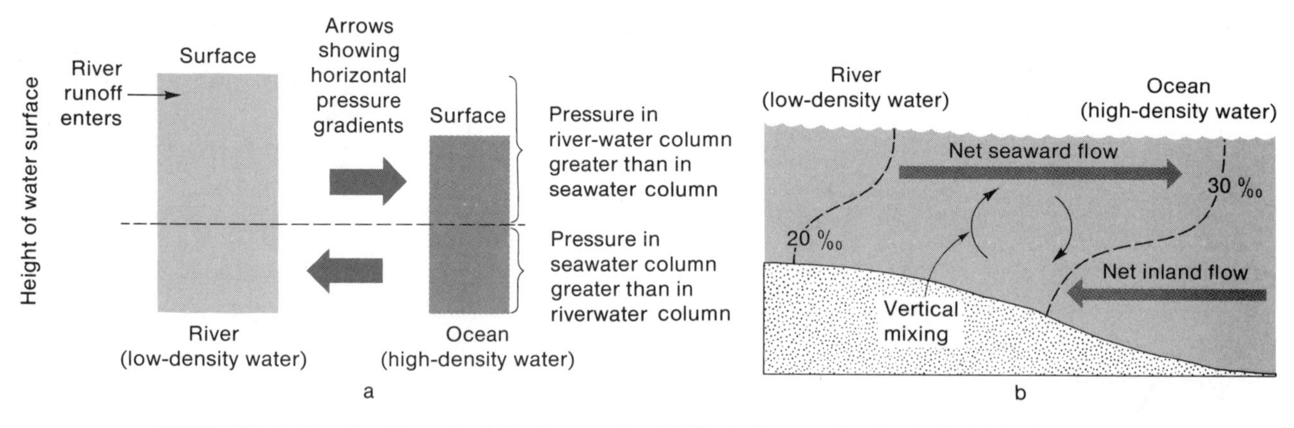

FIGURE 21-1 Density-driven circulation in estuaries. (a) Block diagram showing horizontal pressure gradient. When a column of river water and a column of seawater from the adjacent ocean are compared, the horizontal pressure gradients between the two are found to reverse at depth as shown. (b) Schematic flow diagram with isohalines.

The tides are the second major factor influencing estuarine circulation. In **tidally driven** circulation, pressure gradients in the estuary respond to the periodic rise and fall of the ocean waters along the coast (see Exercise 12). Thus water will flow toward the ocean during ebb tide and landward during flood tide. The seaward flow, however, will last longer and be of slightly greater velocity than the landward flow, resulting in a net seaward transport for a given tidal cycle. The "sloshing" back and forth of the estuary in response to tides is effective in mixing the water column vertically and also produces a diffusion of seawater into the estuary.

The relative importance of the tidal and density-driven components in an estuary determines its salinity structure and circulation type. If the density-driven component is more influential than the tidal component a sharp **halocline** will separate two well-mixed layers. The intrusion of seawater into the estuary appears as a tongue moving up the channel, and this circulation type is termed a **salt wedge** (Figure 21-2a). A classic example is the lower Mississippi River. At the other extreme of the estuarine circulation types is the **well-mixed** estuary (Figure 21-2b), in which tidal fluctuations dominate the density-driven component and the water column is mixed vertically. Portions of Delaware Bay are well mixed. However, most estuaries are intermediate between these two types and are classed as **partially mixed.** Whether density-driven or tidally driven circulation prevails will depend upon the relative magnitudes of freshwater input from the river and of tidal flow from the sea. Since runoff from the land often shows considerable seasonal variation, the characteristics of circulation in

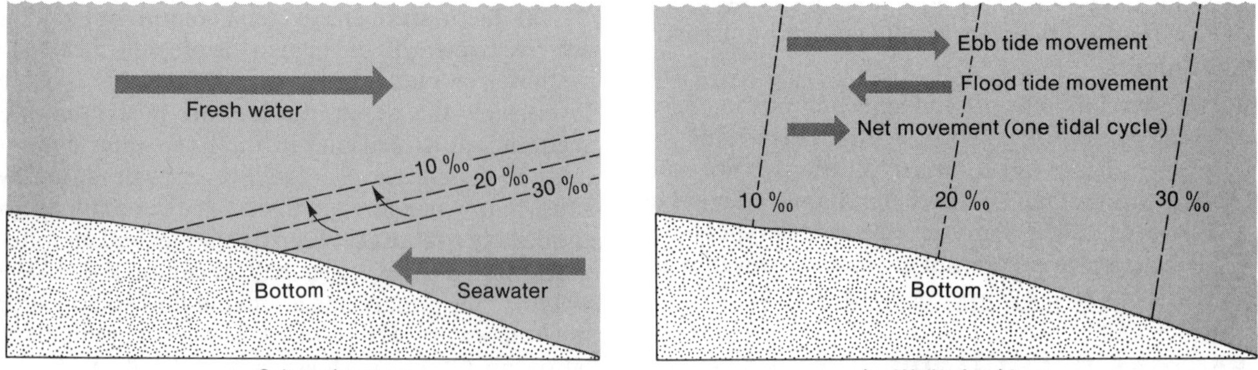

a Salt-wedge type b Well-mixed type

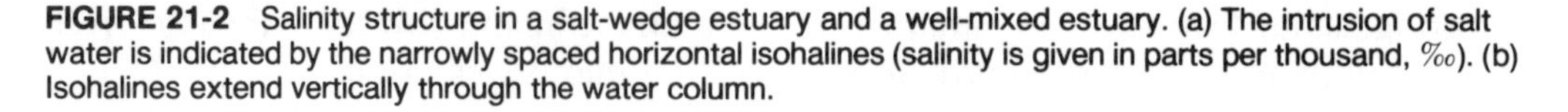
FIGURE 21-2 Salinity structure in a salt-wedge estuary and a well-mixed estuary. (a) The intrusion of salt water is indicated by the narrowly spaced horizontal isohalines (salinity is given in parts per thousand, ‰). (b) Isohalines extend vertically through the water column.

an estuary may vary from season to season. Runoff may vary in response to seasonal variations in precipitation, melting of ice and snow in spring, and increased evaporation of rainfall in summer.

THE IMPORTANCE OF ESTUARIES

Estuaries contain flourishing biological communities. Their high nutrient content stimulates phytoplankton productivity (productivity may be limited by the availability of light in many estuaries). These phytoplankton support a rich and diverse fauna. Extensive shellfish beds may develop; many species of fish use estuaries as a breeding ground; and tidal marshes that develop along estuarine shores support abundant waterfowl.

Estuaries are frequently subjected to conflicting demands. The Hudson River, for example, once supported large fish and shellfish industries, but these have drastically declined in the last few decades. It has long been a major transportation and shipping corridor, as well as a recreational area for owners of small boats. It is used as an emergency water supply for the metropolitan New York area. Finally, it is a receptacle for the sewage effluent of approximately 15 million people. About 30 percent of the total is raw sewage discharged from Manhattan. While construction of treatment facilities for this portion has been in progress for over a decade, they may not be complete until the twenty-first century. Another 30 percent comes from northern New Jersey and receives partial secondary treatment but contains many industrial wastes. The remainder comes mainly from Brooklyn, Queens, and the Bronx and receives secondary treatment. Together these discharges represent about 5 percent of the total United States sewage production and have resulted in elevated levels of bacteria, metals, and organic compounds while reducing dissolved oxygen levels in the lower Hudson. Clearly, careful management of the Hudson River and its estuary on a regional basis is needed to minimize the conficting demands of the diverse uses of this resource.

DEFINITIONS

Density-driven circulation. Variations in salinity create variations in density in estuaries. These variations in density create horizontal pressure gradients, which drive estuarine circulation.

Halocline. A zone in which salinity changes rapidly.

Partially mixed estuary. An estuary that shows a small to moderate salinity change with depth. Both density-driven and tidally driven circulation are important.

Salt-wedge estuary. In this circulation type, the density-driven component dominates, and two well-mixed layers are separated by a sharp halocline. The seawater entering the channel appears as a tongue or wedge.

Tidally driven circulation. Fresh water and seawater are mixed by the sloshing back and forth of the estuary in response to ocean tides.

Well mixed estuary. In this circulation type, tidal fluctuations dominate, and the water column is mixed vertically.

EXERCISE 21

REPORT

ESTUARIES

NAME

DATE

INSTRUCTOR

1. The surface salinity data in Table 21-1 on page 220 were obtained on cruises up the Hudson River estuary during March and August 1974. The bottom salinity during these months was approximately 1.2 times the surface salinity, indicating that the Hudson is a partially mixed estuary. In the table and on the map in Figure 21-3, the symbol "mp" is an abbreviation for miles north of the southern tip of Manhattan, and "mp—" means miles south of the same point. The isohaline (a contour connecting points of equal salinity) is given for 1‰ only. Note that in the Hudson the isohaline of a particular surface salinity will extend horizontally from bank to bank at one mp only.

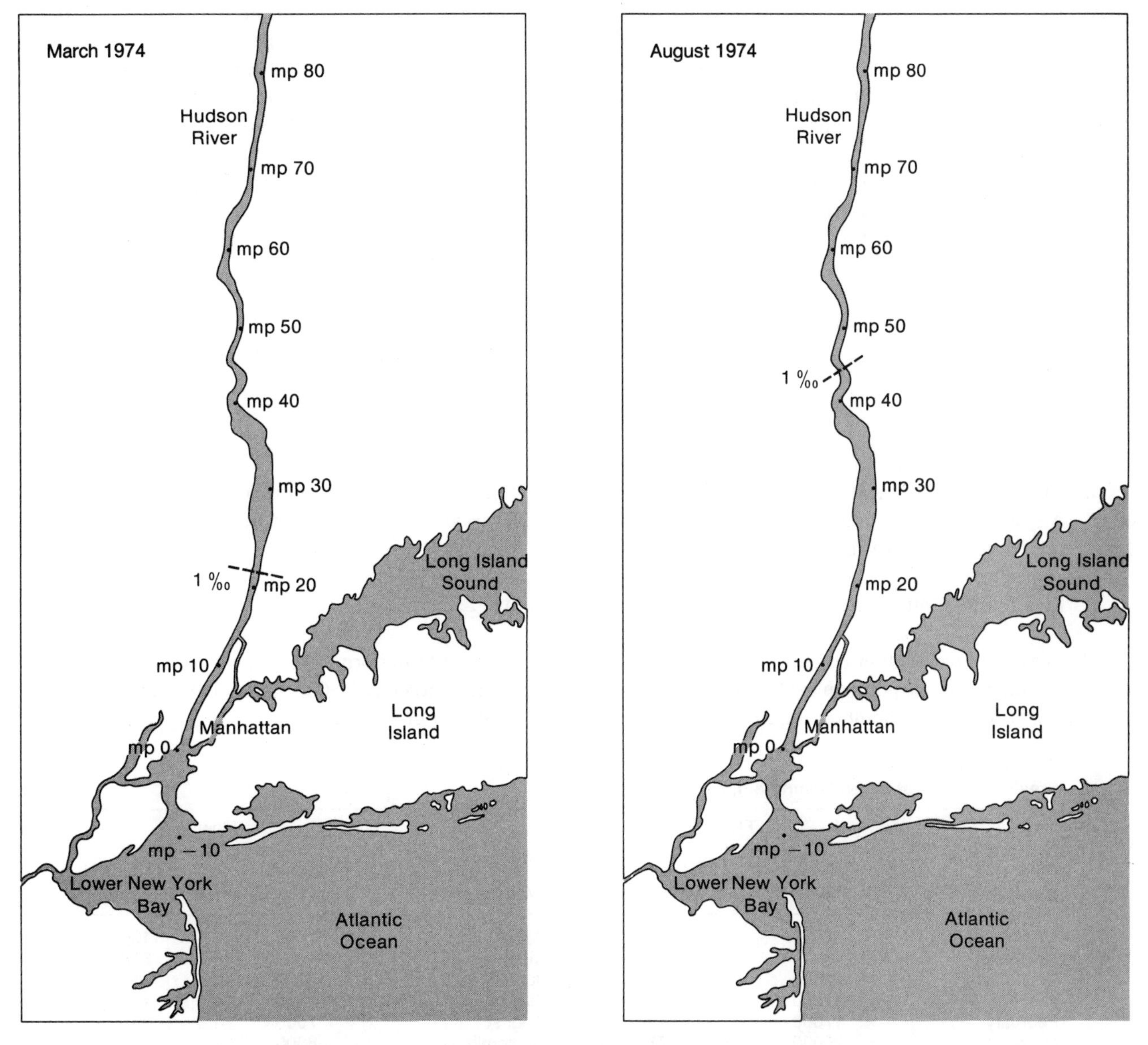

FIGURE 21-3 Surface salinity map of the Hudson River and its estuary, with the contour given for 1‰.

TABLE 21-1
1974 Salinity data on the Hudson River estuary

mp	Surface salinity (‰) March	August
67	<0.1	<0.1
53	<0.1	0.5
41	<0.1	1.5
34	<0.1	4
25	0.5	7
14	2	12
1	9	19
−2	13	20
−8	16	23
−18	28	30

Using these data, plot the isohalines at 25‰, 15‰, and 0.1‰ for surface salinity in March and August. (Hint: Interpolate from the table to find the mp of the desired salinity. For example, if you were to locate the 1‰ isohaline for the month of March—already given on the map—it would be necessary to interpolate between 0.5 and 2‰. Inasmuch as 1‰ is one-third the difference between 0.5 and 2‰, the desired mp is one-third the distance from mp 25 to mp 14. Take one-third the difference of 11 miles, which is 3.6 miles—say, 4 miles—and the 1‰ isohaline would lie at mp 21. Follow the same procedure for the month of August and you would find that the 1‰ isohaline lies at mp 47.)

2. Keep in mind that the data in Table 21-1 are time-averaged salinities. The salinity at mp 20, for example, will change with the phase of the tide, increasing as the tide floods and decreasing as the tide ebbs. Answer the following questions about the contours you have plotted.

   (a) During what month does salt water extend upstream the farthest?

   (b) Briefly state why.

   (c) The average movement of water in a tidal cycle is 8 miles. If the time-averaged surface salinity at mp 20 is 9‰ in August, give the maximum and minimum salinities over a tidal cycle at this point. (Hint: Between mid-tide and high tide, water will move about 4 miles upstream.)

   Maximum salinity ______________ ‰. Minimum salinity ______________ ‰.

3. The seasonal precipitation pattern in the Hudson drainage basin is quite uniform. That is, each month has similar precipitation. The last point on the Hudson at which the river flow is monitored is the Green Island Dam at mp 154. The freshwater flow at this point (averaged over one-month periods) is shown in Table 21-2.

TABLE 21-2
Freshwater flow at Green Island Dam, mp 154

Water year 1974	Flow (cubic feet per second)	Flow (cubic meters per second)	Water year 1974	Flow (cubic feet per second)	Flow (cubic meters per second)
October 1973	4000	145	April	30000	1100
November	9000	330	May	24000	870
December	24000	870	June	10000	365
January 1974	17000	620	July	11000	400
February	17000	620	August	7000	255
March	20000	730	September	8000	290

(a) On the graph provided, plot the monthly average flow as a bar graph.

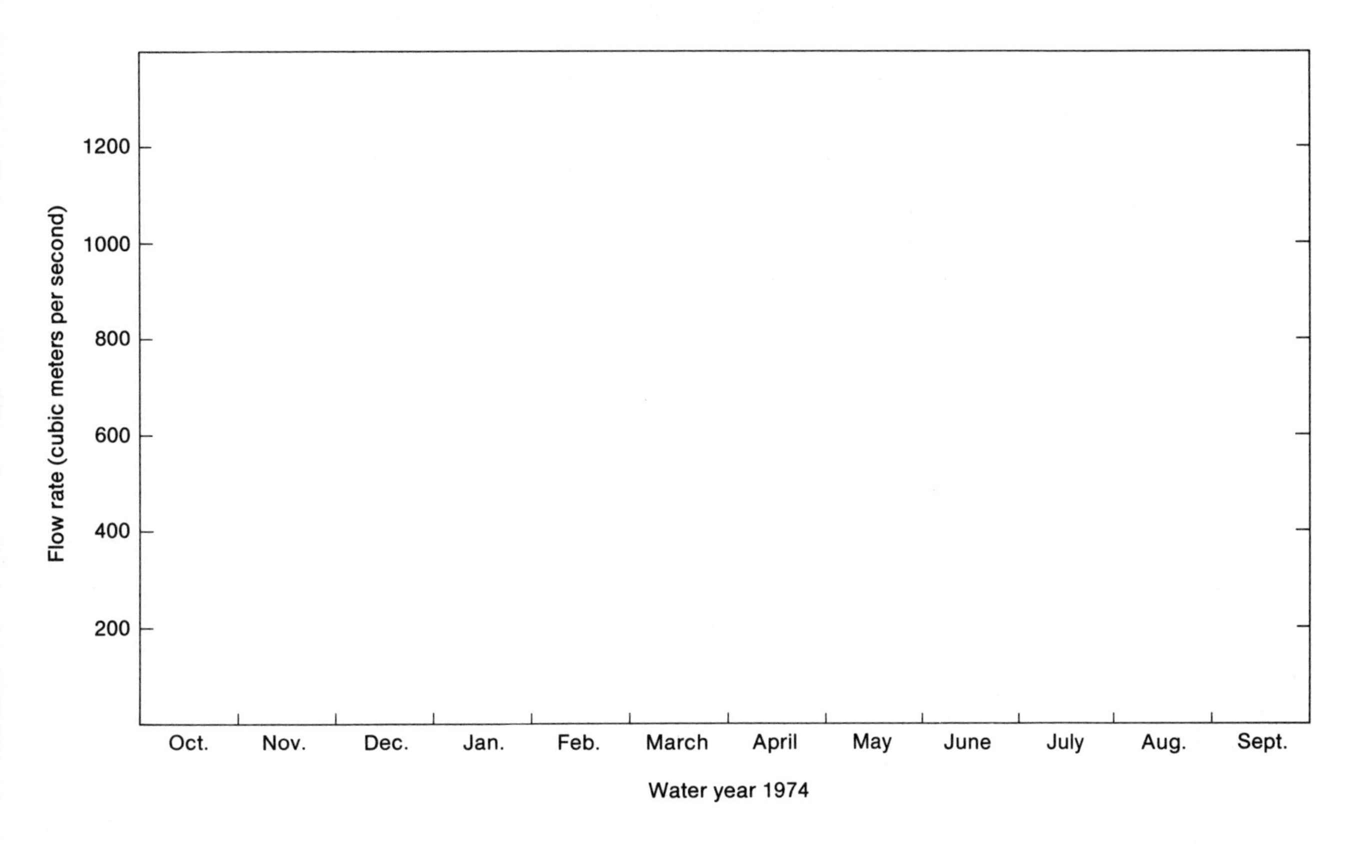

(b) Why is there minimum flow in August, September, and October?

(c) Why is there maximum flow in March through May?

4. On the average, each person in the New York metropolitan region uses more than 150 gallons of water per day. The total demand is nearly 2×10^9 gallons per day, much of which is met by tapping the waters in the Catskill Mountains (which are just to the west of the portion of the Hudson shown on the map). In dry years the rainfall in the Catskills is not sufficient to meet this demand, and the Hudson is the only alternative water source. To maintain water quality, it is desirable to draw water that has a salinity of less than 100 parts per million (0.1‰). Assuming that the position of the 0.1‰ isohaline is linearly related to the rate of freshwater flow over the Green Island Dam (this is approximately true), it would be important to determine the best point along the Hudson to place a water intake so that the water quality would be high enough to make the Hudson water source usable in a drought that had reduced Green Island flow to 100 cubic meters per second.

(a) To determine what this ideal point would be, first plot on the following graph the rate of flow over the Green Island Dam for the months of March and August against the mp of the 0.1‰ isohaline for the same months.

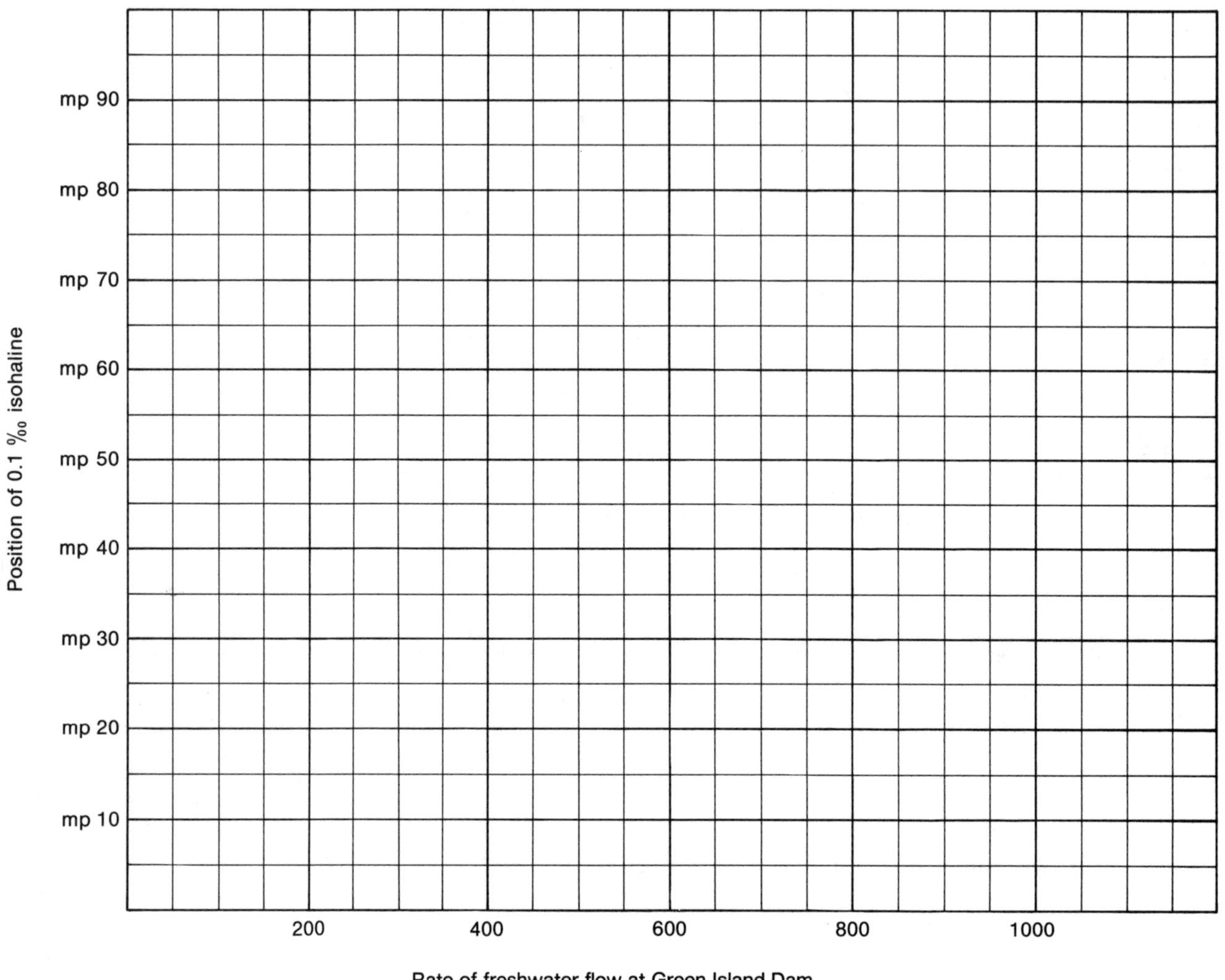

(b) Now state where you would place the water intake. Remember that the farther north this intake is, the more expensive it will be to pipe water to New York City. Note that the present water intake is at mp 67.

5. Most of the sewage generated by New York City and northern New Jersey is discharged into the lower Hudson. Is it likely that these discharges may affect water quality in regions upstream, to the north of New York? Explain why.

EXERCISE 22

OIL SPILLS

In previous exercises various forms of human pollution (thermal, sewage, pesticide, and oil) of the marine environment have been mentioned; in this exercise we are concerned only with petroleum. The major sources of this kind of contamination of the marine environment are tanker accidents and spills at dock facilities, offshore drilling rigs, and coastal refineries. Some noteworthy spills include the 1969 Santa Barbara oil platform spill, which resulted in the death of many marine birds that had become coated with oil; the spill from the tanker *Torrey Canyon,* which in 1967 ran aground off England (the oil on the water surface there was bombed and set afire, and the beaches were cleaned with detergent, which is probably more toxic to marine organisms than oil); and the *Burmah Agate* and *Olympic Glory* spills off the Texas coast in 1979 and 1981, respectively. The largest single oil spill to date is the 1979 blowout of the Ixtoc 1 oil platform in the Gulf of Mexico.

In this exercise we will focus on, and you will use data collected during, the three Gulf of Mexico oil spills just mentioned. Some of the data we will consider were collected by students in an oceanography course like this one.

The spills under discussion occurred in and affected three different oceanographic environments: (1) an open-ocean and open-marine-shelf environment (Ixtoc 1); (2) an open-ocean nearshore marine environment (*Burmah Agate*); and (3) an estuarine environment (*Olympic Glory*). (See Figure 22-1.) The responses of microplankton and microbenthon were recorded following each spill. The data are presented, and you are asked to interpret these responses by using the laboratory and lecture material you have had during this term.

In general, when oil spills occur in the open ocean there may be immediate effects on seabirds and planktonic organisms, but the dispersion of the oil usually has little observable effect on benthic organisms except in regions where the oil might be concentrated, as in nearshore sediments and environments. Spills in nearshore environments (or offshore spills concentrated in nearshore environments) usually have a greater observable impact on a greater number of organisms occupying different habitats. Great harm may be done not only to nearshore birds and planktonic organisms, but also to benthic invertebrates, fish, and other organisms in estuaries, lagoons, and bays, where the oil may remain for a considerable time. Significantly, nearshore regions are

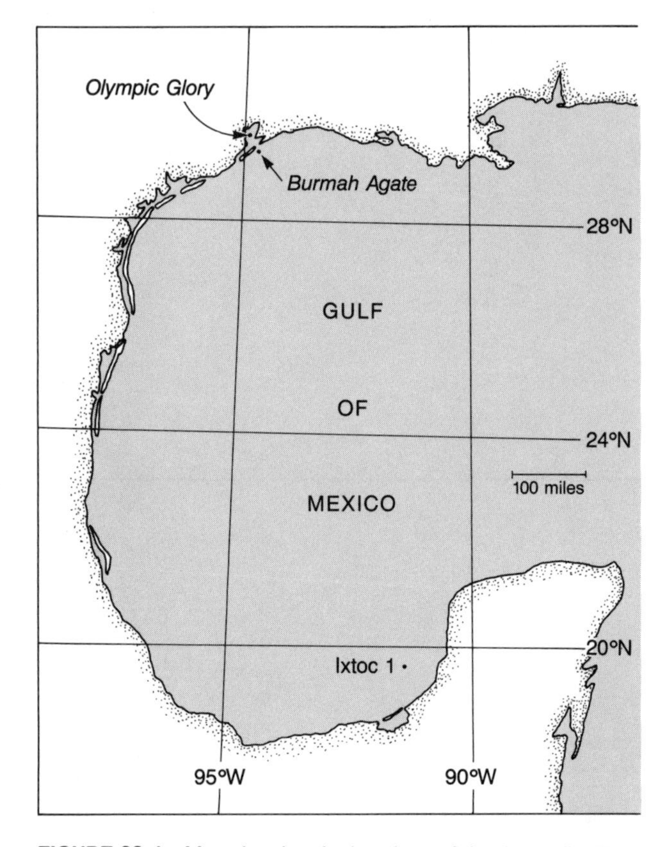

FIGURE 22-1 Map showing the locations of the three oil spills discussed in this exercise. Ixtoc 1 was an oil platform spill in the Bay of Campeche; the *Burmah Agate* and *Olympic Glory* were tankers that spilled oil off the coast of Galveston, Texas, and in the upper reaches of Galveston Bay, respectively. [From R. Casey et al., *Proceedings of the Offshore Technology Conference,* 1982, pp. 449–459.]

some of the most important to human beings and to marine organisms that use them as sanctuaries and nurseries.

FIELD AND LABORATORY TECHNIQUES

Various methods have been used to study the impact of oil spills, and there are no standard techniques for such investigations. In this study sediment samples were collected via a bottom grab lowered over the side of a small boat or from a fishing pier; the upper few centimeters of the sediment were removed at sea or on the pier; and each sample was preserved in a buffered formalin solution containing rose Bengal, a vital dye that stains living and recently dead organisms red. In the laboratory a known volume of this sediment was washed over a 63-micrometer screen. The living organisms larger than 63 micrometers (microbenthon and microplankton) retained by the screen were identified and counted. Throughout this exercise these data are presented as number of organisms per 10 cubic centimeters.

Plankton samples were collected with a 63-micrometer mesh net or water bottle; they were preserved in a buffered formalin solution containing rose Bengal, and the living microplankton were identified and counted. Throughout this exercise these data are presented as number per liter.

Some physical oceanographic and some meteorological data were also collected and are presented where appropriate in the exercise.

IXTOC 1 SPILL

On June 3, 1979, Ixtoc 1 (an oil production platform in the Bay of Campeche) blew out, began spilling oil into the Gulf of Mexico, and continued to do so until March 24, 1980 (Figure 22-2). This prolonged spill affected the shelf region of Texas. Previously (1974–1977) the outer continental shelf of southern Texas had undergone an extensive study in order to determine the taxonomic makeup, distribution, and density of organisms on and over the shelf so that there would be a baseline of information to use in the event of an oil spill or other perturbation. The oil from Ixtoc 1 was contained in Mexican waters by a current moving south along the Texas coast (the Texas Current, labeled 2 in Figure 22-2) except during the summer months, when a current from the south (the Mexican Current, labeled 3 in Figure 22-2) that is usually contained south of the Mexican border by the Texas Current moved up along the Texas coast because of a switch in wind conditions.

The only postspill anomaly involving microplankton or microbenthon that was recorded that might be related to the spill was the presence of recently dead shelled microzooplankton (labeled 4–9 in Figure 22-2) from off the south Texas shore. Recently dead shelled microzooplankton (radiolarians, planktonic foraminiferans, and pteropods) are very rarely observed in the sediment, and their presence there in this case is believed to have been a result of the spill.

BURMAH AGATE SPILL

On November 1, 1979, the oil-laden tanker *Burmah Agate* was involved in an accident 4 miles (about 7 kilometers) outside the Galveston breakwater (Figure 22-3) and sank in about 13 meters of water, spilling Nigerian crude oil that burned for 69 days. The day after the spill started, plankton and benthon collections were made near the burning vessel, on a transect south of the ship, and on a transect toward shore (to

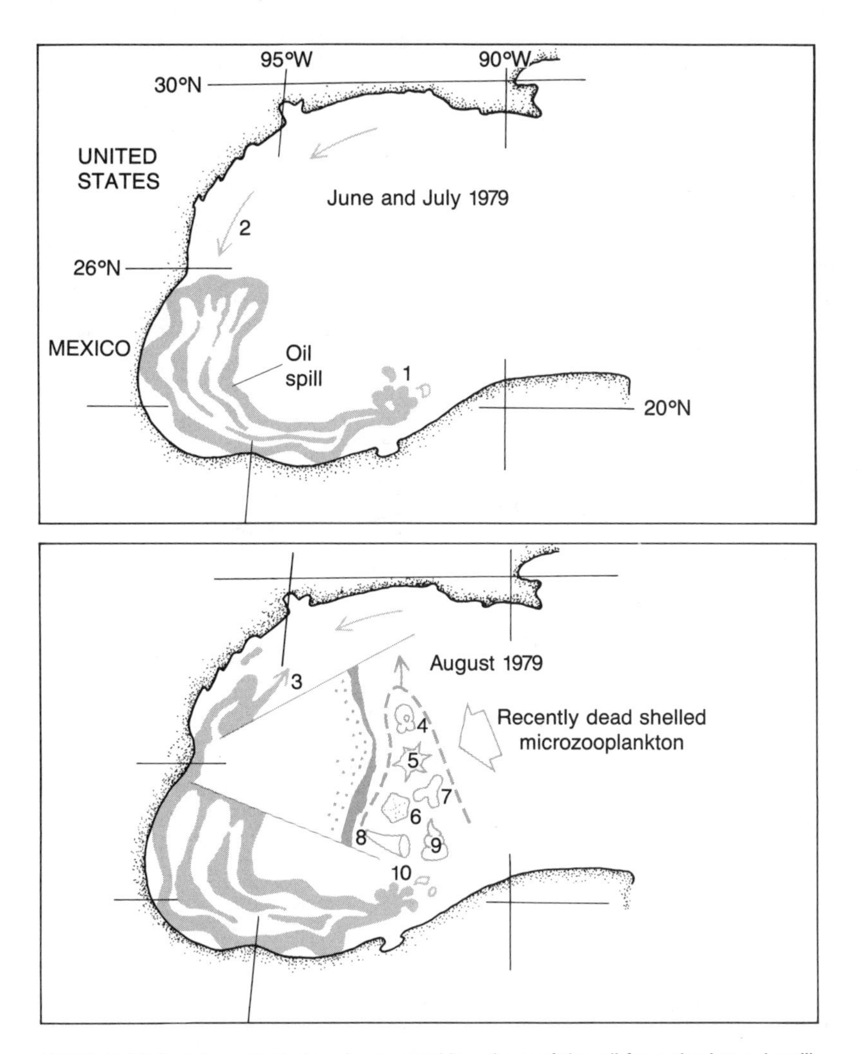

FIGURE 22-2 Maps depicting the general locations of the oil from the Ixtoc 1 spill. During June and July of 1979 the oil from the Ixtoc 1 oil platform (1) remained in Mexican waters, held there by the Texas Current (2) flowing to the south. During August of 1979 the development of the Mexican Current (3) brought oil into Texas waters. The shelf sediments off south Texas contained the remains of recently dead shelled microzooplankton (planktonic foraminiferans, 4; radiolarians, 5, 6, 7, and 9; and pteropods, 8). The oil platform (10) was still spilling some oil. [From R. Casey et al., *Proceedings of the Offshore Technology Conference,* 1982, pp. 449–459.]

TABLE 22-1
Densities of organisms (in number per 10 cubic centimeters) in sediments from various locations after the *Burmah Agate* spill

	1 day (early winter)	1 month (winter)	6 months (spring)	1 year (early winter)	1.5 years (spring)	2 years (early winter)
Nearshore						
Benthonic foraminiferans	40	300	60	20	0	20
Nematodes	1300	200	320	180	480	340
Near ship						
Benthonic foraminiferans	8	24	30	22	30	30
Nematodes	22	56	120	48	260	90
Offshore						
Benthonic foraminiferans	20	30	0	30	60	0
Nematodes	40	0	100	120	310	220

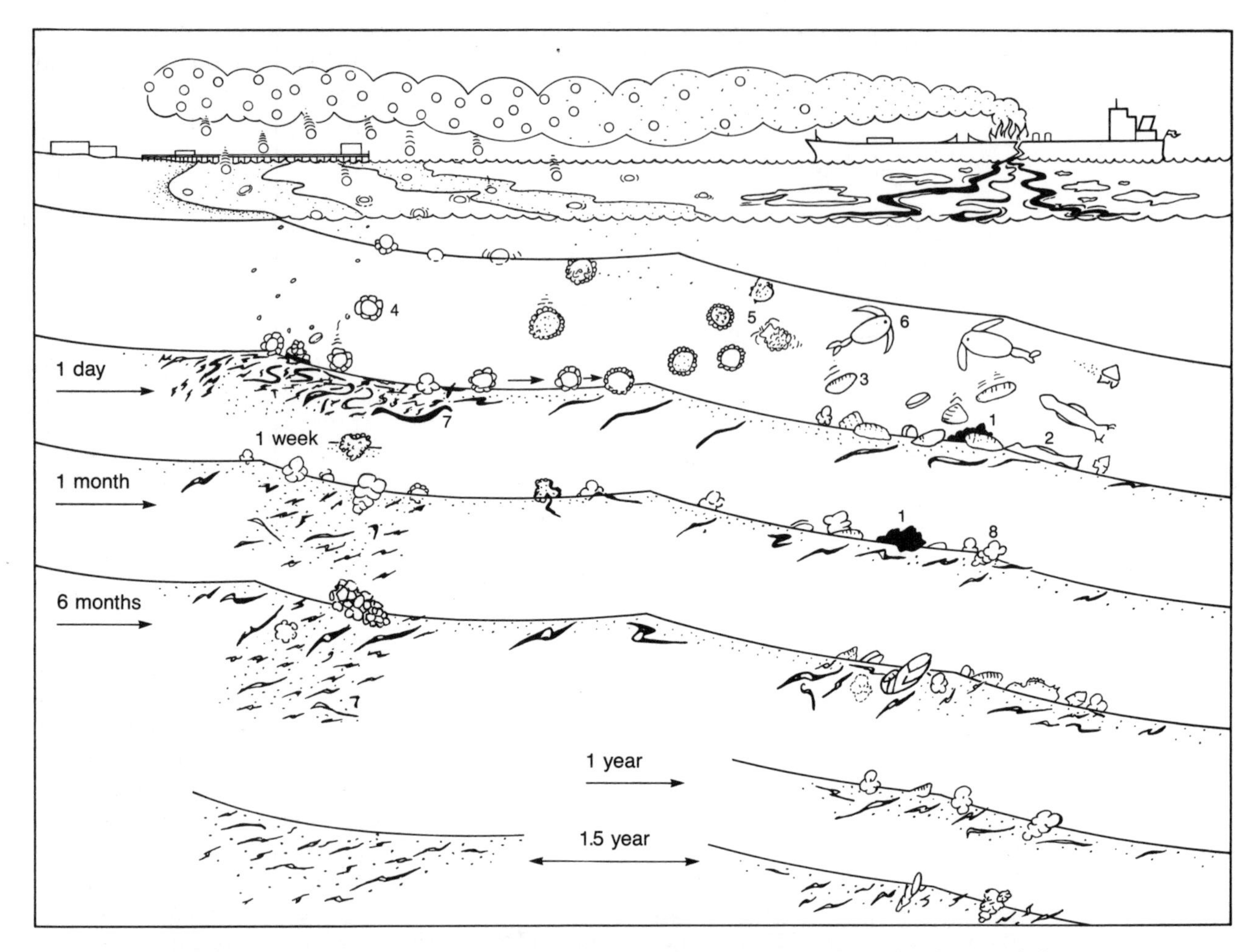

FIGURE 22-3 Drawing of the *Burmah Agate* spill depicting the movement of the oil and suspended sediments and the effects on organisms. The *Burmah Agate* is shown at the upper right spilling oil that moves on the ocean surface toward the Galveston beach. Oil drops in the smoke from the burning oil are also carried toward the beach. From the Galveston breakwater at the upper left a plume of sediment- (clay-) laden water moves parallel to the Galveston beach. Nearshore (off the beach) waves stir up and suspend silt throughout the nearshore waters. Findings after 1 day, 1 week, 1 month, 6 months, 1 year, and 1.5 years are shown. For these intervals 1 denotes oil balls; 2, recently dead chaetognaths; 3, copepod fecal pellets; 4, oil drops coated with sand; 5, oil and water spheres coated with silt; 6, copepods feeding on oil; 7, nematodes; and 8, the benthonic foraminiferan *Brizalina lowmani.* [From R. Casey et al., *Proceedings of the Offshore Technology Conference,* 1982, pp. 449–459.]

TABLE 22-2
Standing crops of microbenthon and microplankton from sediment and water samples, respectively, at Sylvan Beach after *Olympic Glory* spill*

	2/26/81	3/5/81	3/21/81	3/27/81	3/31/81	5/23/81	1/11/82	1/24/82	1/26/82†	1/26/82*	2/1/82	2/5/82
Sediment samples												
Benthonic foraminiferans	290	410	2540	1100	nc	1800	200	nc	nc	nc	nc	nc
Nematodes	300	1000	600	980	nc	250	290	nc	nc	nc	nc	nc
Water samples												
Diroflagellates	16	23	48	6	0.1	0.3	0.07	11.5	4.2	2.3	1.4	0.3
Diatoms	0.1	17	0.05	0.02	0.4	0.06	0.07	1.3	1.9	1.0	0.3	0.5

Source: R. Casey et al., *Proceedings of the Offshore Technology Conference,* 1982.

* Number of benthonic foraminiferans and nematodes per 10 cubic centimeters from sediment samples; millions of dinoflagellates and diatoms per liter from water samples. Sampling period: 1 year; nc means not counted.

† Sampling was done at 0900 and at 1430 hours on this date.

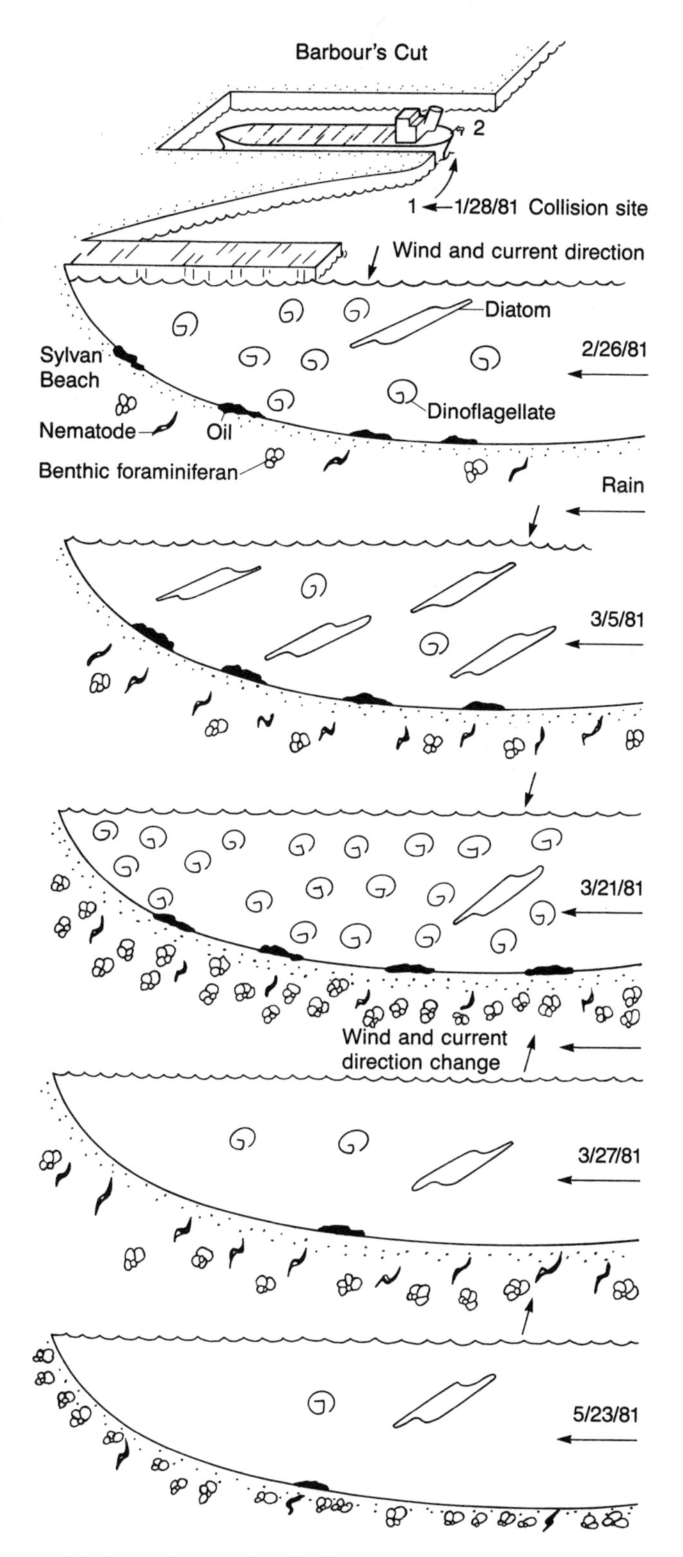

FIGURE 22-4 The *Olympic Glory* oil spill and its observed effects on microplankton and microbenthon. Shown are the collision site (1) and the location of the leaking tanker after the spill (2). Below the sketch of the fishing pier from which the samples were taken are drawings depicting the oil, microplankton, and microbenthon in a time series from 1 month until about 4 months after the spill. Important meterorological and oceanographic events—the rain, and the wind and current direction change—are also shown. [From R. Casey et al., *Proceedings of the Offshore Technology Conference,* 1982, pp. 449–459.]

within 200 meters of the beach at Galveston, Texas) on an anticipated spill track. A few days later the oil took this anticipated track to the Galveston beach, and subsequently sampling was done at 1 week, 1 month, 6 months, 1 year, 1.5 years, and 2 years after the spill to determine the effects of, and recovery from, the spill. The results are presented in Table 22-1.

OLYMPIC GLORY SPILL

On January 28, 1981, the tanker *Olympic Glory* was involved in an accident and began to spill oil in upper Galveston Bay, Texas (1 in Figure 22-4). The ship was maneuvered to Barbour's Cut (2 in the figure), where most of the spilling oil was contained. Samples of the plankton and benthon were taken for a year from a fishing pier at Sylvan Beach, which is near the original accident site and a few miles south of Barbour's Cut. Results are presented in Table 22-2.

Along with the collision site and the *Olympic Glory* in Barbour's Cut, Figure 22-4 shows the most important biological events noted off Sylvan Beach. Diatoms and dinoflagellates in the water column and nematodes, benthonic foraminiferans, and oil on the bottom are shown in their relative group abundances at different times after the spill. In other words, twice as many benthonic foraminiferans drawn means that twice as many were collected at the time indicated. Similarly, twice as many diatoms drawn means that twice as many were collected at that time. Note that these relative indications do not refer to an exact density of organisms, nor to relative densities between different groups (such as between benthonic foraminiferans and diatoms).

Oil spills, which occur off and affect all oil-producing and oil-transporting regions and routes, are not good, and every attempt should be made to avoid them; but it should be borne in mind that considerable efforts are usually made (and were made during the three oil spills considered in this exercise) to minimize their effects. The *Olympic Glory,* for example, was taken into Barbour's Cut and oil curtains (devices to stop the movement of oil at and just below the surface of the water) were placed so that most of the spilling oil was contained in a small region and collected there. Oil curtains were also deployed for the *Burmah Agate* and Ixtoc 1 spills, as were oil skimmers, which are specially designed vessels that skim the oil from the surface of the sea. Commonly some absorbant, such as hay, is spread on beaches that are in danger of being affected by spills. After the oil has washed ashore, or when the danger has passed, the hay (occasionally oil soaked) is removed.

EXERCISE 22

REPORT

OIL SPILLS

NAME ______________________

DATE ______________________

INSTRUCTOR ______________________

1. Contour (as best you can) on the "map" below the densities of recently dead shelled microzooplankton from the Ixtoc 1 spill at the 10- and 20-per-10-cubic-centimeter intervals.

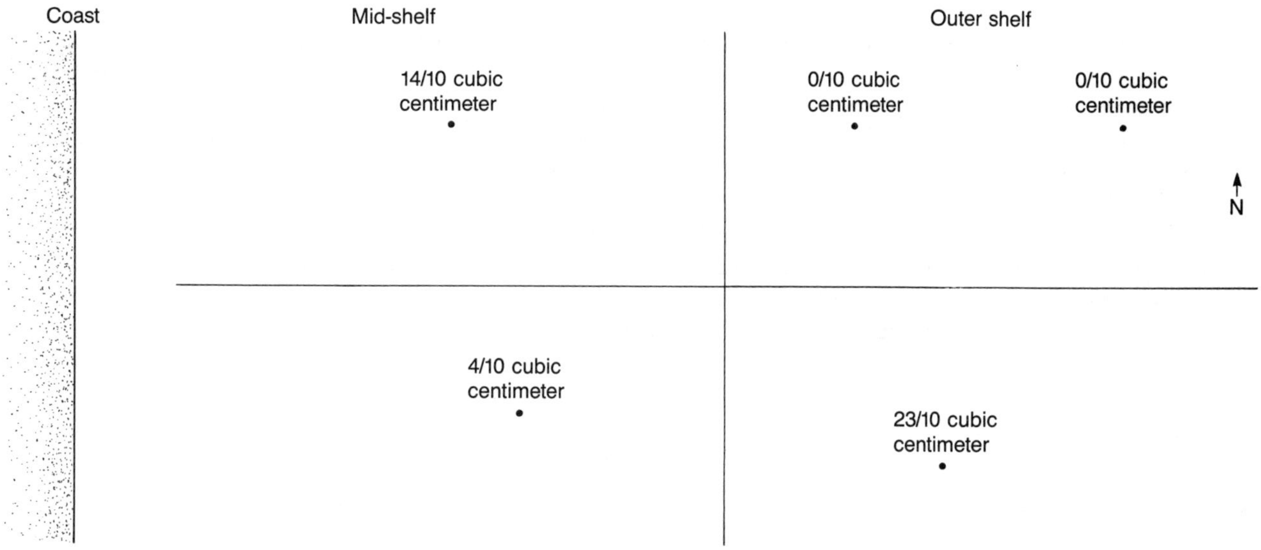

Coast

Mid-shelf

Outer shelf

14/10 cubic centimeter

0/10 cubic centimeter

0/10 cubic centimeter

N

4/10 cubic centimeter

23/10 cubic centimeter

2. Since the Ixtoc 1 spill had no apparent effect (no observable effect in this study) on the microbenthon, but had an obvious effect on the shelled microzooplankton, in the geographic area off south Texas, construct a simple oceanographic model of what might have affected microzooplankton while not noticeably affecting the underlying microbenthon.

3. Is there any evidence of the Mexican Current from the data you contoured in Question 1 for the Ixtoc 1 spill? If so, what is it?

4. Using Figure 22-3 and the data in Table 22-1, answer this and the following questions concerning the *Burmah Agate* spill. On the day after the spill (1 day in Figure 22-3) tar balls (oil balls) were found on the bottom near the spill site (1 in the figure), as were soft-bodied plankton such as chaetognaths (labeled 2 in Figure 22-3), and abundant copepod fecal pellets (3 in the figure). In the water above, however, the only abundant plankton collected in a net were copepods (6). Apparently some plankton in the immediate vicinity of the oil were adversely affected and some were not. Give two pieces of evidence from the illustration that suggest that at least one plankter was not immediately adversely affected by the spill.

5. Two other regions where oil was noticed on the sediments were nearshore, where small (20-micrometer) oil drops coated with silt-sized sediment were found on the sand bottom (4 in Figure 22-3); and at the middle of the Galveston transect, where larger (80-micrometer) spheres of oil plus water (an emulsion) were covered with clay-sized particles and located on a mud bottom (5 in Figure 22-3). Using the illustration, describe the most likely avenues of the oil drops and spheres to the bottom. That is, what two different sequences of events made oil drops or spheres sink to the bottom? Also, give your evidence for each avenue and means you select. In constructing your answers remember the following: (1) oil is lighter than water, as shown by the leaking oil floating on the sea surface near the *Burmah Agate;* (2) there was a plume of sediment-laden (clay sized) water located at middle transect that had exited the Galveston Breakwater; (3) nearshore the waves were suspending silt-sized particles throughout the water column; and (4) there were two avenues for the oil to get to these regions, via the air (drops in smoke) and on the sea surface.

6. The only biological microbenthonic response noted on the first day was the very high density of nematodes in the nearshore sands (7 in Figure 22-3). Assuming that nematode rapid reproduction cannot account for this higher than normal standing crop, how can it be accounted for? Remember how the samples were collected, and that nematodes are infaunal organisms that in this region are in a sand bottom.

7. One month after the spill a shelled microbenthonic organism (the foraminiferan *Brizalina lowmani,* 8 in Figure 22-3) was found at all stations, although it had not been noted at any station in previous samplings. How do you suppose *B. lowmani* got there? (Hint: Refer back to Exercise 15, especially the description of the habitats of the pelagic environment in the section "The Marine Life Zones.")

8. *Olympic Glory* spill data for the Sylvan Beach area are given in Table 22-2. Samples were taken at intervals over a period beginning a month after the spill and ending about a year after it. Plot the densities specified in parts (a) and (b) on the accompanying graphs.

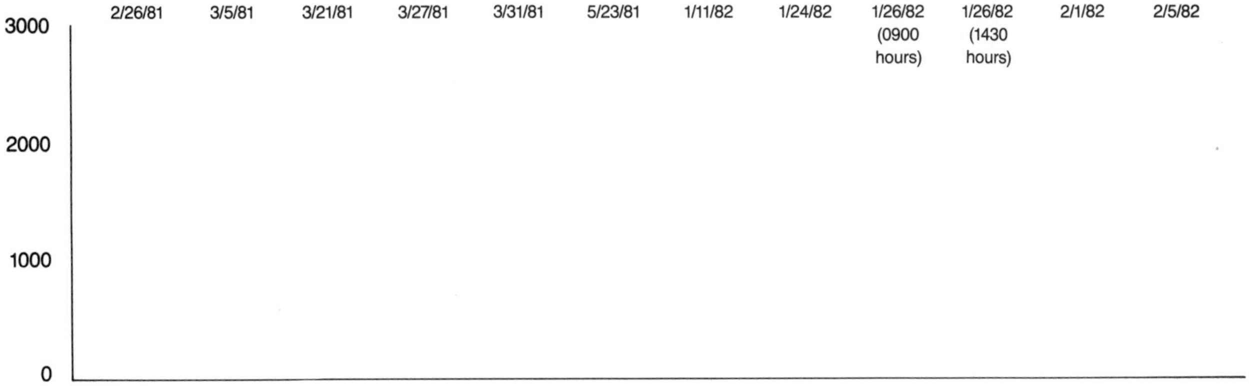

3000
2000
1000
0

2/26/81 3/5/81 3/21/81 3/27/81 3/31/81 5/23/81 1/11/82 1/24/82 1/26/82 (0900 hours) 1/26/82 (1430 hours) 2/1/82 2/5/82

(a) Densities of benthonic foraminiferans and nematodes from the sediment. (Use solid lines to connect the data points for the benthonic foraminiferans and dashed lines for the points for the nematodes.)

(b) Densities of dinoflagellates and diatoms from the water. (Use solid lines to connect the data points for the dinoflagellates and dashed lines for the points for the diatoms.)

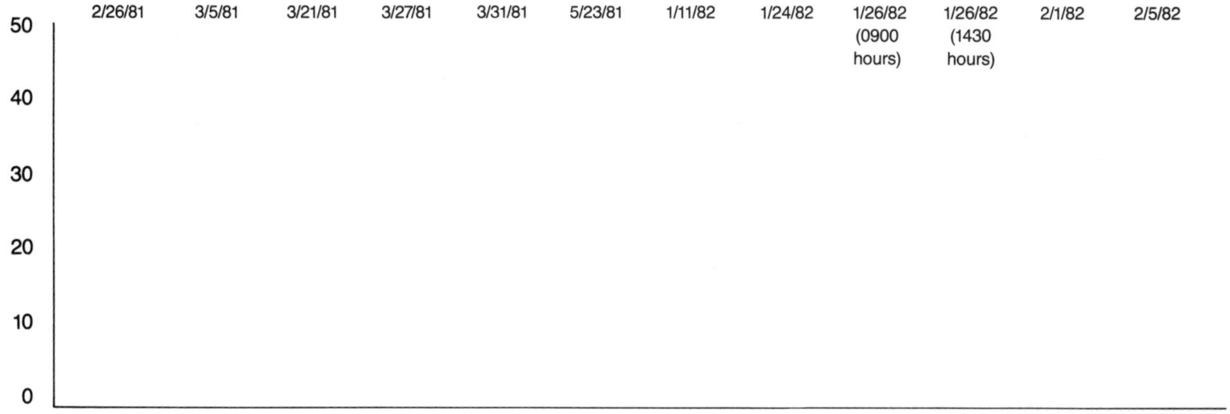

50
40
30
20
10
0

2/26/81 3/5/81 3/21/81 3/27/81 3/31/81 5/23/81 1/11/82 1/24/82 1/26/82 (0900 hours) 1/26/82 (1430 hours) 2/1/82 2/5/82

9. (a) Red tide conditions (dinoflagellate densities greater than 1,000,000 organisms per liter of water) existed off Sylvan Beach from 1 month after the *Olympic Glory* spill until March 27, 1981. Can you think of any way that the oil spill might have been responsible for these prolonged red tides, which are not natural in the area at this time of the year? (Red tides normally occur off Sylvan Beach around the end of January and last for about a week, but would not usually last for over a month, as happened in 1981.)

(b) Why do you think these oil-related red tide conditions had ended by March 27, 1981?

10. (a) The red tide sampled at Sylvan Beach after the *Olympic Glory* spill (on March 5, 1981) was a diatom–dinoflagellate bloom rather than a dinoflagellate bloom. By March 21, 1981, it had reverted to a normal dinoflagellate red tide. How might these observations be accounted for? (Hint: What are diatom skeletons made of, and what happened between February 26 and March 5, 1981?)

(b) From the data presented on Table 22-2 there appears to be one natural red tide represented during the period of data collection. When was that natural red tide? What do you think the high diatom concentrations during that time mean, and why?

(c) From the information you have been asked to evaluate here, what are the conditions that tend to cause natural red tides at Sylvan Beach, and why?

11. By May 23, 1981, the oil-related red tide had been gone from Sylvan Beach for 2 months but the benthonic foraminiferan population there was still high. How could the microbenthonic population still be higher than during normal (prespill) conditions and microplankton be at near-normal prespill levels?

OPTIONAL QUESTIONS

12. Using all the figures and tables in this exercise, tell how you would biologically monitor the recovery of the environments in question from a future oil spill. In other words, what biological conditions or responses could you look for or cite that would indicate that each environment was returning (or had returned) to normal, or prespill, conditions? (Hints: Are there densities of certain organisms, or are certain organisms present or absent, that might reflect the recovery? Also, in any of the data you would collect, can you detect a return to natural (prespill) biological rhythms such as those mentioned in the discussion, or those shown in Figure 16-3, which is for an area similar to that affected by the *Burmah Agate* spill? The offshore area profiled in Table 22-1 was not significantly affected by the spill, and so the data there can be regarded as representing natural, or prespill, conditions for the site near the *Burmah Agate* spill.)

13. The Ixtoc 1 spill was the world's largest single oil spill; the *Burmah Agate* spill was a major one, but of much smaller magnitude; and the leak from the *Olympic Glory* was much smaller than the *Burmah Agate* spill. Each affected a different environment: Ixtoc 1 damaged the open ocean and mid and outer shelf; the *Burmah Agate* the ocean and inner shelf; and the *Olympic Glory* an enclosed estuarine system. Considering the relative amounts of oil spilled and the different environments affected, what would you say are the important factors that determine how great an impact an oil spill will have on a given environment? (Compare the environments.)

14. With Question 13 and your answers to it in mind, do you think it would be more environmentally acceptable to build a major oil port in a bay or offshore, and why?

PALEOCEANOGRAPHY

Paleoceanography, the effort to know ancient oceans, was greatly benefited by the initiation of the Deep Sea Drilling Project in the 1960s. The *Glomar Challenger* (namesake of HMS *Challenger* and drilling ship of the Deep Sea Drilling Project) and its successor, the *Joides Resolution* (drilling ship for the Ocean Drilling Program) are capable of drilling cores through the sediments and sedimentary rocks to the basalt "basement" and retrieving them. On board the *drilling ship* micropaleontologists, sedimentologists, petrologists, and other scientists examine and direct the drilling operations; on shore, they study the historical record revealed by these cores at their laboratories. Much of this ancient oceanographic record is revealed by the microfossils contained in the samples, and this exercise will use actual microfossil assemblages to reconstruct paleoceanographic conditions. You will have the opportunity to perform many of the same tasks that micropaleontologists do: date fossil sediments, determine the paleogeographic region of deposition of fossil sediments, and suggest some oceanographic conditions prevailing at that region at the time of deposition.

AGE DETERMINATION OF FOSSIL SEDIMENTS

Throughout geologic time (the very long period dealt with by historical geology) many species have evolved and become extinct. **Relative geologic time** can be determined by ascertaining from fossil sediments the sequence of first occurrences (evolutions) and last occurrences (extinctions) of certain forms. Radioactive isotopes (discussed in Exercise 24) can be used to determine **absolute geologic time.** In this exercise we will determine the relative geologic time of an actual microfossil assemblage by using the first occurrences and last occurrences of radiolarians.

Table 23-1 illustrates the geologic ranges of some hypothetical radiolarian species.

In it, species A is shown to be restricted to the middle Miocene, whereas species B existed from late-late Miocene through the Pliocene, and species C was extant from early Miocene to the Miocene–Pliocene boundary. Species D is limited to the late Miocene and species E ranges from early Miocene through the Pliocene. If you had a fossil assemblage that contained

TABLE 23-1
Geologic ranges of some hypothetical radiolarian species

Epoch (time)		Approximate age in millions of years	Hypothetical radiolarian species				
			A	B	C	D	E
Pleistocene		1					
Pliocene		2		|			|
		3		|			|
		4		|			|
		5		|			|
Miocene		6		|	|	|	|
		7		|	|	|	|
	Late	8			|	|	|
		9			|	|	|
		10			|	|	|
		11	|		|		|
		12	|		|		|
		13	|		|		|
	Middle	14	|		|		|
		15	|		|		|
		16	|		|		|
	Early	17			|		|

species C, D, and E, that assemblage would have to be a late Miocene assemblage (the geologic range of D). If, however, an assemblage containing C, D, and E also contained species B, you would have to say that its age is late-late Miocene, because the only period in which species B, C, and D coexisted is the very latest Miocene time (the only period in which their geologic ranges overlap). This is the method you should use in this exercise to determine the relative geologic time of the radiolarian assemblage you will date. Figure 23-1 illustrates, names, and gives the geologic ranges and environments of some radiolarian species that will be used in the report.

RECONSTRUCTION OF PALEOCEANOGRAPHIC ENVIRONMENTS

In order to reconstruct paleoceanographic conditions or environments from microfossil data it is important to understand the relationship between the modern relatives of microfossils and their environment. If we can correlate living (or extant) forms with specific environmental parameters (such as their geographic distribution, or whether they occur in eutrophic or oligotrophic environments), we can then, with some confidence, assume that fossil relatives (extinct forms) reflect the same environmental parameters for their fossil environments of deposition. Radiolarians are planktonic protozoans with siliceous skeletons about 100 or 200 micrometers in greatest dimension. There are about 500 extant species of radiolarians in the ocean today. About one-half of these live in the epipelagic region (see Exercise 15) of the tropical and temperate seas; about 50 species live in the epipelagic regions of the polar seas; and about 200 live at mesopelagic and greater depths. Figure 23-2 is a photomicrograph of a radiolarian assemblage taken from the tropical Pacific Ocean (close to area C in Figure 16-2). These radiolarians were collected from a core of sediments taken from the tropical Pacific, and the sediment sample was treated to concentrate the radiolarians. The photomicrograph shows a great diversity of forms, including some that are restricted to warm waters.

Figure 23-3 is a photomicrograph of radiolarians concentrated from polar (Pacific) sediments (close to area D in Figure 16-2). It shows less diversity when compared to the tropical assemblage in Figure 23-2; some species in it are restricted to cold polar waters, and many diatoms are evident. Figure 23-4 illustrates radiolarians from sediments collected under two temperate regions of the Pacific Ocean. Figure 23-4a shows radiolarians concentrated from sediments collected under the open-ocean low-productivity (oligotrophic) region in the middle of the North Pacific gyre (in a region near A in Figure 16-2), and Figures 23-4b and 23-4c represent radiolarians concentrated from sediments underlying the high-productivity (eutrophic) California Current region of the temperate

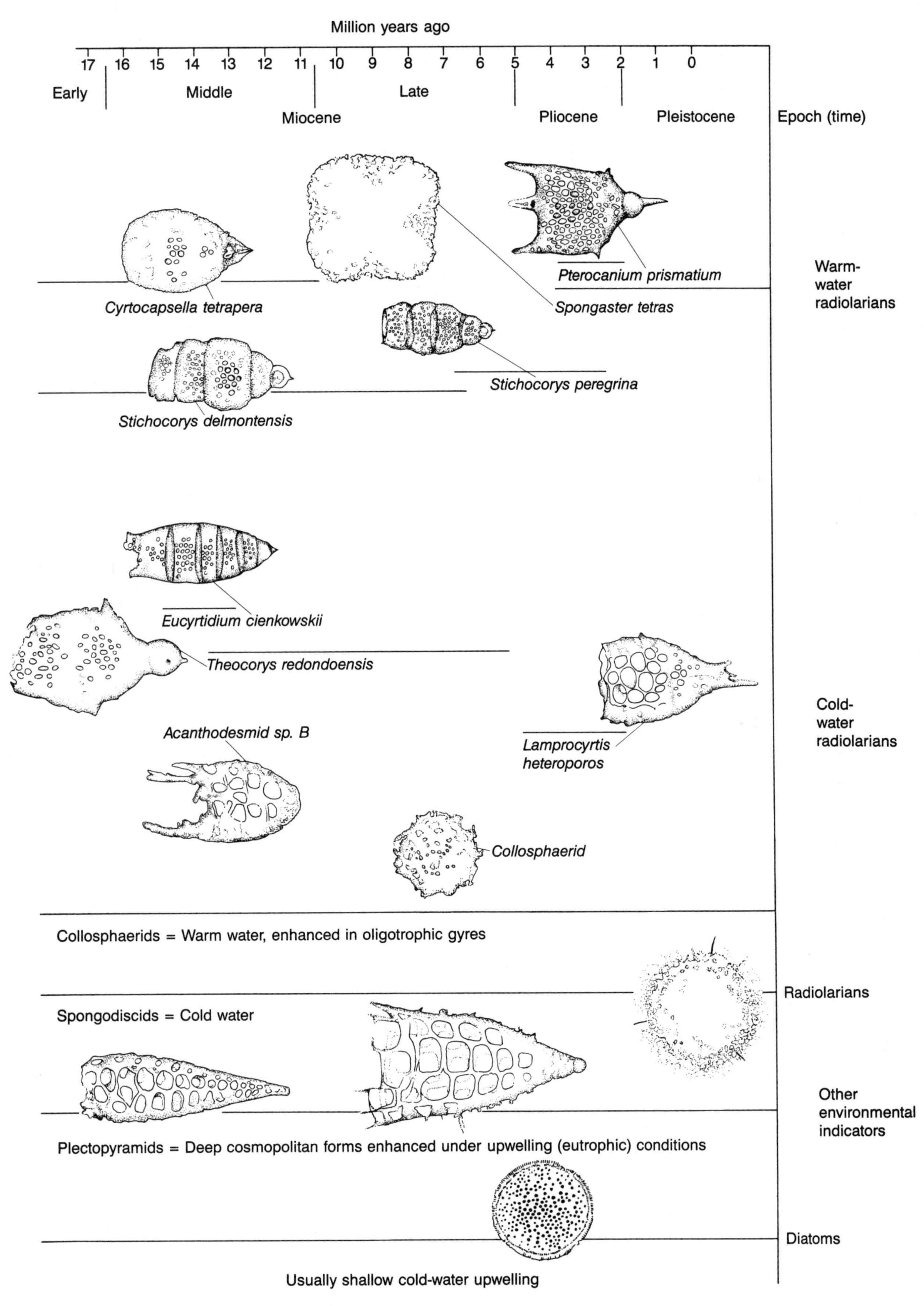

FIGURE 23-1 Radiolarian geologic ranges and environments.

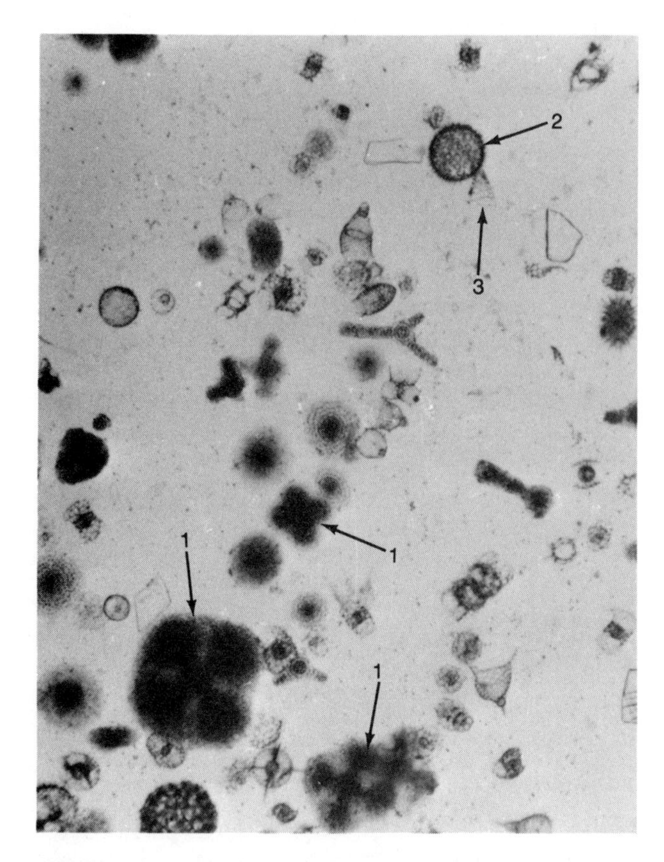

FIGURE 23-2 Among the many species in this highly diverse tropical radiolarian assemblage are such warm-water forms as *Spongater tetras* (1) and collosphaerids (2). A deep-water plectopyramid (3) is also present.

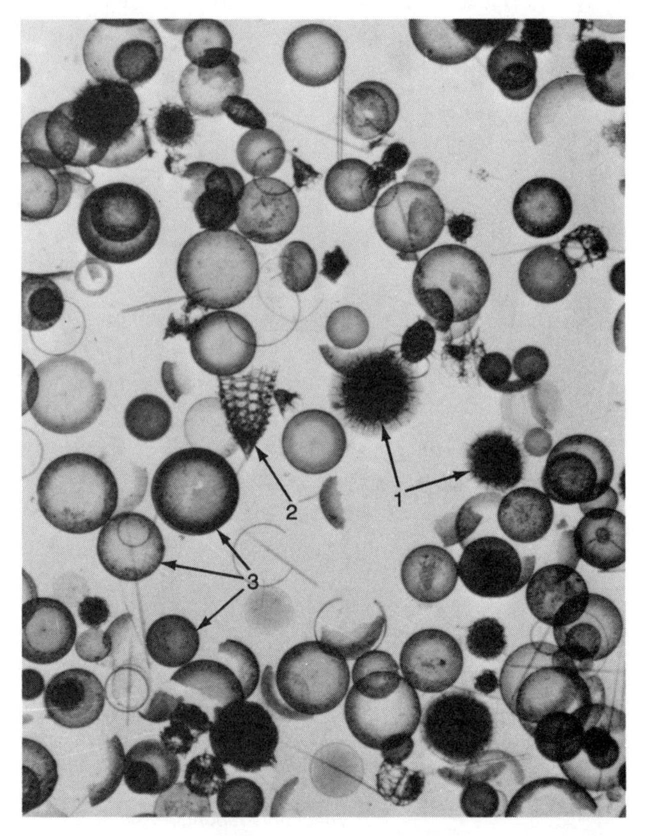

FIGURE 23-3 This low-diversity polar radiolarian assemblage includes such cold-water forms as spongodiscids (1) and deep-water forms such as plectopyramids (2); there are also many diatoms (3).

North Pacific (in regions near B in Figure 16-2). Figure 23-4b contains some terrigenous sands, showing that the sediment was collected near a continent, and Figure 23-4c contains some angular sands that are volcanic shards, showing that the sediment was collected near a volcanic source.

Even though the three samples in Figure 23-4 are from a temperate region, they differ in radiolarian content in that the California Current samples contain some polar cold-water radiolarians and diatoms brought down by the current and are less diverse in radiolarians than the oligotrophic sample. (Recall the differences between eutrophic and oligotrophic regions in Question 9 of Exercise 16.) Present in all samples are deep-water radiolarians (plectopyramid radiolarians from mesopelagic and deeper waters) that are so labeled on Figures 23-2, 23-3, and 23-4. The physical and chemical oceanographic conditions at depth (mesopelagic and deeper; see Table 20-1) are very uniform (especially in the bathypelagic), and therefore the radiolarians (and many other organisms) are widely distributed at these depths and show up in sediment samples from polar to tropical regions.

Although deep-water radiolarians occur in all the environments shown, some occur as higher percentages of the fauna in some samples when compared to other samples. Such is the case when we compare the percentage of deep-water radiolarians from one of the temperate California Current samples (Figure 23-4c) with that from the temperate open-ocean North Pacific sample (Figure 23-4a). The reason for the enhancement (high percentage) of deep-water radiolarians accumulating under this California Current sample when compared to those accumulating under the open-ocean North Pacific region appears to be the enhancement of deep radiolarians under upwelling environments. The higher percentage associated with upwelling may be due to the deep radiolarians' being carried to (or toward) the surface, dying, and raining down on the sea floor, or to there being a larger standing crop of deep-living radiolarians under high-productivity regions (California Current) than under low-productivity regions (mid-North Pacific), as noted for other deep-living organisms (Exercise 20).

The oligotrophic North Pacific region (Figure 23-4a) appears to be enhanced in collosphaerid radio-

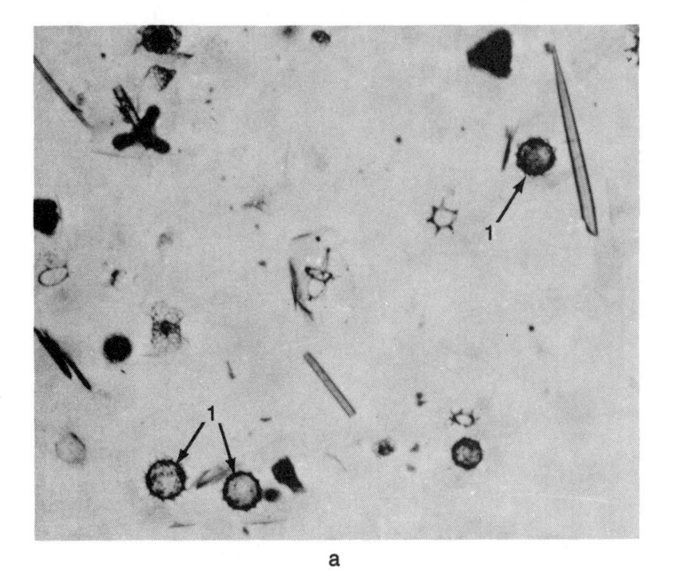

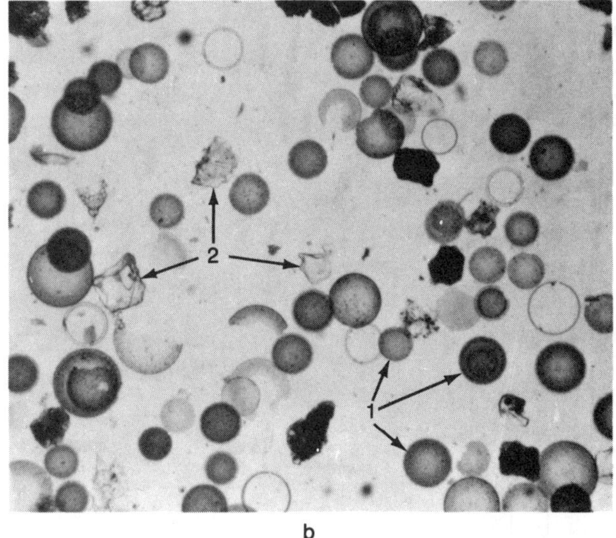

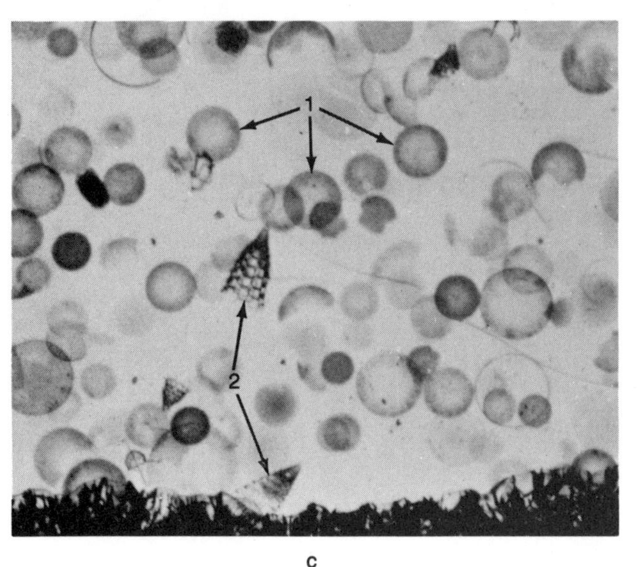

FIGURE 23-4 Temperate radiolarian assemblages. (a) An oligotrophic-gyre radiolarian assemblage (mid-ocean North or South Hemisphere, low-productivity region) showing high radiolarian diversity and enhancement of collosphaerid radiolarians (1). (b) California Current radiolarian assemblage exhibiting low radiolarian diversity, abundant diatoms (1), and terrigenous sediment (2), showing that it was taken near a continent. (c) California Current radiolarian assemblage exhibiting low radiolarian diversity, abundant diatoms (1), and enhanced deep-water radiolarians (2), showing it is from a more intense upwelling region than the assemblage in (b).

larians (labeled in Figure 23-4a). The latter are a colonial type of radiolarian—a number of individual radiolarians living in close association (symbiosis), with algae. The protozoan animal and the plant (algae) both appear to benefit by their relationship, which seems to allow for high standing crops of collosphaerid radiolarians in these temperate oligotrophic waters and the enhancement of such forms in the sediments accumulating underneath.

These present-day characteristics of radiolarian distribution and enhancement can be used to infer that similar geographies and environmental conditions prevailed when their fossil counterparts were deposited in the sediments.

DEFINITIONS

Absolute geologic time. Geologic time in years (in absolute time); for example, radioactive dating yields absolute geologic time.

Assemblage. A collection of organisms (living or dead) found in the natural environment.

Relative geologic time. Geologic time in the context of earlier or later geologic events or times, for example, a fossil or rock unit is determined to be relatively older or younger than another fossil or rock unit based on its position in sequence (in earlier- or later-deposited rocks).

Symbiosis. A close association of two different species to the mutual benefit of each.

EXERCISE 23

REPORT

PALEOCEANOGRAPHY

NAME ______________________

DATE ______________________

INSTRUCTOR ______________________

1. From the information in Figure 23-1 and what you learned from the text and Table 23-1, what is the geologic epoch (or portion of, such as late-late Miocene), and the range in millions of years for the radiolarian assemblage illustrated below? Explain why you think this assemblage was deposited within that time range. (Give the names of species you used for your determination, circle and label these species in the figure, and tell why you used them, instead of other species in the assemblage, to date the assemblage correctly.)

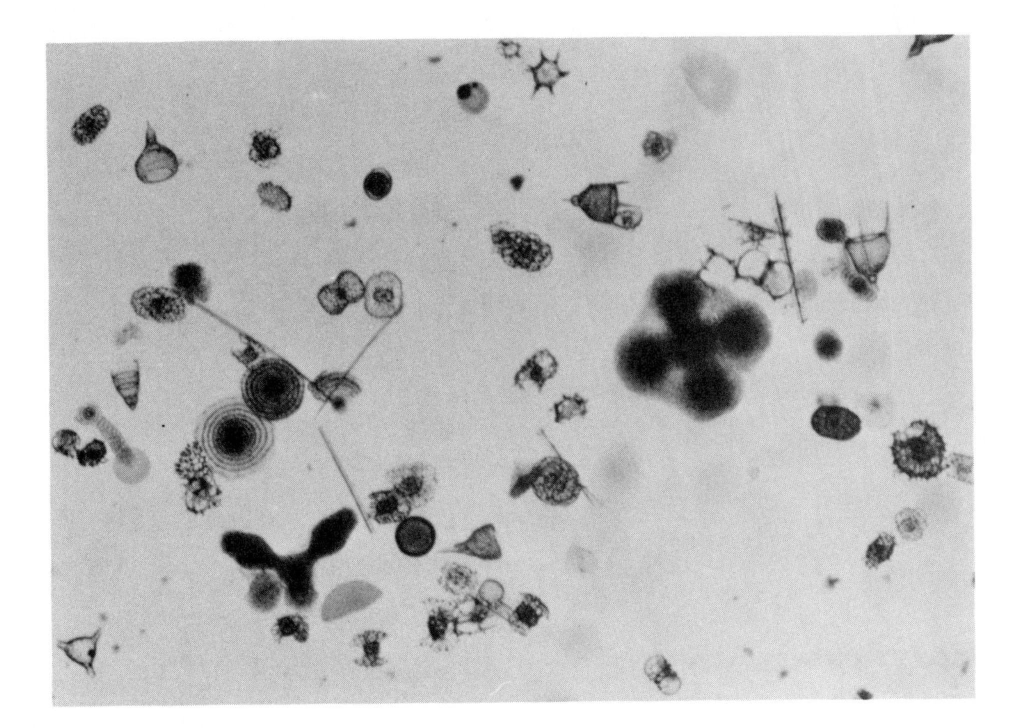

2. Using the information from Figures 23-2, 23-3, and 23-4 and what you learned in the text, interpret the paleoceanographic conditions illustrated by the radiolarian assemblage that you dated in response to Question 1. What paleoenvironment was it deposited from, and why? (For example, it might be from an oligotrophic gyre because it shows high radiolarian diversity and enhanced collosphaerids).

3. Date and determine the paleoceanographic conditions of the following assemblage, and defend these conclusions as you did for Questions 1 and 2.

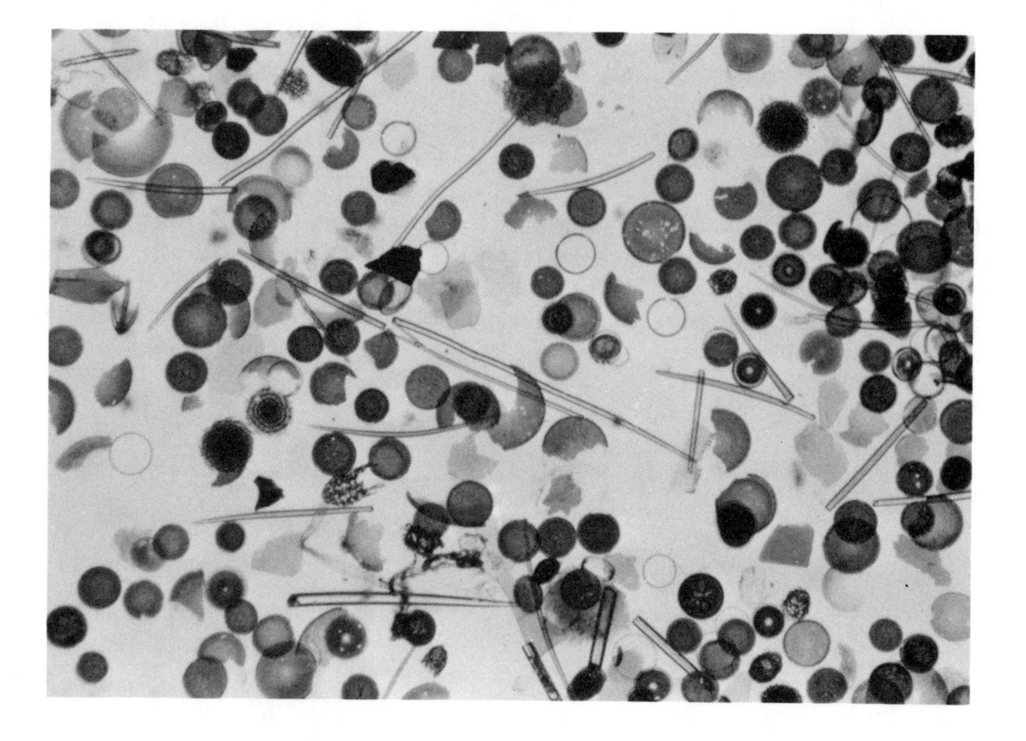

OPTIONAL QUESTIONS

4. Using the information in Table 23-1, give one good explanation for the occurrence of a few A species in a fossil assemblage that is late-late Miocene and contains common occurrences of species B, C, D, and E.

5. Referring to Table 23-1, explain how during the same exact time period (late-late Miocene) one fossil assemblage may contain species B, C, and D but no E, whereas a fossil assemblage a few hundred miles away may contain species B, C, D, and E.

6. Why might deep-living fossil radiolarians be better for dating fossil sediments than shallow-living fossil radiolarians?

7. The characteristics that make a fossil a good time indicator (make it good for fine-resolution dating of rocks) are often the opposite of the characteristics that make a fossil a good environmental indicator (make it helpful for determining the environment of deposition, such as polar or tropical, eutrophic or oligotrophic, deep or shallow). List below two good characteristics for a time indicator, two good characteristics for an environmental indicator, and one characteristic that is both a good time indicator and a good environmental indicator.

Time indicator		Environmental indicator
1. ______	different	______
2. ______	different	______
3. ______	in common	______

EXERCISE 24

THE USE OF RADIOACTIVE ISOTOPES IN OCEANOGRAPHY

A **radioactive isotope** is an atom with an unstable nucleus, which decays into another **isotope.** The rate of decay depends only on the nuclear properties of the unstable atom, and is independent of the identity of its surrounding atoms. Because of this and because the rate of decay of a radioactive isotope can be measured in the laboratory, these isotopes have been used as "clocks" in many branches of earth science, including oceanography.

The ocean is a dynamic system, and to understand its operation we must have techniques that can be used to measure the rates at which oceanic processes occur. It has been shown that radioactive isotopes can be used to measure the rates of some of these processes.

Many processes, such as the accumulation of sediment on the sea floor, occur so slowly that thousands of years are required to produce a measurable change. For these, naturally occurring radioactive isotopes offer the only "clock" for measuring rates. Other processes, such as large-scale oceanic mixing, are subject to local and transient fluctuations, which make observation of their average rates by current-meter measurements impossible. Radioactive isotopes, added to the ocean by fallout of debris from nuclear weapons, have been useful for observing average rates of these large-scale phenomena.

Among the oceanographic processes whose rates can be successfully measured by the use of naturally occurring and man-made radioactive isotopes are the following:

1. sediment accumulation on the sea floor;
2. mixing of ocean waters;
3. removal of highly reactive substances from seawater to the sediments; and
4. exchange of dissolved gases across the air–water interface.

Although the details of many of these processes are not completely understood, radiochemical tools have provided some basic information about them—the average rate at which they occur. In this exercise, we will illustrate the way in which one radioactive isotope, carbon 14 (^{14}C), can be used to estimate the rate of sediment accumulation in a deep-sea core.

CARBON AND ITS RADIOACTIVE ISOTOPE, ^{14}C

First, it is important to understand what we mean by an isotope of a particular **element.** Any atom of an element has in its nucleus a specified number of protons (particles with a positive electrical charge). An equal number of electrons (negatively charged particles) orbit around the nucleus. The number of protons in the nucleus dictates the chemical behavior of the atom. For example, carbon has six protons. An atom also contains another type of particle in its nucleus, the neutron (an uncharged particle of about the same mass as a proton). Although the number of protons is the same for any atom of an element, the number of neutrons can vary, producing different isotopes of the same element. Thus, whereas carbon has six protons, it has three isotopes with six, seven, or eight neutrons. To distinguish these, we write a superscript to the left of the chemical symbol. The value of the superscript equals the sum of the protons and neutrons (approximately the atomic weight).

The isotopes of carbons are ^{12}C, ^{13}C, and ^{14}C. They have identical chemical behavior, but their nuclear stability depends on the relative numbers of protons and neutrons. The isotopes ^{12}C and ^{13}C are stable, but ^{14}C is not; and eventually a ^{14}C nucleus will decay by emitting a negatively charged beta particle (a high-energy nuclear electron, expressed as β^-), leaving behind a neutron that has turned into a proton. The remaining nucleus has seven protons and seven neutrons, and will now behave like the element nitrogen. The reaction is written as

$$^{14}_{6}C \longrightarrow {}^{14}_{7}N + \beta^- + \bar{\nu}$$

(where $\bar{\nu}$ represents an antineutrino that is also lost in this decay but is not important for this discussion). For clarity, the number of protons in the parent atom (^{14}C here) and its daughter (^{14}N here) has been included as a subscript.

Radioactivity is a statistical phenomenon. A radioactive atom has a certain probability of decaying in a given period of time. The more unstable the atom, the greater the probability that it will decay in this time period. For example, ^{14}C is less stable than another well-known radioactive isotope, ^{238}U. Thus, in a 1-year period, a ^{14}C atom has a greater probability of decaying than does a ^{238}U atom. If we deal with large numbers of atoms, we find that they follow the laws of probability.* Suppose that we start with a large number of ^{14}C atoms, equal to N. Instead of specifying the probability for the decay of one atom in 1 year, we frequently specify another parameter that can be related to this probability, the time that is required for 50 percent of the N atoms to decay. This period is called the **half-life** of ^{14}C and is about 5700 years. The more stable ^{238}U isotope has a half-life of 4.5×10^9 years. Suppose we start with 1000 ^{14}C atoms. In 5700 years, about 500 will remain. In the next 5700 years another 50 percent will decay, leaving approximately 250.

This concept is illustrated in the graph in Figure 24-1, in which the fraction of atoms remaining, N/N^0, is plotted against time. The number of half-lives that have elapsed is also indicated. If we can measure the number of ^{14}C atoms, N, currently present in a sample and can deduce the number of ^{14}C atoms it had when it was formed, N^0, we can calculate the time elapsed since formation of the sample by use of Figure 24-1 (assuming the sample has not exchanged ^{14}C atoms with its surroundings since it was formed). For example, if $N/N^0 = 0.25$, two half-lives, or 11,400 years, have elapsed since a sample was formed. If $N/N^0 = 0.125$, three half-lives of ^{14}C, or 17,000 years, have passed since the sample was formed. Note that N/N^0 decreases exponentially with time.

THE USE OF CARBON-14 TO DETERMINE SEDIMENTATION RATES

Carbon 14 atoms are continuously produced in the ionosphere as an end product of the interaction of cosmic rays and the solar wind with the upper atmosphere. These atoms are oxidized to carbon dioxide ($^{14}CO_2$) and mix into the lower atmosphere within a few years after their formation. Most $^{14}CO_2$ will dissolve in the ocean before the ^{14}C atom decays. The consequence of the continuous production and transport of ^{14}C is that the ratio of ^{14}C to ^{12}C (the major form of stable carbon) has remained fairly constant with time (certainly for the last 10,000 years and probably much longer).

Foraminifera are small animals that live in the ocean and build calcium carbonate ($CaCO_3$) tests. Because ^{14}C and ^{12}C have the same chemistry, the ratio of ^{14}C to ^{12}C (hereafter expressed as $^{14}C^{\star}$) in foraminifera shells will be the same as the $^{14}C^{\star}$ of dissolved carbon in the water they grow in. Growth of a foraminiferan "sets the clock." When the foraminiferan dies, it sinks to the sea floor (see Exercise 6 on the materials of the sea floor). The ^{14}C in the shell decays as it is continuously buried under the younger shells. Thus, $^{14}C^{\star}$ in foraminifera shells decreases with increasing depth in the sediment. Assuming that $^{14}C^{\star}$ in surface ocean waters has remained constant through time, $^{14}C^{\star}$ of buried foraminifera tests can be used to find the rate of sediment accumulation.

---

* Note that an atom has a decay probability, and not a decay rate. It will decay completely or not at all in a given period of time. If we deal with large numbers of atoms, then we can refer to the decay rate of the group. The decay rate is equivalent to the decay probability times the number of radioactive atoms.

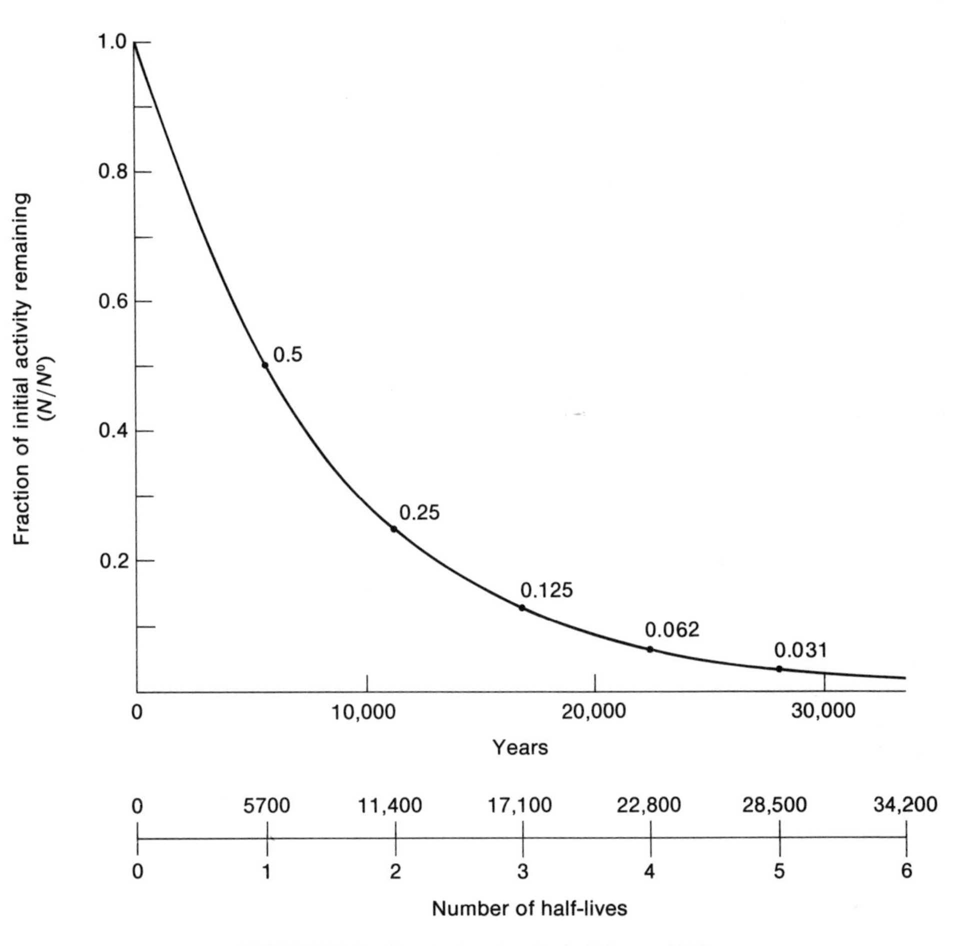

FIGURE 24-1 Graph showing the half-lives of ^{14}C.

Measurements of $^{14}C\star$ require rather sophisticated analytical procedures. In one commonly used technique, shells are returned to the laboratory, cleaned, and heated to a high temperature, which causes the release of CO_2 gas. The CO_2 that is released is collected, purified, and placed in a gas proportional counter, which detects the beta decay of ^{14}C. The length of time required to count a sample depends on $^{14}C\star$, but one day or more may be needed for deep-sea sediment samples.

The data given in Table 24-1 are similar to data that might be obtained from a core of carbonate-rich sediments in the South Atlantic. The value $^{14}C\star$ has been normalized to the ratio at the top of the core.

TABLE 24-1
Ratio of $^{14}C\star$ to surficial sediments in a hypothetical deep-sea core

Depth in core (centimeters)	$^{14}C\star$
0	1.00
5	0.60
10	0.36
15	0.22
20	0.13
30	0.05
40	0.02

DEFINITIONS

Element. An atom with a specified number of protons. The number of protons controls the chemical behavior of the atom.

Half-life. The period of time required for the amount of a particular isotope in a sample to decay 50 percent.

Isotope. An atom with a specified number of protons and a specified number of neutrons. Note that the element carbon (which has six protons) has three isotopes: ^{12}C (six neutrons), ^{13}C (seven neutrons), and ^{14}C (eight neutrons). The superscript indicates the sum of protons and neutrons.

Radioactive isotope. An atom with an unstable nucleus that will eventually disintegrate into another nucleus. The initial nucleus is called the *parent,* and the product nucleus is called the *daughter.*

REPORT

THE USE OF RADIOACTIVE ISOTOPES IN OCEANOGRAPHY

NAME ______________________

DATE ______________________

INSTRUCTOR ______________________

1. On the graph below, plot $^{14}C^*$ against depth in sediment for the data in Table 24-1.

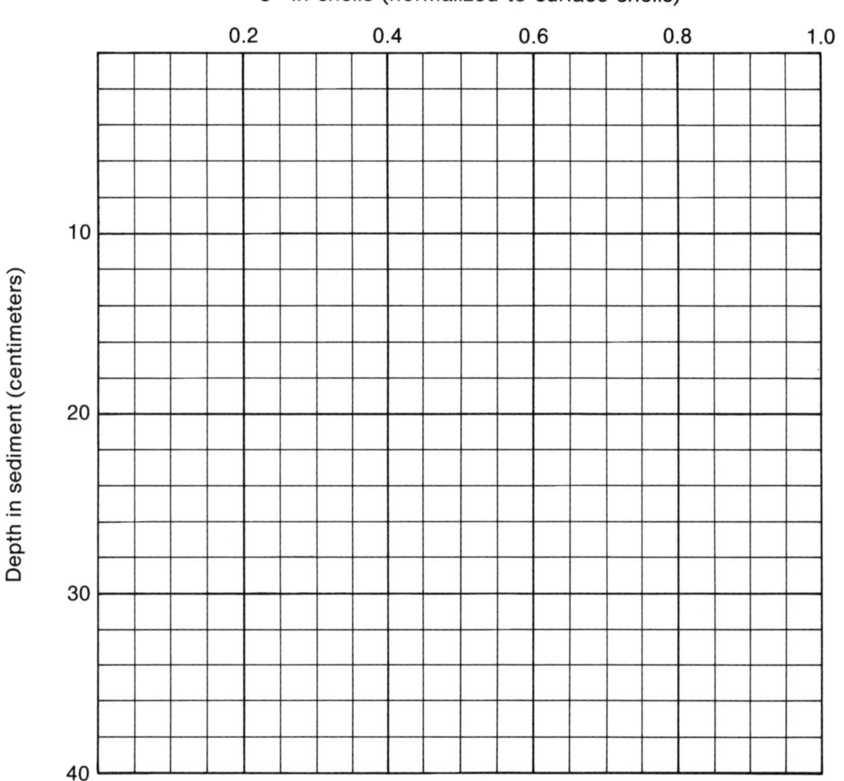

What do you notice about the shape of this curve?

2. If the sediment at the surface was formed today, how many years ago was the sediment at 10 centimeters formed? (Hint: Look at the curve of N/N^0 for ^{14}C shown earlier.)

At 20 centimeters?

3. What is the average rate of accumulation (in centimeters per thousand years) of sediment at this location between 0 and 10 centimeters?

Between 10 and 20 centimeters?

4. If a sample of sediment that was 1×10^6 years old was contaminated with 10 percent atmospheric CO_2 during preparation of the sample for counting, how old would the sample appear to be?

Why do you think ^{14}C age dates are considered to be minimum ages on very old samples? (Hint: Assume atmospheric CO_2 has the same $^{14}C^\star$ as the surface ocean.)

CONVERSION FACTORS AND NUMERICAL DATA

CONVERSION FACTORS

Length	
1 micrometer (μm)	0.001 millimeter
1 millimeter (mm)	1000 micrometers 0.1 centimeter 0.001 meter
1 centimeter (cm)	10 millimeters 0.394 inch 10,000 micrometers
1 meter (m)	100 centimeters 39.4 inches 3.28 feet 1.09 yards
1 kilometer (km)	1000 meters 1093 yards 3280 feet 0.62 statute mile 0.54 nautical mile
1 inch (in.)	25.4 millimeters 2.54 centimeters
1 foot (ft)	30.5 centimeters 0.305 meters
1 yard	3 feet 0.91 meter
1 fathom	6 feet 2 yards 1.83 meters
1 statute mile	5280 feet 1760 yards 1609 meters 1.609 kilometers 0.87 nautical mile
1 nautical mile	6076 feet 2025 yards 1852 meters 1.15 statute miles

Speed	
1 statute mile per hour	1.61 kilometers per hour 0.87 knot
1 knot	1.15 miles per hour 1.85 kilometer per hour
1 kilometer per hour	0.62 mile per hour 0.54 knot
Velocity of sound in sea water at 34.85 parts per thousand (‰)	4945 feet per second 1507 meters per second 824 fathoms per second

Mass	
1 kilogram (kg)	2.2 pounds 1000 grams
1 metric ton	2205 pounds 1000 kilograms 1.1 tons
1 pound	16 ounces 454 grams 0.45 kilogram
1 ton	2000 pounds 907.2 kilograms 0.91 metric ton

Pressure	
1 atmosphere (sea level)	14.7 pounds per square inch 33.9 feet of water (fresh) 29.9 inches of mercury 33 feet of seawater

CONVERSION FACTORS (continued)

Time	
1 hour	3600 seconds
1 day	24 hours 1440 minutes 86,400 seconds
1 calendar year	31,536,000 seconds 525,600 minutes 8760 hours 365 days

Volume	
1 cubic inch (in.^{3})	16.4 cubic centimeters
1 cubic foot (ft^3)	1728 cubic inches, 28.32 liters, or 7.48 gallons
1 cubic centimeter (cc; cm^3)	0.061 cubic inch
1 liter	1000 cubic centimeters, 61 cubic inches, 1.06 quarts, or 0.264 gallon
1 cubic meter (m^3)	10^6 cubic centimeters, 264.2 gallons, or 1000 liters
1 cubic kilometer (km^3)	10^9 cubic meters, 10^{15} cubic centimeters, or 0.24 cubic mile

Area	
1 square inch (in.^{2})	6.45 square centimeters
1 square foot (ft^2)	144 square inches
1 square centimeter (cm^2)	0.155 square inch or 100 square millimeters
1 square meter (m^2)	10^4 square centimeters or 10.8 square feet
1 square kilometer (km^2)	247.1 acres, 0.386 square mile, or 0.292 square nautical mile

NUMERICAL OCEANOGRAPHIC DATA

Scaling Ratios for Charts	
Number of nautical miles per inch on the chart	Reciprocal of the natural scale divided by 72,913
Number of statute miles per inch on the chart	Reciprocal of the natural scale divided by 63,360
Number of inches on the chart per nautical mile	Natural scale times 72,913
Number of inches on the chart per statute mile	Natural scale times 63,360

Equivalences in Concentration of Seawater	
Seawater with 35 grams of salt per kilogram of seawater	3.5 percent 35.00 parts per thousand (‰) 35,000 parts per million (ppm)

Area, Volume, and Depth of the World's Oceans

Body of water	Area (10^6 km^2)	Volume (10^6 km^3)	Mean depth (m)
Atlantic Ocean	82.4	323.6	3926
Pacific Ocean	165.2	707.6	4282
Indian Ocean	73.4	291.0	3963
All oceans and seas	361	1370	3795

THE GEOLOGIC TIME SCALE

Geologic column and time scale

Era	System or Period (rocks) (time)	Series or Epoch (rocks) (time)	Approximate age in millions of years (beginning of unit)
Cenozoic *(recent life)*	Quaternary (an addition to the 18th-century scheme)	Holocene *(entirely recent)*	0.01
		Pleistocene *(most recent)*	1.6
	Tertiary (Third, from the 18th-century scheme)	Pliocene *(very recent)*	5.3
		Miocene *(moderately recent)*	24
		Oligocene *(slightly recent)*	37
		Eocene *(dawn of recent)*	58
		Paleocene *(early dawn of the recent)*	66
Mesozoic (*intermediate life*)	Cretaceous (*chalk*)		144
	Jurassic (Jura Mountains, France)		208
	Triassic (from threefold division in Germany)		245
Paleozoic (*ancient life*)	Permian (Perm, a Russian province)		286
	Pennsylvanian		320
	Mississippian		360
	Devonian (Devonshire, England)		408
	Silurian (an ancient British tribe, Silures)		438
	Ordovician (an ancient British tribe, Ordovices)		505
	Cambrian (the Roman name for Wales, Cambria)		570
Precambrian	Many local systems and series are recognized, but no well-established worldwide classification has yet been delineated.		

Source: Adapted from J. G. Gilluly, A. C. Waters, and A. O. Woodford, *Principles of Geology,* 4th ed. (New York: W. H. Freeman and Company, 1975). Information of the approximate ages based on A. Palmer, 1983, "The Decade of North American Geology 1983 Geologic Time Scale," *Geology,* 11:503–504.
Notes: Terms in italics indicate the Greek derivations of some names. Many provincial series and epochs have been recognized in various parts of the world for Mesozoic and older strata. Most of the systems have been divided into Lower, Middle, and Upper series, to which correspond Early, Middle, and Late epochs as time terms.

BIBLIOGRAPHY

General References

Barnes, H., *Apparatus and Methods of Oceanography.* London: Allen & Unwin, 1959, 341 pp.

Bascom, W., *Waves and Beaches.* New York: Anchor Books, Doubleday, 1964, 267 pp.

Bowditch, N., *American Practical Navigator.* Hydrographic Office Publication No. 9, U.S. Navy. Washington, D.C.: Government Printing Office, 1943, 387 pp.

Broecker, W., *Chemical Oceanography.* New York: Harcourt Brace Jovanovich, 1974, 214 pp.

Davis, R., *Principles of Oceanography.* Reading, Massachusetts: Addison-Wesley, 1972, 434 pp.

Drake, C., Imbrie, J., Knauss, J., and Turekian, K., *Oceanography.* New York: Holt, Rinehart and Winston, 1978, 447 pp.

Duxbury, A., and Duxbury, A., *An Introduction to the World's Oceans.* Reading, Massachusetts: Addison-Wesley, 1984, 549 pp.

Fairbridge, R., *The Encyclopedia of Oceanography.* New York: Van Nostrand Reinhold, 1966, 1021 pp.

Gross, M. G., *Oceanography, A View of the Earth.* Englewood Cliffs, New Jersey: Prentice-Hall, 1972, 581 pp.

Hedgpeth, J. (ed.), *Treatise on Ecology and Paleoecology,* Vol. 1 (*Ecology*), Geological Society of America Memoir 67; Baltimore: Waverly Press, 1957, 1296 pp.

Heezen, B. C., and Hollister, C. D., *The Face of the Deep.* New York: Oxford University Press, 1971, 659 pp.

Hopkins, T. L., *A Survey of Marine Bottom Samplers: Progress in Oceanography,* Vol. 2, pp. 213–256. London: Pergamon Press, 1964.

Komar, P., *Beach Processes and Sedimentation.* Englewood Cliffs, New Jersey: Prentice-Hall, 1976, 429 pp.

Libby, W. F., *Radiocarbon Dating.* Chicago: University of Chicago Press, 1965, 175 pp.

McCormick, J., and Thiruvathukal, J., *Elements of Oceanography.* Philadelphia: W. B. Saunders Co., 1976, 346 pp.

Mero, J. L., *The Mineral Resources of the Sea.* New York: Elsevier, 1965, 312 pp.

Pirie, R. G., *Oceanography, Contemporary Readings in Ocean Sciences.* New York: University Press, 1973, 529 pp.

Ricketts, E. F., and Calvin, J., *Between Pacific Tides,* 4th ed. Stanford: Stanford University Press, 1968, 614 pp.

Sears, M., and Merriman, D., *Oceanography: The Past.* New York: Springer-Verlag, 1980, 812 pp.

Shepard, F. P., *Submarine Geology.* New York: Harper & Row, 1973, 517 pp.

Shepard, F. P., *Geological Oceanography.* New York: Crane, Russak, 1977, 214 pp.

Snead, R., *Coastal Landforms and Surface Features.* Stroudsburg, Pennsylvania: Hutchinson Ross, 1982, 247 pp.

Sverdrup, H. U., Johnson, M. W., and Fleming, R. H., *The Oceans, Their Physics, Chemistry and General Biology.* Englewood Cliffs, New Jersey: Prentice-Hall, 1942, 1087 pp.

Thurman, H., *Essentials of Oceanography.* Columbus, Ohio: Merrill, 1983, 374 pp.

U.S. Navy Oceanographic Office, *Glossary of Oceanographic Terms.* Special Publication SP-35, U.S. Navy. Washington, D.C.: U.S. Government Printing Office, 1966, 204 pp.

van Andel, T., *Science at Sea, Tales of an Old Ocean.* New York: W. H. Freeman and Company, 1981, 186 pp.

Weyl, P., *Oceanography, An Introduction to the Marine Environment.* New York: Wiley, 1970, 535 pp.

Wilson, T. R. S., "Salinity and the Major Elements of Sea Water." In *Chemical Oceanography* (J. P. Riley and G. Skirrow, eds.), 2d ed., Vol. 1, pp. 365–413. New York: Academic Press, 1975.

Scientific American References

Arnon, D. I., "The Role of Light in Photosynthesis," *Scientific American,* November 1960.

Astin, A. V., "Standards of Measurement," *Scientific American,* June 1968.

Bailey, H. S., Jr., 1953, "The Voyage of the *Challenger,*" *Scientific American,* May 1953.

Baker, D. J., Jr., "Models of Oceanic Circulation,' *Scientific American,* January 1970. Offprint No. 890, W. H. Freeman and Company, New York.

Bascom, W., "Beaches," *Scientific American,* August 1960. Offprint No. 845, W. H. Freeman and Company, New York.

Bascom, W., "Ocean Waves," *Scientific American,* August 1959. Offprint No. 828, W. H. Freeman and Company, New York.

Bernstein, J., 1954, "Tsunamis," *Scientific American,* August 1954.

Borgese, E., "The Law of the Sea," *Scientific American,* March 1983.

Bullard, Sir E., "The Origin of the Oceans," *Scienfitic American,* September 1969. Offprint No. 880, W. H. Freeman and Company, New York.

Carr, A., "The Navigation of the Green Turtle," *Scientific American,* May 1965. Offprint No. 1010, W. H. Freeman and Company, New York.

Chapman, S., "Tides in the Atmosphere," *Scientific American,* May 1954.

Clark, J. R., "Thermal Pollution and Aquatic Life," *Scientific American,* March 1969. Offprint No. 1135, W. H. Freeman and Company, New York.

Courtillot, V., and Vink, G., "How Continents Break Up," *Scientific American,* January 1983.

Covey, C., "The Earth's Orbit and the Ice Ages," *Scientific American,* February 1984.

Dietz, R. S., and Holden, J. C., "The Break-Up of Pangaea," *Scientific American,* October 1970. Offprint No. 892, W. H. Freeman and Company, New York.

Donaldson, L., and Joyner, T., "The Salmonid Fishes as a Natural Livestock," *Scientific American,* January 1983.

Edmond, J., and Von Damm, K., "Hot Springs on the Ocean Floor," *Scientific American,* April 1983.

Eisley, L. C., "Charles Darwin," *Scientific American,* February 1956. Offprint No. 108, W. H. Freeman and Company, New York.

Emery, K. O., "The Continental Shelves," *Scientific American,* September 1969. Offprint No. 882, W. H. Freeman and Company, New York.

Emiliani, C., "Ancient Temperatures," *Scientific American,* February 1958. Offprint No. 815, W. H Freeman and Company, New York.

Ericson, D. B., and Wollin, G., "Micropaleontology," *Scientific American,* July 1962. Offprint No. 856, W. H. Freeman and Company, New York.

Fisher, R. L., and Revelle, R., "The Trenches of the Pacific," *Scientific American,* November 1955. Offprint No. 814, W. H. Freeman and Company, New York.

Francis, P., and Self, S., "The Eruption of Krakatoa," *Scientific American,* August 1983.

Gamov, G., "Gravity," *Scientific American,* March 1961.

Harbron, J., "Modern Ice Breakers," *Scientific American,* June 1983.

Heezen, B. C., "The Origin of Submarine Canyons," *Scientific American,* August, 1956.

Heirtzler, J. R., "Sea-Floor Spreading," *Scientific American,* December 1968. Offprint No. 875, W. H. Freeman and Company, New York.

Heiskanen, W. A., "The Earth's Gravity," *Scientific American,* September 1955. Offprint No. 812, W. H. Freeman and Company, New York.

Hurley, P. M. "The Confirmation of Continental Drift," *Scientific American,* April 1968. Offprint No. 874, W. H. Freeman and Company, New York.

Hutner, S. H., and McLaughlin, J. J. A., "Poisonous Tides," *Scientific American,* August 1958.

Isaacs, J. D., "The Nature of Oceanic Life," *Scientific American,* September 1969. Offprint No. 884, W. H. Freeman and Company, New York.

King-Hele, D., "The Shape of the Earth," *Scientific American,* October 1967. Offprint No. 873, W. H. Freeman and Company, New York.

Kort, V. G., "The Antarctic Ocean," *Scientific American,* September 1962. Offprint No. 860, W. H. Freeman and Company, New York.

Kuenen, P. H., "Sand," *Scientific American,* April 1960.

Kurten, B., "Continental Drift and Evolution," *Scientific American,* March 1969. Offprint No. 877, W. H. Freeman and Company, New York.

Lack, D., "Darwin's Finches," *Scientific American,* April 1963. Offprint No. 22, W. H. Freeman and Company, New York.

Lynch, D., "Tidal Bores," *Scientific American,* April 1982.

MacIntyre, F., "Why the Sea Is Salt," *Scientific American,* November 1970. Offprint No. 893, W. H. Freeman and Company, New York.

McDonald, J. E., "The Coriolis Effect," *Scientific American,* May 1952. Offprint No. 839, W. H. Freeman and Company, New York.

McVay, S., "The Last of the Great Whales," *Scientific American,* August 1966.

Menard, H. W., "The Deep-Ocean Floor," *Scientific American,* September 1969. Offprint No. 883, W. H. Freeman and Company, New York.

Molnar, P., and Tapponnier, P., "The Collision Between India and Eurasia," *Scientific American,* April 1977.

Munk, W., "The Circulation of Oceans," *Scientific American,* September 1955. Offprint No. 813, W. H. Freeman and Company, New York.

Murphy, R. C., "The Oceanic Life of the Antarctic," *Scientific American,* September 1962.

Oliver, J., "Long Earthquake Waves," *Scientific American,* March 1959. Offprint No. 827, W. H. Freeman and Company, New York.

Opik, E. J., "Climate and the Changing Sun," *Scientific American,* June 1958. Offprint No. 835, W. H. Freeman and Company, New York.

Pequegnat, W. E., "Whales, Plankton and Man," *Scientific American,* January 1958. Offprint No. 853, W. H. Freeman and Company, New York.

Plass, G. N., "Carbon Dioxide and Climate," *Scientific American,* July 1959. Offprint No. 823, W. H. Freeman and Company, New York.

Price, D. J. de S., "An Ancient Greek Computer," *Scientific American,* June 1959.

Rabinowitch, E. I., "Photosynthesis," *Scientific American,* August 1948.

Ritchie-Calder, Lord, "Conversion to the Metric System,"

Scientific American, July 1970. Offprint No. 334, W. H. Freeman and Company, New York.

Sauer, E. G. F., "Celestial Navigation by Birds," *Scientific American,* August 1958. Offprint No. 133, W. H. Freeman and Company, New York.

Schmidt-Nielsen, K., "Salt Glands," *Scientific American,* January 1959.

Sokal, R. R., "Numerical Taxonomy," *Scientific American,* December 1966. Offprint No. 1059, W. H. Freeman and Company, New York.

Stewart, R. W., "The Atmosphere and the Ocean," *Scientific American,* September 1969. Offprint No. 881, W. H. Freeman and Company, New York.

Stommel, H., "The Anatomy of the Atlantic," *Scientific American,* January 1955. Offprint Nol. 810, W. H. Freeman and Company, New York.

Wenk, E., "The Physical Resources of the Ocean," *Scientific American,* September 1969. Offprint No. 885, W. H. Freeman and Company, New York.

Wiebe, P., "Rings of the Gulf Stream," *Scientific American,* March 1983.

Woodcock, A. H., "Salt and Rain," *Scientific American,* October 1957. Offprint No. 850, W. H. Freeman and Company, New York.

Woodwell, G., "The Carbon Dioxide Question," *Scientific American,* January 1978.

Wooster, W. S., "The Ocean and Man," *Scientific American,* September 1969. Offprint No. 888, W. H. Freeman and Company, New York.